Ultrananocrystalline Diamond Coatings for Next-Generation High-Tech and Medical Devices

This book presents a comprehensive review on the fundamental and applied materials science and technology development for a transformational ultrananocrystalline diamond (UNCD™) thin-film technology enabling a new generation of industrial products and high-tech and external and implantable medical devices and prostheses. Edited and co-authored by a co-originator and pioneer in the field, it describes the synthesis and material properties of UNCD coatings and integration with multifunctional oxide/nitride thin films and nanoparticles, to enable the development and commercialization of new generations of high-tech and external and implantable medical devices and prostheses, and treatments of human biological conditions. Bringing together contributions from experts around the world, this book covers a range of clinical applications, including ocular implants to restore partial vision to blind people, glaucoma treatment devices to protect people from becoming blind due to glaucoma, UNCD-coated implantable prostheses (e.g., dental implants, artificial hips and knees) to eliminate current failures of metal prostheses, scaffolds for stem cell growth and differentiation into other human cells to provide biological treatment of many conditions, Li-ion batteries for defibrillators and pacemakers, and drug delivery and sensor devices. Technology transfer and regulatory issues are also covered.

This is essential reading for researchers, engineers, and practitioners in the field of high-tech and medical device technologies across materials science and biomedical engineering.

Orlando Auciello is a distinguished endowed chair professor at the University of Texas at Dallas, and a co-founder of Advanced Diamond Technologies, Inc., Original Biomedical Implants (OBI-USA), LLC, and OBI-México, SRL, which are commercializing the UNCD coating technology in industrial, high-tech, and biomedical products worldwide.

Ultrananocrystalline Diamond Coatings for Next-Generation High-Tech and Medical Devices

Edited by

ORLANDO AUCIELLO

University of Texas, Dallas

Original Biomedical Implants (OBI-USA)

Original Biomedical Implants (OBI-México)

CAMBRIDGE
UNIVERSITY PRESS

University Printing House, Cambridge CB2 8BS, United Kingdom

One Liberty Plaza, 20th Floor, New York, NY 10006, USA

477 Williamstown Road, Port Melbourne, VIC 3207, Australia

314–321, 3rd Floor, Plot 3, Splendor Forum, Jasola District Centre, New Delhi – 110025, India

103 Penang Road, #05–06/07, Visioncrest Commercial, Singapore 238467

Cambridge University Press is part of the University of Cambridge.

It furthers the University's mission by disseminating knowledge in the pursuit of education, learning, and research at the highest international levels of excellence.

www.cambridge.org
Information on this title: www.cambridge.org/9781107088733
DOI: 10.1017/9781316105177

First published 2022

A catalogue record for this publication is available from the British Library.

Library of Congress Cataloging-in-Publication Data
Names: Auciello, Orlando, 1945– editor.
Title: Ultrananocrystalline diamond coatings for next generation high-tech and medical devices / edited by Orlando Auciello, University of Texas, Dallas.
Description: United Kingdom ; New York, NY : Cambridge University Press, 2022. | Includes bibliographical references and index.
Identifiers: LCCN 2021029135 (print) | LCCN 2021029136 (ebook) | ISBN 9781107088733 (hardback) | ISBN 9781316105177 (epub)
Subjects: LCSH: Medical instruments and apparatus–Materials. | Diamond thin films–Industrial applications. | Nanodiamonds–Industrial applications. | Diamond powder–Industrial applications. | Protective coatings.
Classification: LCC R857.M3 U58 2022 (print) | LCC R857.M3 (ebook) | DDC 610.284–dc23
LC record available at https://lccn.loc.gov/2021029135
LC ebook record available at https://lccn.loc.gov/2021029136

ISBN 978-1-107-08873-3 Hardback

Contents

Contributors

Jesus J. Alcantar-Peña
Division of Microtechnologies, Engineering and Industrial Development Centre (CIDESI), Mexico

Orlando Auciello
University of Texas at Dallas, USA

Alejandro Berra
Ocular Research Laboratory, University of Buenos Aires, Argentina

Elida de Obaldia
Technological University of Panama, Panama, and University of Texas at Dallas, USA

Pablo Gurman
Research Scientist, Department of Materials Science and Engineering, University of Texas at Dallas, USA

Karam Kang
Bioengineering, University of Texas at Dallas, USA

Gilberto López-Chávez
Bioingeniería Humana Avanzada, Iturbide Esquina Moctezuma, San José Iturbide, México

Geunhee Lee
BTI Solutions, dba Blue Telecom, Inc., USA

Enio Lima, Jr.
Bariloche Atomic Centre and CONACIT, Argentina

Daniel G. Olmedo
School of Dentistry, Department of Oral Pathology, University of Buenos Aires, Argentina

Mario J. Saravia
Buenos Aires Macula-Clinical Research, Argentina

Bing Shi
X-ray Science Division, Argonne National Laboratory, USA

Debora R. Tasat
Environmental Bio-Toxicology Laboratory, National University of General San Martín, Argentina

Yonhua Tzeng
National Cheng Kung University (NCKU), Taiwan

Martin Zalazar
Faculty of Engineering, Bioengineering, National University of Entre Ríos, Argentina

Roberto D. Zysler
Bariloche Atomic Centre, Argentina

Preface

This book provides detailed descriptions of the materials science, materials integration strategies, materials properties, and design and development of a new generation of implantable and external medical devices based on the unique biocompatible ultrananocrystalline diamond (UNCD™) coating and some key oxide films and nanoparticles, to improve the quality of life for people worldwide.

The target audience for this book includes undergraduate students, graduate students, postdocs, researchers, and practitioners in industry in different fields of science and technology, including materials scientists, mechanical engineers, bioengineers, applied physicists, medical device designers and manufacturers, and even medical doctors and surgeons, who by knowing more about the biomaterials and devices they are implanting in people may be able to make better decisions when selecting a device based on the appropriate material for implantation. The book will also be useful to manufacturers of medical implants, so they become aware of the new revolutionary UNCD coating that can be grown directly on current bare metal-based implantable devices (e.g., dental implants, hips, knees, and more) to eliminate failure due to mechanical degradation and chemical attack by bodily fluids.

1 Fundamentals on Synthesis and Properties of Ultrananocrystalline Diamond (UNCD™) Coatings

Orlando Auciello, Jesus J. Alcantar-Peña, and Elida de Obaldia

1.1 Background on UNCD Film Synthesis, Properties, and Applications

This chapter serves as the introduction of the focus of this book on the science and technological applications of a new material paradigm provided by a novel material in thin-film (coating) form named ultrananocrystalline diamond (UNCD™), which has enabled new generations of industrial, high-tech, and electronic-related products. In this respect, this chapter focuses on presenting a review of the status of the synthesis and properties of the UNCD film technology, originally developed and patented by a group of scientists (O. Auciello, D. M. Gruen, A. R. Krauss, and J. A. Carlisle), with a view to industrial and high-tech applications, and subsequently under R&D by Auciello's group for biotechnological and medical devices/prostheses and medical treatment applications, which is the main topic of this book.

Early research on the growth and characterization of properties of diamond films provided valuable information for understanding the underlying physical [1, 2], chemical [1, 3], and structural [1, 4, 5] properties of the various diamond films, from single crystalline diamond (SCD) to microcrystalline diamond (MCD; ≥ 1 μm grain sizes) to nanocrystalline diamond (NCD; ~10–1000 nm grain sizes) films, which were synthesized by various different processes reported by several different groups [6–15]. The technique used to grow the SCD, MCD, and NCD films involved mainly the microwave plasma chemical vapor deposition (MPCVD) process, whereby a mixture of H_2 (99%) and CH_4 (1%) gases were flown in a chamber evacuated from air to relatively high vacuum (around $\leq 10^{-6}$ Torr), followed by coupling of microwave power to the gas mixture to create a plasma, producing CH_x^{o} and CH_x^{+} ($x = 1, 2, 3$), C^+ and C^o species, and electrically neutral atomic H^o atoms and electrically charged H^+ ions, which, interacting on the surfaces of substrates, produced the SCD, MCD, and NCD films. The large quantity of H inserted into the gas mixture was to produce neutral H^o atoms and H^+ ions, which exhibit strong chemical reaction with open chemical bonds of C atoms on the surface of the graphite impurity phase that was observed to grow concurrently with the growth of the diamond phase [6–15], thus providing the means to minimize or practically eliminate the undesirable impurity graphite phase when growing the diamond films. However, the atomic neutral H^o atoms and H^+ ions may also induce chemical etching of diamond nanograins (≤ 10 nm grain size), which are nucleated in the initial phase of MCD and NCD film growth,

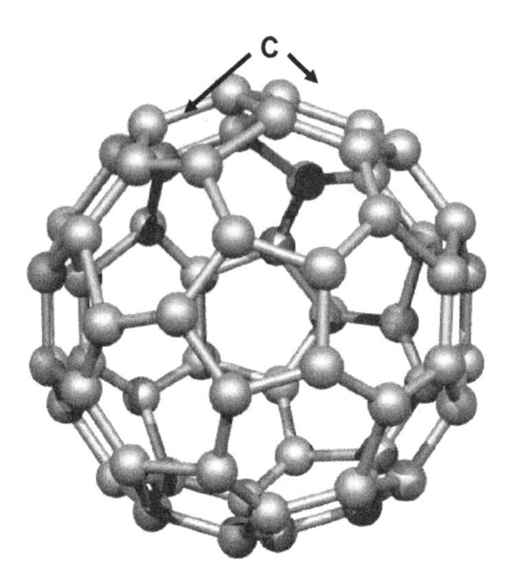

Figure 1.1 Schematic of a fullerene molecule showing the football-type geometry of the molecule with C atoms on the corners of pentagon-type cells.

thus eliminating the possibility of growing polycrystalline diamond films with grain sizes $\leq$10 nm.

The next materials science breakthrough in relation to growing polycrystalline diamond films happened with the discovery of a new process to grow polycrystalline diamond films with grain sizes $\leq$10 nm. Gruen et al. [16–18] demonstrated that by flowing Ar gas through an oven heated to ~900 °C to evaporate C_{60} fullerene molecules from powder, the Ar gas induced flow of C_{60} fullerene molecules (Figure 1.1) into an evacuated MPCVD chamber, where coupling of microwave power to the Ar/C_{60} fullerene molecules mixture induced cracking of the C_{60} fullerene molecules, releasing C atoms that upon landing on the surfaces of substrates produced the growth of polycrystalline diamond films with $\leq$10 nm grain size. The C_{60} fullerene molecule had been discovered previously in 1985 by H. W. Kroto, R. E. Smalley, and R. F. Curl, Jr. (1996 Nobel Prize Winners in Chemistry for this discovery, see review in [19]).

Although the use of C_{60} fullerene molecules produced the growth of polycrystalline diamond films with $\leq$10 nm grain size, this process was too expensive to produce polycrystalline diamond films for industrial applications, and in addition the need for an oven producing the gas-phase C_{60} fullerene molecules induced the formation of substantial amounts of carbon soot, which was not appropriate for being interconnected with a clean vacuum system, where the films were grown.

The next breakthrough in growing polycrystalline diamond films with grain sizes $<$10 nm happened through the discovery led by Gruen, Krauss, and Auciello [20], which was based on replacing the H_2 gas in the H_2/CH_4 gas mixture, used previously to grow from SCD to NCD films [6–15], with Ar gas (the less expensive inert gas on the market today) to produce an Ar (vol. 99%)/CH_4 (vol. 1%) gas mixture in the

MPCVD system, which resulted in the growth of polycrystalline diamond films with grain size in the range 3–5 nm, the smallest grain size of any polycrystalline diamond film today (see [20], patent [21], and reviews [22, 23], and Figure 1.4a,b). The R&D described in [20–23] superseded prior work that showed that diamond films grown by MPCVD, using mixtures of Ar (80–97%)/CH_4 (1%)/H_2 (19 to 2%), produced diamond films with grain sizes in the range 50–10 nm [24], respectively, as the H_2 flux volume was reduced. Gruen, Krauss, and Auciello coined the name ultrananocrystalline diamond (UNCD) to designate the new polycrystalline diamond film with a grain size of 3–5 nm.

Auciello and Carlisle trademarked the name UNCD™ when founding Advanced Diamond Technologies, Inc. (founded in 2003, profitable in 2014, and sold for profit to a large company in 2019) [25]), the first and only company worldwide marketing UNCD-coated industrial products, as described below.

Auciello then led the founding of Original Biomedical Implants (OBI-USA 2013–present [26] and OBI-México 2016–present), which are currently developing revolutionary new generations of external and implantable medical devices (new long-life Li-ion batteries for a new generation of defibrillators/pacemakers with >10 times longer life and safer than the current technology – see Chapter 7), prostheses (UNCD-coated dental implants, hips, knees, and more – see Chapter 8) with superior performance to current metal-based devices and prostheses, which are corroded by body fluids sooner or later, resulting in the need for replacement earlier than desired, and other devices for medical applications, as reviewed in various chapters in this book. The UNCD coating provides the best biocompatible material for insertion in the human body, since UNCD is made of C atoms (the element of life in the DNA, cells, and molecules of every human body), and it exhibits the strongest resistance to mechanical wear [27] and chemical attack by body fluids [28], and the lowest coefficient of friction compared to any biocompatible material [27], a condition very favorable for operation of any prostheses involving friction (hips, knees, and others).

Other breakthroughs in material-related science and development of the UNCD film technology include the following:

1. The discovery that incorporation of nitrogen atoms in grain boundaries of UNCD films, grown flowing a mixture of Ar/CH_4/N_2 gases for the MPCVD process, produced the first electrically conductive diamond film, via N atoms' incorporation in UNCD grain boundaries, satisfying C atoms' dangling bonds and providing electrons for electrical conduction through the large grain boundary network of the structure of the as defined N-UNCD film [29, 30].
2. The discovery that B atoms replacing C atoms in the UNCD lattice of the nanograins provides electrons to the electron energy conduction band, inducing true semiconductor-type doping, which yielded high electrical conductivity B-UNCD films that were inserted as a corrosion-resistant coating on metal electrodes in the first worldwide transformational electrolysis-based/ozone-generation/water purification system (DIAMONOX-Advanced Diamond Technologies) marketed by ADT [25]. The B-doping of UNCD coatings done by ADT, including the recent

demonstration of the first process for low-temperature (460–600 °C) growth of B-UNCD by hot filament chemical vapor deposition (HFCVD) [31], a world-first for UNCD coatings, was recently reproduced by independent groups using the MPCVD process [32]. The development of the B-UNCD films followed the early R&D on B-doping of crystalline and MCD and NCD films using MPCVD and HFCVD processes, based on H_2/CH_4 gas mixture chemistry (see [33–36] and a recent review [37]). The problem with all MPCVD and HFCVD growth processes used to produce B-UNCD or B-SCD, B-MCD, B-NCD films is that the B-containing molecular precursors flown into the vacuum systems during diamond film growth result in contamination of expensive systems used for growing any type of diamond films, thus limiting those systems to growing only B-doped diamond films. In relation to this issue, Auciello's group recently developed a post-grow diamond film doping with B atoms using an independent, relatively low-cost/dedicated rapid thermal annealing system, which produces B-UNCD films or any other B-diamond doped films with similar quality and high electrical conductivity as the B-doped diamond films produced by MPCVD and HFCVD growth processes, by simply inserting the diamond film with a spin-on-dopant (SOD) solution (Boroflim 100®) dispersed on the UNCD surface via a spinning speed of 3000 rpm and a process time of 20 s, and subjecting the surface-doped UNCD film to a rapid thermal annealing (RTA) process for ~10 s, which diffuses the B atoms into the UNCD film lattice to produce B-doped B-UNCD films [38].

3. Development of a new process to grow UNCD films at the lowest temperature (~350–400 °C) [39, 40] demonstrated for any diamond films at that time (early 2000s) and today, to enable integration with microelectronic devices based on complementary metal-oxide semiconductors (CMOS) technology requiring a thermal budget processing ≤425 °C. The integration of UNCD films with Si-based microchips was demonstrated by developing a hermetic/biocompatible/humor eye corrosion-resistant UNCD coating to encapsulate an Si-based microchip implantable in the eye as the main component of an artificial retina to restore partial vision to people blinded by genetically induced degeneration of photoreceptors (see [28] and Chapter 2).
4. Discovery of basic materials/chemistry/physical processes to integrate dissimilar materials like high-dielectric constant (*k*) dielectrics and UNCD and MCD (e.g., UNCD or MCD/HfO_2 [41]), ferroelectric/piezoelectric oxides/UNCD (e.g., $PbZr_xTi_{1-x}O_3$/UNCD [42]; piezoelectric nitrides/UNCD (e.g., AlN/UNCD [43]), all enabling new generations of micro/nanoelectronic and microelectromechanical systems/nanoelectromechanical systems (MEMS/NEMS) devices, as described below.

The fundamental and applied material science on UNCD films, reviewed above, provides valuable information revealing that UNCD films exhibit a remarkable synergistic combination of exceptional mechanical, tribological, chemical, electrical, thermal, electron emission, and biocompatibility properties, which are being used to enable a new generation of multifunctional devices from the macro- to nanoscale:

1. The outstanding mechanical properties of UNCD films enable their application to produce a new generation of MEMS devices [44, 45], including the first demonstrated UNCD/piezoelectric oxide films integration to produce piezoelectrically actuated UNCD-based MEMS actuators and sensors [42]; see also a review [46].
2. The chemistry and atoms bonding configuration of UNCD coating surfaces induce unique nanotribological properties resulting in outstanding resistance to mechanical wear and the lowest coefficient of friction demonstrated today in commercial products requiring low wear and friction, such as UNCD-coated mechanical pump seals and bearings and AFM tips (NaDiaProbes^{TM}), providing practically wear-free/high-resolution AFM nanolithography and imaging capabilities, which are currently on the market commercialized by Advanced Diamond Technologies [25].
3. R&D has demonstrated that UNCD films exhibit excellent dielectric properties (as dielectric layers for RF MEMS switches [45]), which enable for the first time RF MEMS switches without failure as occurred in prior switch technologies involving dielectric layers made of SiO_2 or Si_2N_3, which failed due to electrical charging (~80 μs) of the oxides and nitrides and long discharging times (hundreds of seconds), leading to sticking of the movable switch membrane to the dielectric layer on top of the bottom electrode, and thus failure [45]. In the case of the UNCD dielectric layer used in RF MEMS switches there is fast charging and discharging, both in ~80 μs, due to the large grain boundary network of the UNCD layer, thus eliminating the electrical charging failure [45].
4. The outstanding electrical conductivity properties of nitrogen grain boundary incorporated UNCD (named N-UNCD) films [22, 29, 30] enables a new generation of metal electrodes for the new generation of implantable neural stimulation and Li-ion batteries (LIBs) with $\geq$10 times longer life and which are safer than current LIBs due to the corrosion-resistant N-UNCD coating of anodes and cathodes [47–48]. Alternatively to electrical conduction by nitrogen incorporation into the grain boundaries, highly conductive UNCD films have been achieved by doping of UNCD films, in the true sense of semiconductor doping, by inserting boron (B) atoms to replace C atoms in the diamond lattice, providing electrons to the conduction band as discussed above.
5. Excellent electric field-induced electron emission properties based on electric field-induced electron emission from UNCD film surfaces [49, 50–52] enables field emitter cold cathodes [53] and field emitter flat panel displays [54].
6. Excellent chemical properties, demonstrated via surface functionalization of the UNCD films by electrochemical reduction of aryl-diazonium salts to enable growth of biomolecules on the UNCD surface [55].
7. Excellent biological properties, demonstrated by DNA-induced modification of UNCD thin-film surfaces [56] that provide stable, biologically active scaffolds for biological cell growth and differentiation [57, 58].
8. Excellent biocompatibility, demonstrated in the application of UNCD coatings for encapsulation of a microchip implantable in the eye to restore partial vision to

people blinded by genetically induced death of photoreceptors [28], and application to a new generation of implantable medical devices such as dental implants, artificial hips and knees, and many more prostheses that are implantable in the human body [59].

All applications of the UNCD film (coating) technology require growing the films with the appropriate nanostructure mentioned above and described in detail in the following sections, which discuss the two main growth techniques: MPCVD and HFCVD.

1.2 Fundamentals of UNCD Film Synthesis via MPCVD and Properties

To grow diamond films on nondiamond substrates using MPCVD or HFCVD, or any other CVD method, it is necessary to induce a nucleation step. Two main methods have been developed and used over the years to condition the surface of substrates to grow diamond films:

1. Surface "seeding," embedding diamond particles (with micro- or nanoscale dimensions) on the substrate surface via polishing with diamond-based polishing material or immersing the substrate in a container with a solution of micro- or nano-size diamond particles in methanol in an ultrasound wave-generating system such that the sound waves shake the diamond particles, embedding them on the substrate surface as "seeds" to induce the nucleation and subsequent growth of diamond films [1–18, 20–24]. Following the seeding process, standard thin-film deposition methods based on MPCVD and HFCVD processes have been and are still used to grow diamond films, as discussed below.
2. The other process to nucleate and grow diamond films, which has been used more recently and is still being optimized to nucleate and grow films without using the wet chemical process, is the so-called bias enhanced nucleation-bias enhanced growth (BEN-BEG) process, whereby a negative voltage is applied to an electrically conductive substrate to attract positive ions of C^+, CHx^+, and other species that interact jointly on the substrate's surface, inducing nucleation and growth of diamond films. A review of the synthesis of diamond films using MPCVD and HFCVD processes with both the chemical seeding and the BEN-BEG processes is discussed in this chapter.

1.2.1 Fundamentals on the Synthesis of MCD, NCD, and UNCD Thin Films via MPCVD with the Chemical Seeding Process

1.2.1.1 MPCVD Growth Process for MCD and NCD Films with the Wet Seeding Process

Growth of diamond films, using conventional wet seeding process, plus film growth using the MPCVD technique, involves several gas flow chemistries, resulting in different grain sizes of the diamond films. The hydrogen-rich chemistry (H_2 (99.9 to

Figure 1.2 Scanning electron microscope (SEM) micrographs showing grain sizes and surface morphology of diamond films grown with different $CH_4/Ar/H_2$ gas mixtures in the MPCVD process: (a) MCD film with rough surface morphology; (b) MCD film with less rough surface morphology than in (a); (c) and (d) NCD films with different grain size and surface morphology, depending on the increased Ar and decreased H_2 percentage, as indicated in each subfigure.

96%)/CH_4 (1%)) [15, 60, 61] results in MCD (1–5 μm grains with columnar microstructure for ~1% CH_4/2% Ar/97% H_2 – see Figure 1.2a; and 0.5–1 μm grains for 1% CH_4 /80% Ar /19% H_2 – see Figure 1.2b) and NCD films (50–100s nm grains for ~1% CH_4 /90% Ar/9% H_2 – see Figure 1.2c; and 10–50–nm grains for 1% CH_4/97% Ar/2% H_2 – see Figure 1.2d).

High-quality NCD films can also be grown with relatively low methane percentages (0.3%) [61]. The MCD and NCD films grown on surfaces seeded with conventional solutions of diamond micro- or nanoparticles, without proper functionalization to reduce agglomeration, experience the drawback of relatively low initial nucleation density ($<10^{10}/cm^2$). The MCD film surface coarsens with thickness, exhibiting a rough, highly faceted surface morphology with a root mean square (RMS) roughness generally ~10% of the film thickness. When an optimized seeding process and CH_4 (0.3%) is used in the H_2/CH_4 gas mixture, at substrate temperatures in the range of 450–900 °C, very high nucleation densities are achieved ($>10^{12}/cm^2$) jointly with relatively smooth, high-quality NCD structure of film with various thicknesses [60]. In any case, there is no report in the literature of successful integration of NCD films with CMOS devices, which is a critical proof of low-temperature growth of diamond films, as demonstrated for UNCD films encapsulating an Si microchip implanted inside the eye as the main component of the Argus II device, which returns partial vision to people blinded by retinitis pigmentosa [28]. The growth process related to the H_2/CH_4 chemistry is driven by $CH_3^{\cdot}$ radicals interacting on the substrate surface, involving hydrogen extraction from the radicals and ultimately resulting in C atoms' chemical reaction with the surface to induce the growth of the NCD films. The atomic neutral H^0 atoms and H^+ ions generated in the plasma induce preferential etching of a graphitic phase that co-deposits with the diamond phase. Unfortunately, the H^0 atoms and H^+ ions also induce chemical etching of the diamond phase, specifically nano-grains that try to continually nucleate. However, the etching of the diamond nano-grains by the H^0 atoms and H^+ ions occurs at a much lower rate (~50 times) than for graphite, resulting in the formation of intergranular voids and columnar morphology with large grains ($\geq$1 μm). The grains of MCD (1–5 μm) and NCD (10–100s nm) films are much and a medium step larger, respectively, than those of the UNCD films (3–5 nm), which are the main topic of this chapter (compare Figures. 1.2a–d and Figure 1.4).

The MCD films produced by the MPCVD process exhibit high residual compressive stress, poor intergranular adhesion, and very rough surfaces (see Figures 1.2a,b). Consequently, MCD films are not well suited, for example, to producing MEMS/NEMS structures with smooth surfaces and sharply defined geometries. In addition, MCD films exhibit high coefficient of friction due to high roughness, which makes them inappropriate for applications such as coating of prostheses (e.g., hips, knees, and others), which requires a low coefficient of friction. The grain size can be reduced to 10–100 nm, characteristic of films typically known as nanocrystalline diamond, by inserting Ar gas into the mixture and reducing the H_2 percentage, thus increasing the CH_4/H_2 ratio in the plasma. This process produces a smoother surface than for MCD films, but at the cost of increased nondiamond components at the grain boundaries [62]. There is also another class of NCD film with high sp^3 content which is grown with the CH_4/H_2 gas chemistry with relatively low CH_4 content (0.3%) using a special diamond seeding/nucleation process [63], but these films exhibit the NCD structure only when the film thickness is limited to $<$100 nm [64], while for thicker films grain coarsening dominates due to practically no renucleation and film growth with columnar structure, resulting in increasing grain size and roughness as the film thickness increases.

1.2.1.2 MPCVD Growth Process for UNCD Films with the Conventional Wet Seeding Process

Growth, Structure, and Chemical Characterization of Insulating UNCD Films

Growth Process. Contrary to the growth process for MCD and NCD films described above, the UNCD films, discussed in this chapter, were first produced by the MPCVD process, but in recent years the HFCVD process was also developed to produce high-quality UNCD films, as discussed in this chapter. The original MPCVD process used to grow UNCD films was based on gas mixture flowing into an air evacuated chamber, and involved – and still involves – a novel Ar-rich chemistry (Ar [99%]/CH4 [1%]) with no H_2 gas flow in the system [20–23, 50] and in more recent years including also extremely small H_2 gas flow ($\leq 1\%$). In both cases, it was determined, using *in situ*/real-time optical emission spectroscopy imaging of the MPCVD plasma, generated with the Ar/CH_4 chemistry, that carbon dimers (C_2) are produced in the plasma (inducing a green color to the plasma from the light emission from the excited C_2 dimers) generated by the MPCVD process, from methane decomposition, via reactions (1.1) and (1.2):

$$2CH_4 \rightarrow C_2H_2 + 3H_2, \tag{1.1}$$

$$C_2H_2 \rightarrow C_2 + H_2. \tag{1.2}$$

Although the Ar-rich/CH_4 plasma induces the formation of a complex mixture of carbon (C_2 dimers) and hydrocarbon molecules (CH_x, with $x = 1$, 2, and 3), the C_2 dimers have been proposed and demonstrated to play a critical role in the UNCD nucleation and growth process (see review in [22]). Calculations predict that C_2 dimers have low activation energy (~6 kcal/mol) for insertion into the surface of substrates to induce nucleation and subsequent growth of UNCD films, thus establishing the unique growth process of the UNCD films. However, recent modeling indicated that while the C_2 population in the plasma is high, the population near the surface may be lower, and other hydrocarbon radicals (e.g., C_2H_2) may also contribute to the growth of UNCD films [65, 66]. However, the model [66] is related to the growth of diamond films produced by MPCVD using a mixture of $Ar/CH_4/H_2$ gases, which, as shown in Figure 1.2, does not produce the unique UNCD structure (with 3–5 nm grains) grown by MPCVD with the Ar/CH_4 chemistry [21–23]. In addition, the model could not fully explain the low-temperature growth of UNCD films as demonstrated for UNCD film growth at ~400 °C [39, 40]. Clearly, more experimental and modeling studies are needed. Regardless of the mechanism, the distinctive characteristic of the UNCD film growth process is that the plasma contains very small quantities of hydrogen, which arise mainly from the thermal decomposition of methane to acetylene in the plasma (about 1.5%) and eventual addition of an extremely small amount of H_2 ($\leq 1\%$). The MPCVD process is implemented in small research and, most importantly, industrial-type systems, like the ones operating in Auciello's group laboratory at UTD and the company OBI-México (Figure 1.3a), which grows UNCD, NCD, and MCD films on up to 200 mm diameter substrates (Figure 1.3b) with outstanding uniformity in thickness ($\leq 1\%$) and nanostructure.

A critical outcome of the nucleation and growth processes for UNCD films is that they are the only diamond films that have been demonstrated to grow at temperatures as low as 350–400 °C [39, 40], as determined not only by *in situ* substrate temperature measurements during growth [22, 39, 40], but most importantly by the demonstration that CMOS devices exhibit practically the same performance before and after growing UNCD film on them [28, 67]. Similar demonstration of low-temperature NCD films integration with CMOS devices has not been published in the open literature yet. The demonstrated integration of UNCD films with CMOS devices is paving the way for the integration with CMOS for the development of monolithically

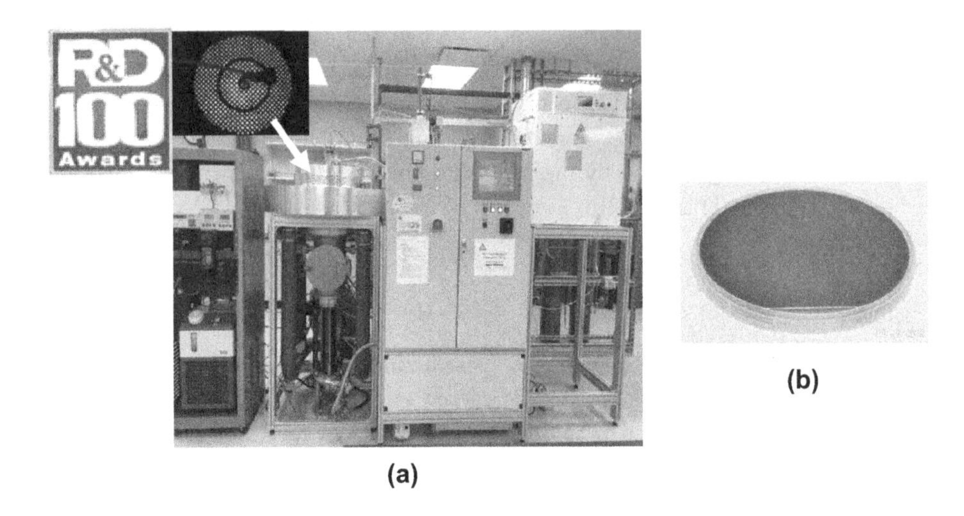

Figure 1.3 (a) Industrial MPCVD system (IPLAS-Germany); (b) Si wafer (200 mm diameter) coated with outstanding uniform UNCD film.

integrated UNCD-MEMS/NEMS/CMOS devices, as recently demonstrated [67], and for encapsulation of Si microchips implantable inside the eye [28] or other parts of the human body.

Structure Characterization. The nucleation and growth process described above produce the UNCD films with the name based on the equiaxed 3–5 nm grains dimension and 0.4 nm wide grain boundaries for undoped insulating UNCD films grown both at 800 °C (Figure 1.4a) and 400 °C (Figure 1.4b), which exhibit extremely smooth as-grown surfaces (~3–5 nm) (Figure 1.4d) when using optimized seeding techniques.

Chemical Characterization. The UNCD films grown by MPCVD were and are characterized by Raman analysis, to provide information on the chemical bonds of the C atoms in the grains and in the grain boundaries. Raman spectroscopy is a nondestructive chemical analysis technique that provides detailed information about the chemical bonds of atoms in a material, distinguishing the atomic bonds in a crystal-type structure such as crystalline diamond and other structures such as the noncrystalline structure of grain boundaries in polycrystalline diamond films. Raman analysis is based on the interaction of light with the chemical bonds of molecules within a material, such that light induces excitation of electrons in the electronic cloud around the atomic nuclei in molecules, producing displacement of the negatively charged electrons with respect to the positively charged nucleus, inducing molecular vibration and causing a "change in polarizability" of the molecule, such that the induced dipole emits or scatters light at the optical frequency of the incident light wave, and the scattered light detected in a detector provides the information on the chemical bonds of the atoms (see a recent review in [69]).

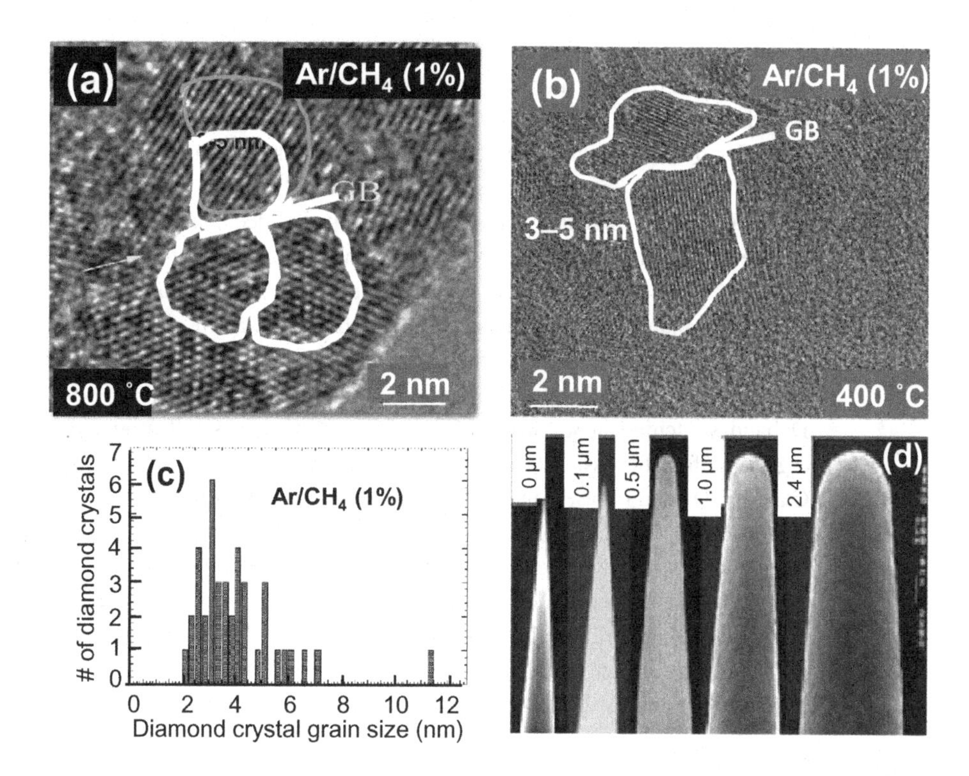

Figure 1.4 High-resolution transmission electron microscopy (HRTEM) images of UNCD films grown by MPCVD at 800 °C (a) and 400 °C (b), showing the unique grain structure (3–5 nm dimensions) of UNCD produced at both temperatures, confirmed by statistical grain size measurements (c). (d) Cross-section SEM images of an Si tip coated with UNCD films grown by the MPCVD process described above, with thicknesses from 0.1 μm to 2.4 μm, showing that the surface roughness remains in the range 3–5 nm independently of the film thickness (reproduced from *J. Appl. Phys.*, vol.88 (11), p. 2958, 2001 (Fig. 2) in [50] with permission from AIP Publisher).

Raman spectroscopy is the most widely used technique to confirm the sp^3 chemical bonds of C atoms in the diamond crystalline lattice and sp^2 C atoms bonds in the grain boundaries of the films. It is a nondestructive technique, often used to determine the nanostructure of carbon thin films due to its ability to discriminate between the presence of sp^2- and sp^3-bonded carbon, as well as the local bonding environment of the carbon itself in some cases. Difficulty arises when Raman spectroscopy is used to examine the structure of carbon films with a mixture of sp^2- and sp^3-bonded carbon in a number of different bonding configurations that possess different short- and long-range order, as is found in nanocrystalline diamond and amorphous carbon films. When a laser with a wavelength in the visible region is used (e.g., 632 nm, 535 nm), the energy of the incident photons is much lower than the energy of the band gap for sp^3-bonded carbon, resulting in a much larger Raman scattering cross-section for sp^2-bonded carbon than for sp^3-bonded carbon due to the well-known resonant Raman effect [69]. The spectra observed are thus completely dominated by Raman scattering from the sp^2-bonded

C atoms, even when a significant amount of the C in the sample, like in UNCD films, is sp^3-C bonded in the small diamond crystalline grains, but with a large network of grain boundaries with sp^2-C atom bonds (Figure 1.5a,b, with a wide peak around ~1560 cm^{-1}) [70]. For polycrystalline diamond films grown with less than 10% H_2 in the plasma (Figure 1.5a), the fingerprint peak of diamond at 1332 cm^{-1} practically disappears, and the visible Raman signal is dominated by the sp^2-bonded C atoms at the grain boundaries. Several distinct broad peaks can be seen for the films: at 1140, 1330, 1450, and 1560 cm^{-1}. The peak at 1560 cm^{-1} corresponds to the G mode Raman peak, which arises from the in-plane stretching modes of the sp^2-bonded C atoms at the grain boundaries of UNCD [70]. On the other hand, the 1330 cm^{-1} broaden peak is attributed to D band-type stretching of C atom bonds in the grain boundaries. In disordered sp^2-bonded C atoms, D band stretching arises from the breathing modes in small aromatic clusters. The argument for the assignment of the broad peak at 1330 cm^{-1} to the sp^2-bonded C atoms' D band is based on the fact that the peak is absent from the UV Raman spectra (Figure 1.5b). The D peak is believed to arise from a double resonant Raman process (incident photon wave vector $k = 0.5q$), where the intensity is highest for low-energy excitation and decreases strongly with increasing excitation energy. In order to see the diamond peak at 1332 cm^{-1}, even for UNCD films, UV laser needs to be used for the Raman analysis (Figure 1.5b). The peaks observed in the UV Raman spectra at 1550 cm^{-1} and 1560 cm^{-1} are mainly due to sp^2-C atoms bonding in the grain boundaries [70].

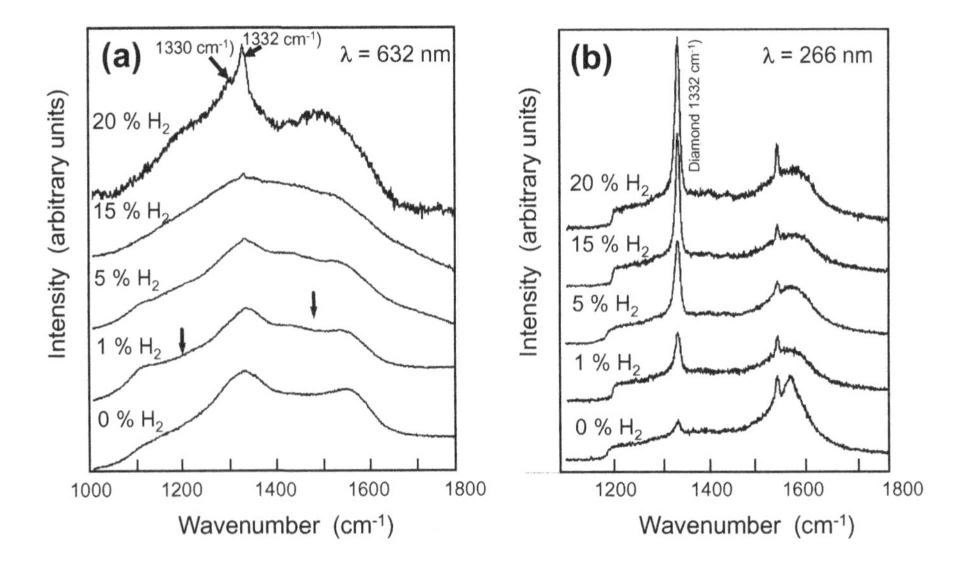

Figure 1.5 (a) Raman spectra obtained with a laser beam wavelength in the visible range, showing the diamond peak at 1332 cm^{-1} only for NCD films grown with H_2 flow $\geq 5\%$. (b) Raman spectra obtained using a laser beam with UV wavelength, for the same polycrystalline diamond films analyzed with the visible laser (a), such that the UV laser enables seeing the diamond peak at 1332 cm^{-1}, even for the UNCD film grown with H_2 flow of 0% (reprinted from *Diam. Relat. Mater.*, vol. 14, p. 86, 2005 (Fig. 4) in [70] with permission from Elsevier Publisher).

The confirmation that the dominant sp^2-C atoms bonds are not from an impurity graphite phase (i.e., parallel planes formed by hexagonal unit lattice with C atoms on the corner) is provided by near-edge X-ray absorption fine structure spectroscopy (NEXAFS) done on UNCD, MCD, and a crystal diamond gem. Core-level photo absorption has been used to characterize the empty electronic states of a wide variety of materials [71]. Specifically, the near-edge electronic energy band region of the photo-absorption process has been used to determine the relative quantity of sp^2- or sp^3-C atoms bonding in several powders and thin films [71]. The technique is known by several acronyms, including NEXAFS and XANES (X-ray absorption near-edge spectroscopy). Its sensitivity to the local atomic bond order in a material arises from the dipole-like electronic transitions from core atoms' electronic states, which have well-defined orbital angular momenta, into empty electronic (e.g., antibonding) states. The symmetry of the final state can be determined, and thus the difference between sp^2 (n-like) or sp^3 (a-like) bonding can be readily observed in covalent, low-z materials such as diamond.

The NEXAFS spectra show the characteristic spectrum of crystalline diamond for UNCD, MCD, and a diamond gem (Figure 1.6a), which is completely different from the NEXAFS spectrum done on a pure graphite film (Figure 1.6a) [72].

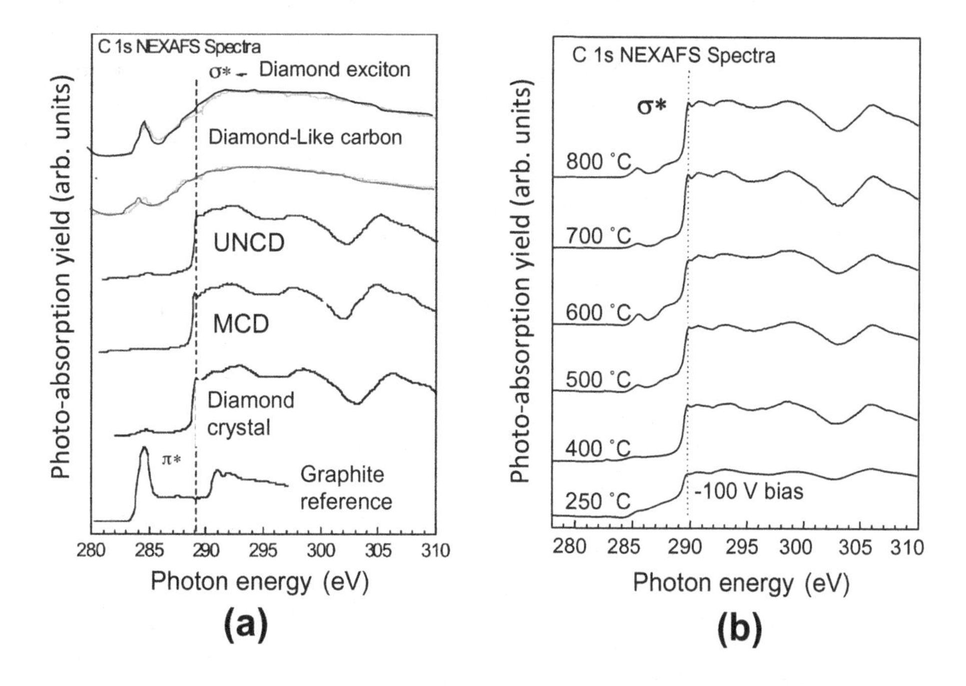

Figure 1.6 (a) NEXAFS spectra from UNCD, MCD, and crystal diamond, showing that they are all identical and do not have graphite impurity, demonstrated by no graphite peak as shown in the graphite NEXAFS spectrum reference. (b) NEXAFS spectra from several UNCD films grown at different temperatures in the range 250–800 °C, showing that all UNCD films have the same chemistry and structure (reprinted from *Mat. Res. Soc. Proc.*, vol. 437, p. 211, 1996 (Fig. 2) in [72] with permission from Cambridge Publisher).

In addition to the NEXAS analysis, electron diffraction from HRTEM done on UNCD films grown at 800 °C and 400 °C (see Figure 1.4a,b) confirm that the UNCD films are made of crystalline diamond nanograins, and **no impurity graphite phase** is present. Relevant information given by the NEXAFS analysis is that UNCD films grown at different temperatures in the range 250–800 °C exhibit the same chemistry and nanostructure (Figure 1.6b). It is important to note that Figure 1.6b shows that UNCD films may be grown by MPCVD at ~250 °C, which may enable coating polymers used in biomedical devices such as glaucoma valves used to drain humor of the eye from people suffering from glaucoma, to keep eye pressure stable, or in polymer soft contact lenses. Both of these case require hydrophobic (liquid-rejecting) surfaces to minimize or eliminate protein and other biomolecule adsorption on normally hydrophilic (liquid-adsorbing) polymer surfaces, to avoid undesirable effects such as clogging of the humor's eye-draining tubes in glaucoma valves, or the need for frequent cleaning of polymer contact lenses' surfaces due protein adsorption (see review in [59]). On the other hand, it has been demonstrated that hydrophilic UNCD film surface provides an excellent scaffold for embryonic cell growth and differentiation for developmental biology [58]. In relation to hydrophilic vs. super-hydrophobic UNCD films' surfaces, recent R&D has demonstrated that the surface of UNCD films can be tailored from super-hydrophilic surfaces, for scaffolds for developmental biology, to super-hydrophobic surfaces for medical devices, as just described.

To finish this section reviewing the growth of UNCD films using the conventional wet chemical seeding plus MPCVD processes, it is relevant to give a brief review of recent research [73], focused on exploring, in a systematic series of studies, the combined effect of total gas pressure (20–80 mbar), precursor gas chemistry (Ar/CH_4/H_2), and microwave power on the grain size, atoms' chemical bonding, and growth rate of polycrystalline diamond films grown by MPCVD [73], with a particular focus on determining the conditions that can produce UNCD films even with the inclusion of H_2 gas flow in the gas chemistry. Figure 1.7 shows a series of SEM images and Raman analysis spectra which provide very revealing information about the mechanism for growing UNCD to MCD diamond films using a tailored combination of gas chemistry and flows, total pressure and power in the MPCVD process. For a gas chemistry of H_2 (98%)/CH_4 (2%), the film grown at 20 mbar total pressure for 2 h was only ~60 nm thick (Figure 1.7a – right) and the surface morphology was characteristic of NCD film structure, with grains ~14 nm, as calculated from the Debye–Scherrer equation using parameters from XRD measurements (Figure 1.7a – left), mainly due to the fact that at low pressure the amount of C-based species is much lower than for the higher pressures, resulting in a film with initial nucleation nanoscale grains. SEM imaging of films grown at 40 mbar pressure (not shown here – see figure 3b in [73]), also for 2 h, showed a thickness of ~850 nm and grain size also of ~14 nm. Figure 1.7b,c shows that films grown at 60 mbar and 80 mbar pressure, with the same H_2 (98%)/CH_4 (2%) chemistry, for 2 h, exhibit the highest thickness (~2500 nm) and grain size of ~230 nm, and 1500 nm thickness and ~230 nm grain size, respectively. The Raman spectra of the films grown at 20 and 40 mbar show the typical low end of NCD (~14 nm) structure, close to UNCD (3–7 nm [22]) (Figure 1.7d), while the

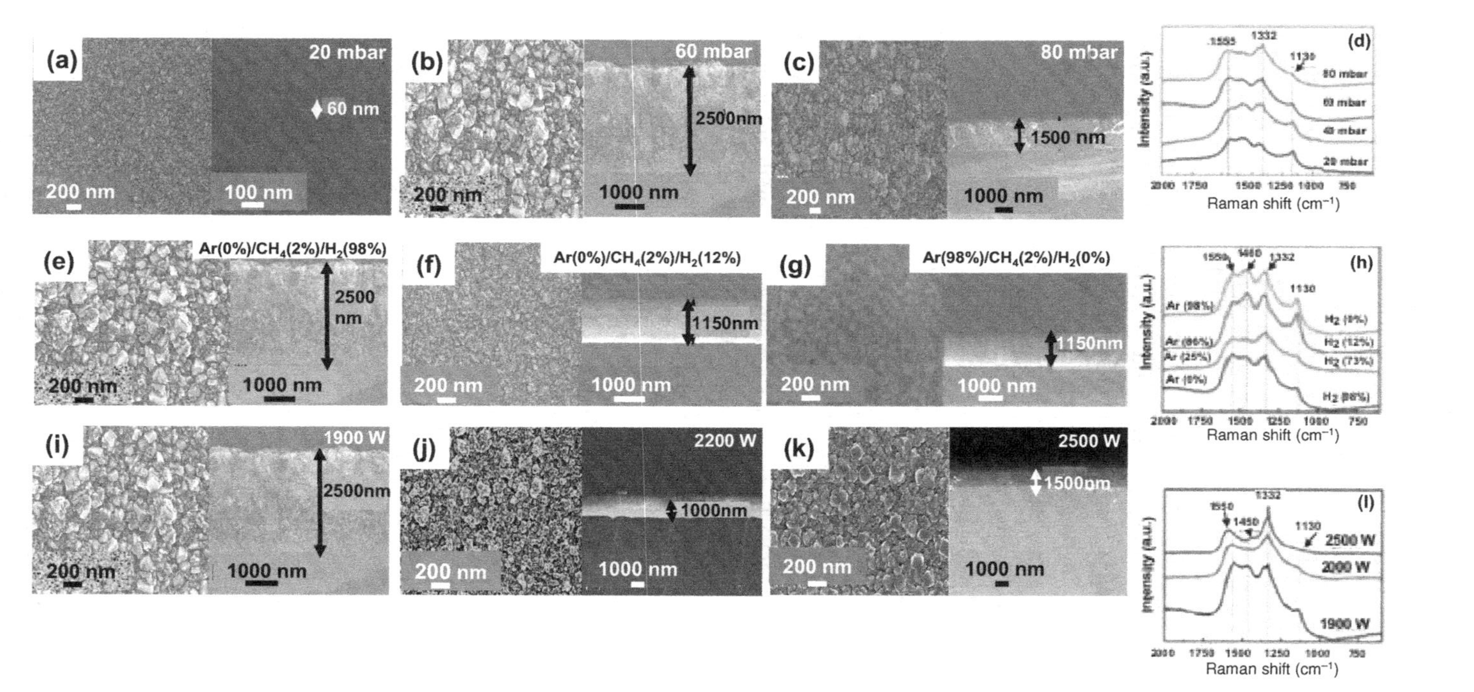

Figure 1.7 SEM top view and cross-section images of MPCVD-processed polycrystalline diamond (PCD) films grown at (a) 20, (b) 60, and (c) 80 mbar using a gas mixture composition of H_2 (98%)/2% CH_4 and microwave power of 1900 W. (d) Raman spectra corresponding to the PCD films, for which SEM images are shown in (a)–(c). Top view and cross-section SEM images of PCD films grown with different Ar/CH_4/H_2 compositions as (e) 0%/2%/98%, (f) 86%/2%/12%, (g) 98%/2%/0%, with total pressure of 60 mbar and 1900 W of microwave power. (h) Raman spectra corresponding to the PCD films, for which SEM images are shown in (e)–(g). Top view and cross-section SEM images of PCD films grown at (i) 1900 W, (j) 2200 W, and (k) 2500 W using gas composition of H_2 (98%)/CH_4 (2%) at a total pressure of 60 mbar. (l) Raman spectra corresponding to the PCD films, for which SEM images are shown in (i)–(k) (reprinted from *IEEE 7th International Engineering, Sciences and Technology Conference*, p. 85, 2019 (Figs. 1.7 (a, b, c) – reprint of Figs. 3 (a, c, d) in [73], respectively; Figs. 1.7 (e, f, g) – reprint of Figs. 5 (a, b, d) in [73], respectively; Figs. 1.7 (I, j, k) – reprint of 2nd Figs. 5 (a, b, d) in [73], respectively; Figs. 1.7 (d), 1.7 (h), 1.7 (l) – reprint of Figs. 2 (a), 4 (a), 6 (a), respectively, with permission from IEEE Publisher).

Raman spectra of the films grown at 60 and 80 mbar (Figure 1.7d) show a small, sharp peak at 1332 cm^{-1}, indicative of the more noticeable sp^3-type C atomic bond from diamond, indicative of grain sizes in the hundreds of nanometer range, as described above (see figures 3 and 4 in [73]). XRD analysis of the polycrystalline diamond films (Figure 2b in [73]) reveal only peaks corresponding to diamond, indicating that no graphite was formed in any of the grown films. The intensity of the peaks increased with increasing pressure, indicative of thicker films with larger grains due to grain growth with film thickness.

The SEM images of films grown with different chemistries, as indicated in Figures 1.7e–g and the corresponding Raman spectra (Figure 1.7h) indicate that films grown with H_2 concentrations of 0–12% (Ar 98–86%) show the typical UNCD/small NCD structures with grain sizes in the range 2–10 nm [22, 23]. On the other hand, in the Raman spectra of the films grown with H_2 concentrations of 73–98% (Ar 25–0%), in addition to the bands discussed above, a peak appears at 1332 cm^{-1}, attributed to sp^3-C bonds of diamond [74, 75], indicating that the grain size increased significantly in the diamond films by increasing the H_2 concentration during growth. XRD analysis shows only peaks attributed to diamond.

The data shown in Figure 1.7 indicate that films with structure close to UNCD can be grown by MPCVD, even when using H_2 gas flow under specific percentages in combination with Ar and CH_4 gases, but only for very thin films, correlated with the initial nucleation of NCD to MCD films, while for films grown with substantially higher thickness the MCD structure is developed. This statement is confirmed by diamond films grown with H_2 (98%)/CH_4 (2%) at 60 mbar and microwave power of 2500 W, but increasing the growth time to 6 h, since it has been reported in the literature that the film grain size increases with growth when using the HFCVD technique, which demonstrated MCD films [74]. Figure 1.8a shows a Raman spectrum of a large grain size MCD film, while Figure 1.8b shows the top surface SEM image of the same MCD film, grown by the MPCVD process for 6 h using a gas mixture of H_2 (98%)/CH_4 (2%) at 60 mbar pressure and 2500 W microwave power. Both figures show clearly an MCD structure.

Figure 1.8 confirms the hypothesis that the growth time, using H_2-rich/CH_4 gas chemistry, induces the development of an MCD film structure due to grain growth.

1.2.1.3 MPCVD Growth Process for UNCD Films via the BEN-BEG Process

Background on MPCVD BEN-BEG of MCD and NCD Films

The new BEN process to nucleate PCD films was investigated and developed, first focused on growing MCD and NCD films. The BEN process provides an alternative to the chemical diamond seeding process discussed in Section 1.2.1.2. The initial research focused on doing BEN followed by film growth without bias [76, 77]. One of the key initial works on BEN for growing MCD/NCD films [77] revealed an enhanced nucleation density of diamond structure (10^{11} cm^{-2}) on virgin silicon wafers vs. 10^7 cm^{-2} on Si wafers seeded with diamond particles via mechanical polishing of the Si surface with diamond particles, both followed by growth of MCD and NCD films without bias. The research described in [76] revealed that if the substrate bias

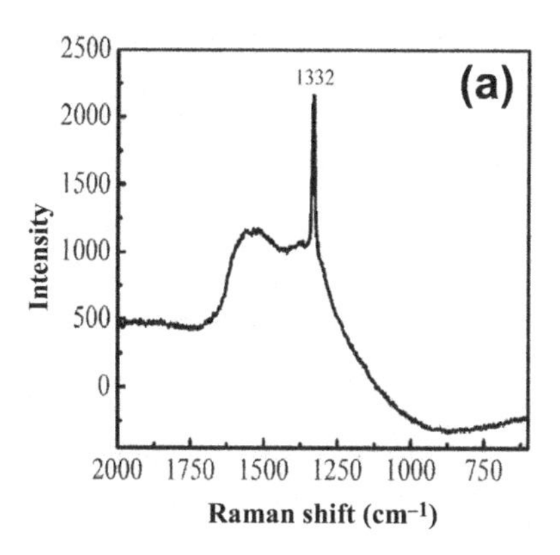

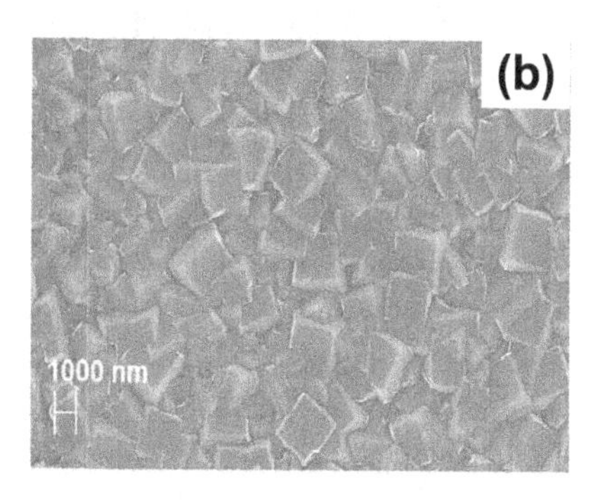

Figure 1.8 (a) Raman spectrum and (b) SEM top view image of an MCD film grown using the MPCVD process with a gas mixture of H_2 (98%)/CH_4 (2%), 60 mbar pressure, 2500 W microwave power, 6 h growth time [73] (reprinted from *IEEE 7th International Engineering, Sciences and Technology Conference*, p. 85, 2019 (Figs. 8 (a, b) in [73] with permission from IEEE Publisher).

was on during diamond film growth, the resulting film exhibited very poor structural quality compared with films grown without bias after the BEN process. Analysis of diamond films, presented in [76], showed that a complete SiC template layer developed on the Si substrate surface before the diamond structure appeared on the film. Research from another group [77] involved growing MCD/NCD films using BEN (2–30 min), accelerating C^+ and CH_x^+ ions from plasmas produced with CH_4/ H_2 gas mixtures (gas pressure of 20 mbar and 800 W microwave power) via substrate biases with voltages in the range –150 V to –350 V. The bias was then turned off and the films were grown by the conventional unbiased MPCVD process for about 30 min, with the substrate temperatures in the range 650–800 °C. The results of the film growth process reported in [77] provided evidence that nucleation density of PCD films reached a maximum at a bias for which the ion energy distribution exhibited a maximum of 80 eV, independent of the substrate temperature. This effect was interpreted as providing evidence that the BEN process involves C^+ ions sub-plantation underneath the substrate surface, leading to C atoms' reaction with the Si substrate to form the SiC layer nucleation to induce the diamond film growth. Cross-section HRTEM imaging showed that the Si interfacial layer involved Si pyramidal structures (see figure 16 in [77]) on top of which SiC interfacial layers were nucleated during the BEN process. However, no explanation was provided for the mechanism for formation of the Si pyramidal structures. In addition, the HRTEM images revealed amorphous-like layers between the pyramidal structures, with thickness in the range 10–100 Å. Raman analysis and XRD presented in [77] indicated that the MCD/NCD films/Si interface produced by BEN consists substantially of sp^2-C atoms bonded to noncrystalline carbon between the Si pyramidal structures. The presence of the (0002) XRD peak from the interlayer between the Si pyramidal structures indicates that the carbon interlayer is primarily graphitic. The early work involving BEN plus BEG processes, using CH_4/H_2 chemistries, produced NCD (30–150 nm grains) cluster and not fully dense films (see figure 15 in [77]), relatively high surface roughness, high compressive stress, delamination of the film, and high content of nondiamond phase [76, 77].

Subsequent relevant research was focused on investigating the effect of key parameters on the nucleation of NCD films using the BEN process [78]. This research [78] revealed a relevant parameter, which is the evolution of electron emission current from the substrate surface during film nucleation, which rises from practically zero to a stable value once the film has been nucleated (see Figure 1 in [78] and Figure 1.9a in this chapter). The results reported in [78] showed that the time for the electron emission current to reach a steady state, and thus BEN of the NCD films, changed from ~82 min for a CH_4/H_2 gas ratio of 3% to ~10 min for a CH_4/H_2 gas ratio of –15% [78]. On the other hand, the electron emission current change from zero to steady state changed from ~70 min for a total gas pressure of 40 Torr to ~10 min for a total gas pressure of 55 Torr [78]. Finally, plasma power is the parameter that induces the smallest change in the time for electron emission current stabilization, from ~30 min for 1000 W to ~6 min for 1500 W. This work indicated that a key parameter in the demonstrated BEN process for MCD and NCD films is the amount of C^+ and H^+ ions contained in the plasma.

MPCVD BEN-BEG of UNCD Films

The early work on BEN-BEG of UNCD films was done first when exploring growing UNCD films with CH_4/N_2/very small H_2 gas mixtures in an MPCVD process [79], but because it was focused on producing UNCD films with N atoms in grain boundaries to develop electron field emission (EFE) devices, it will be described in detail in Section 1.2.1.4, discussing the synthesis of N-UNCD films. Subsequently, two groups demonstrated BEN processes to grow UNCD films using the patented [21] CH_4/Ar gas mixture in the MPCVD process. One group demonstrated that BEN UNCD films, followed by growth without bias on Si substrates, exhibit stronger adhesion to the Si substrate's surface than for UNCD films grown on chemically seeded Si surfaces [80]. Such an effect was attributed to the relatively high kinetic energy of C^+ ions extracted from the plasma and accelerated to the Si substrate surface, which easily form covalent bonding, Si–C, and bond strongly to both the Si and diamond. The work reported in [77] did not include HRTEM, which is critical to understand the nanoscale structures of the nucleation layer and the UNCD films.

Auciello's group [81] first performed R&D to develop a low-pressure BEN-BEG process to grow UNCD films using the MPCVD-based BEN-BEG technique, with the parameters described below. The low-pressure BEN-BEG process described in [81] involved several steps: (1) etching of the natural SiO_2 layer on the surface of an Si (100) substrate for 10 min in a pure H-based plasma, with a substrate biased at –350 V; (2) i*n situ* BEN-BEG using a H_2 (93%)/CH_4 (7%) gas mixture (the idea of using the H_2/CH_4 plasma chemistry was based on the hypothesis that having more H^0 atoms and H^+ ions in the plasma could help etch the impurity graphitic phase that in many cases grows concurrently with the diamond phase when growing NCD films

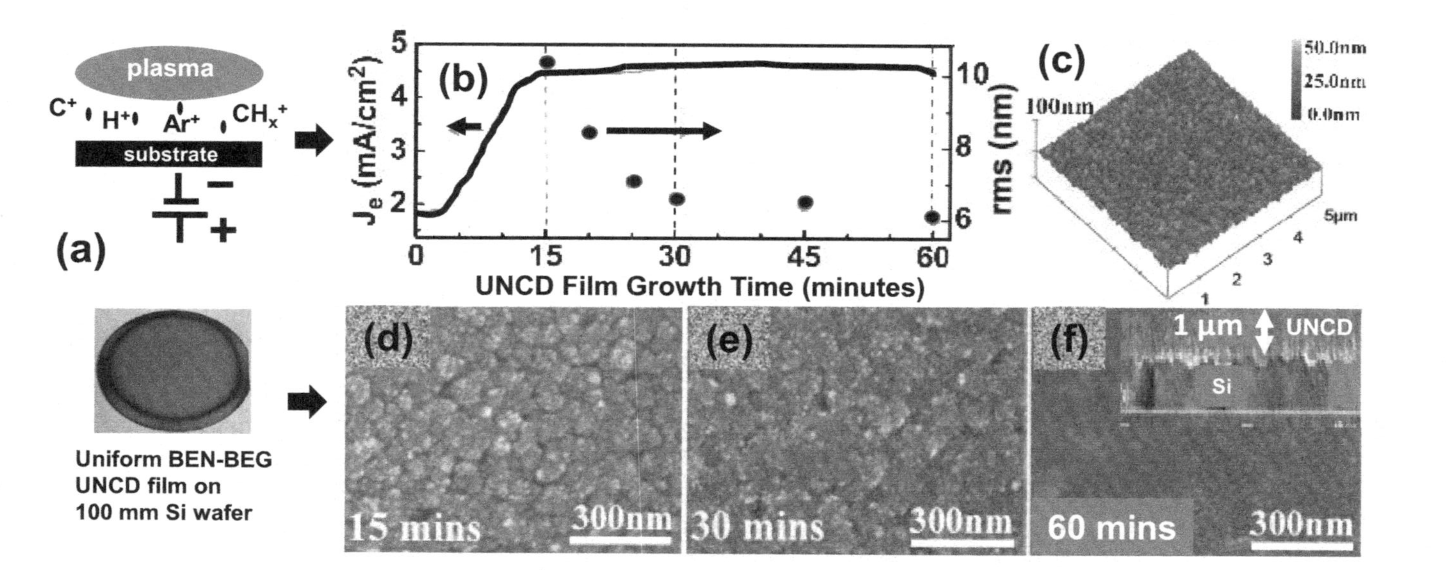

Figure 1.9 (a) Electron field emission current from biased substrate vs. BEN-BEG time for UNCD film growth, showing increase from 0 to stable 4.5 mA/cm^2 value (line curve) and UNCD film grain size reduction from ~10 to 6 nm (dots) as the film reaches dense structure (d). SEM images of UNCD film surface at 15 min (b), 30 min (c), and 60 min (d) BEN-BEG process. (e) AFM measurement of UNCD film surface roughness (~6 nm rms); (f) cross-section SEM image of dense BEN-BEG uniform UNCD film on 100 mm diameter Si surface, showing 1 μm thickness in 60 min growth. (g) Schematic of plasma on negatively biased substrate holder with 100 mm diameter Si wafer, accelerating positively charged atomic and molecular ions toward an Si substrate surface to produce a uniform UNCD film on a 100 mm diameter Si wafer [81] (reprinted from *App. Phys. Lett.*, vol. 92, 133113, 2008 (Fig. 3) in [81] with permission from AIP Publisher).

[76–78]). The UNCD film was grown using a plasma produced by 2.2 kW microwave power at low 25 mbar pressure and applying –350 V bias on a substrate heated to 850 °C in a 2.45 GHz, 6-inch IPLAS CYRANNUS MPCVD system [81]. The BEN-BEG process described here yielded UNCD films with low stress, smooth surfaces (~4–6 nm), high growth rates (~1 μm/h), and uniform grain size (3–7 nm) throughout the film area on 100 mm diameter Si wafers (Figures. 1.9 and 1.10) [81].

Bias current density (J_e) vs. time curves acquired under constant bias during BEN-BEG of UNCD films showed an abrupt increase of J_e after an incubation period of about 5 min and saturation at ~4.5 mA/cm^2 at 15 min after the onset of the BEN process. The abrupt increase and subsequent saturation of J_e can be attributed to the rapid growth of diamond nuclei and subsequent full coverage of the substrate surface by the diamond grains, respectively (see detailed discussion in [81]). The surface roughness of a 1 μm thick BEN-BEG UNCD film was ~10 nm rms over a ~10 μm^2 area after J_e reached steady state, then reduced to ~6 nm rms after 60 min (Figure 1.9f), as revealed by AFM analysis (Figure 1.9c) [81].

Studies of the UNCD–Si interface for a BEN-BEG UNCD film on a Si substrate, using cross-sectional HRTEM, revealed the UNCD grains, an etching effect on the Si surface and the UNCD–Si interfacial layer, revealing the formation of a triangular profile on the surface of the Si substrate (Figure 1.11a) with a peak-to-valley roughness of about 15 nm. The same triangular Si surface morphology was observed previously, although without explanation, on a silicon surface during BEN followed by no-bias growth of NCD isolated grains on an Si surface [77]. The work reported in [81] indicates that H^+ ion bombardment during the SiO_2 layer-cleaning process results in preferential etching of the Si surface along the (100) crystallographic direction, which etches faster than the (111) direction due to a higher physical sputtering rate of the (100) direction, as shown in prior work on physical sputtering of Si surfaces [82]. The physical-dominated etching process may be enhanced by chemical reactions of H atoms with Si atoms on the surface of the Si substrate. This physical-dominated mechanism for etching of the Si surface results in the triangular Si interface profile defined by the (111) planes. The cross-section HRTEM image in Figure 1.11a reveals a preferential nucleation of nanoscale SiC structures (~5 nm thick) on the tip of Si pyramidal structures, which induce the nucleation and growth of the UNCD grains at the peaks of the Si pyramidal structures. Simultaneously, a-C/graphitic phases appear to nucleate at the valleys between the Si pyramidal structures (Figure 1.11a). The UNCD grains nucleated at the peaks of the Si pyramidal structures rapidly extend sideways, covering the a-C and graphite phases grown in the valleys with (0001) planes approximately perpendicular to the Si (100) surface (Figure 1.11a). Figure 1.11e shows a schematic of the mechanism of BEN-BEG growth for UNCD films, which correlates with the observation provided by the cross-section HRTEM, which shows the nucleation of SiC on top of the Si pyramidal structures that induce the subsequent growth of the UNCD film, expanding sideways and covering the graphite areas nucleated between the pyramidal structures. Figure 1.11f shows UNCD films grown by the BEN-BEG process on sharp Si tips for EFE devices. Figure 1.11g shows growth of UNCD films on vertically positioned Ti-alloy-based

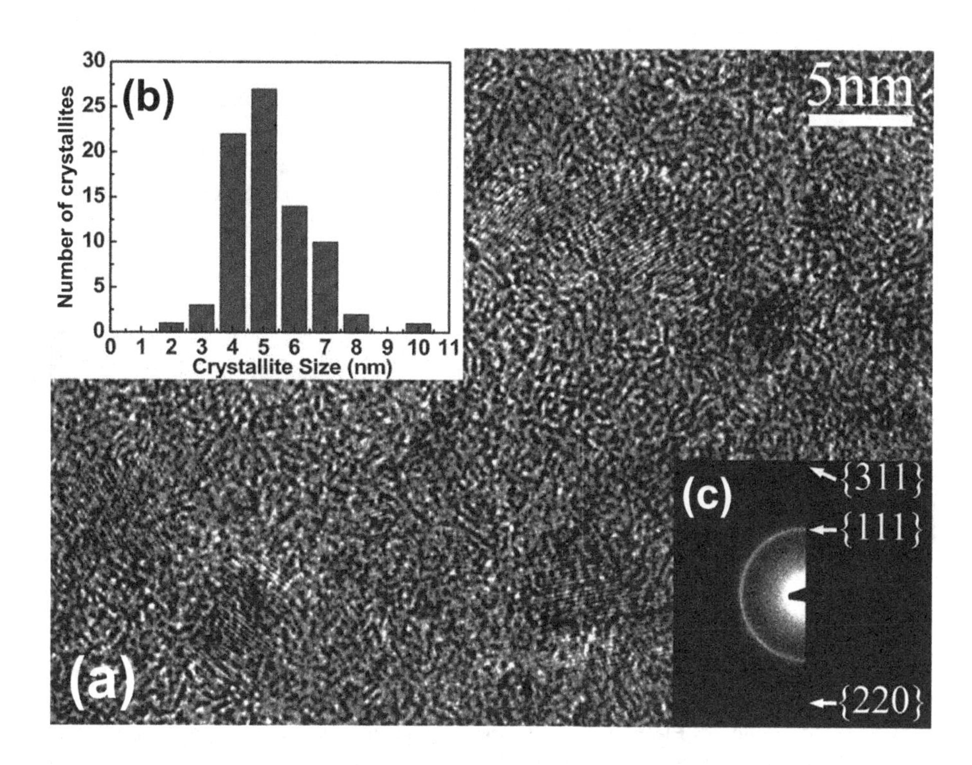

Figure 1.10 (a) HRTEM image of BEN-BEG grown UNCD film shown in Figure 1.9. (b) Statistical number of grains vs. grain size measured on several areas of the UNCD film. (c) Electron diffraction pattern showing the presence of the diamond structure (reprinted from *App. Phys. Lett.*, vol. 92, 133113, 2008 (Fig. 2) in [81] with permission from AIP Publisher).

dental implants in the industrial MPCVD systems shown in Figure 1.3a, using the BEN-BEG process, which provides the means for massive simultaneous coating of many dental implants at low cost because of the elimination of the time-consuming and more expensive chemical seeding process.

Confirmation of the interpretation of the nature of the materials in the areas revealed in the HRTEM images of Figure 1.11a was obtained through two types of systematic studies [83]:

1. Localized electron diffraction in each particular region:

   Figure 1.11b: electron diffraction on Si, showing the Si diffraction peaks;
   Figure 1.11c: electron diffraction on UNCD film, showing the diamond diffraction peaks;
   Figure 1.11d: electron diffraction on the graphite (G) area between Si pyramidal structures, showing the graphite peaks.

2. Systematic studies were performed to confirm the electron diffraction information, using HRTEM *in situ* electron energy loss spectroscopy (EELS) [83]. This technique enables performing EELS spectroscopy along the length extension through the materials to obtain information on atomic chemical bonds along the

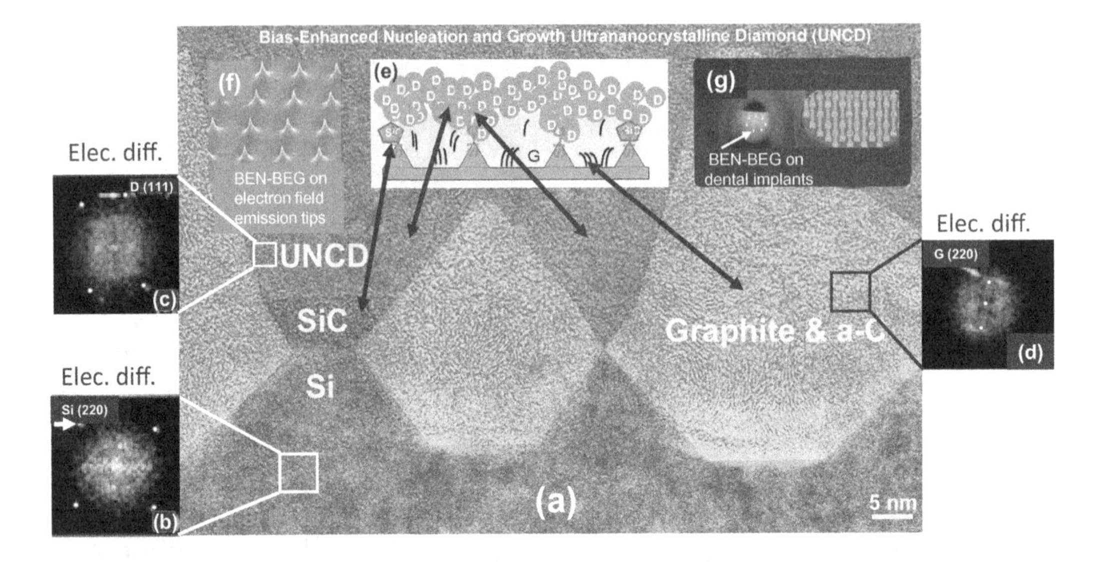

Figure 1.11 (a) Cross-section HRTEM image of the BEN-BEG grown UNCD film, where plain/cross-section SEM and plain HRTEM images are shown in Figures 1.9 and 1.10, respectively. (b) Electron diffraction on the Si substrate, showing the diffraction points characteristic of Si. (c) Electron diffraction on the UNCD film, showing the diffraction points characteristic of diamond. (d) Electron diffraction on the graphitic structured film area, showing the diffraction points characteristic of graphite [79, 80]. (e) Schematic representing the mechanism of BEN-BEG growth for UNCD films. (f) SEM image of UNCD films grown by the BEN-BEG process on sharp Si tips for EFE devices. (g) Optical picture (left) of UNCD films grown simultaneously on several Ti-alloy-based dental implants, positioned vertically on the substrate holder of the industrial MPCVD system shown in Figure 1.3, and schematic of holder under design to hold up to 300 dental implants for industrial low-cost coating by BEN-BEG biocompatible UNCD coating [81, 83] (reprinted from *J. App. Phys.*, vol. 105, 034311, 2009 (Fig. 2 (b)) in [83] with permission from AIP Publisher).

pathway. Initially, general EELS spectra were obtained from the UNCD films grown under different processing condition, such as the BEN-BEG process and UNCD film growth without any bias (see Figure 1.12a). The EELS spectra revealed the two main peaks relevant to characterizing the chemical bonding from the Si substrate to the UNCD film, as shown in the cross-section HRTEM image of Figure 1.11a and 1.12a. The analysis of the EELS spectra in Figure 1.12b revealed two main peaks: (1) a peak at 285 eV, characteristic of the so-called π^* peak, corresponding to the C atoms sp^2 bonds in graphite; and (2) a σ^* peak at 290 eV, characteristic of C atoms sp^3 bonds in diamond. Not shown in the spectra of Figure 1.12b are the Si K edge at 100 eV for Si–Si bonds, and the Si K edge at 103 eV, confirming the existence of the Si–C bonds in the SiC area on top of the Si pyramidal structures, and the Si–Si bond of the Si substrate [83].

Figure 1.12c shows the EELS signal evolution (for the π^* (G) and σ^* (D) peaks, the Si–Si bond signal from Si, and the Si–C signal from SiC) vs. distance along a pathway going from the Si substrate through a pyramidal structure, the SiC layer, and the

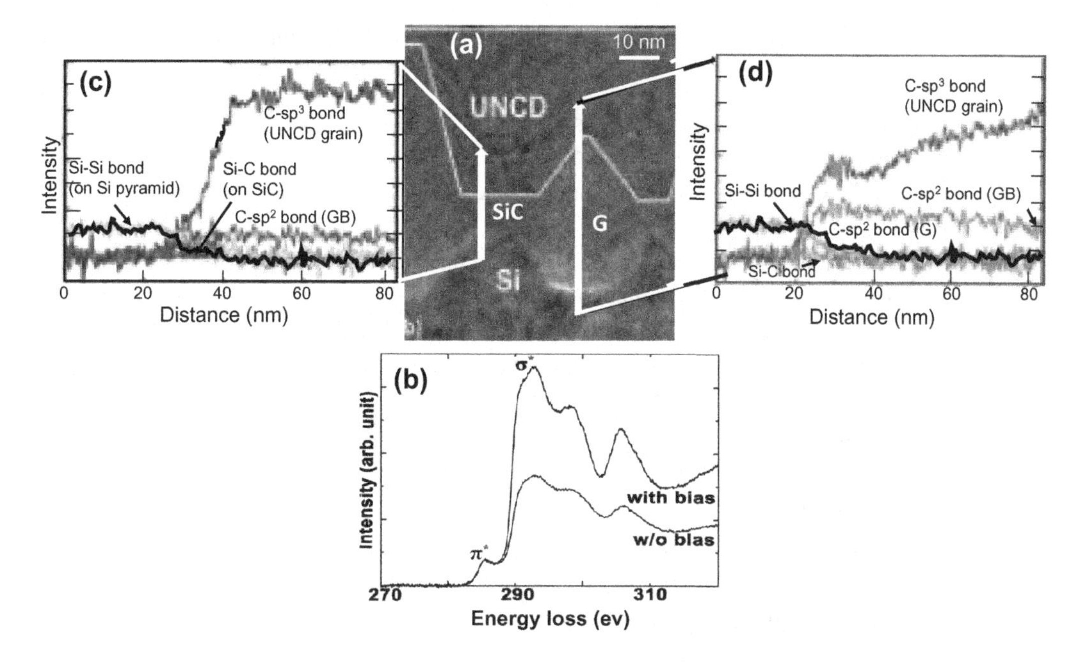

Figure 1.12 (a) Cross-section HRTEM image of the BEN-BEG UNCD film grown on an Si substrate. (b) EELS spectra for a BEN-BEG UNCD film and a UNCD film grown without any bias, showing the key π^* peak (corresponding to the sp^2-C atom bonds in UNCD grain boundaries [GB]), and the other key σ^* peak (corresponding to the sp^3-C atom bonds in the diamond grains of the UNCD film). (c) EELS signal evolution (for the π^* (G) and σ^* [D] peaks, the Si–Si bond signal from Si, and the Si–C signal from SiC) vs. distance along a pathway going from the Si substrate through a pyramidal structure, the SiC layer, and the UNCD film, grown on top of the Si pyramid. (d) The EELS signal evolution (for the π^* [G] and σ^* [D] peaks, the Si–Si bond signal from Si, and the Si–C signal that would correspond to SiC) vs. distance along a pathway going from the Si substrate through the graphitic area between the Si pyramids, and into the UNCD film.

UNCD film, grown on top of the Si pyramid. Four EELS signals evolutions are clear, namely:

- the Si–Si signal goes to zero as the EELS signal goes from the Si to the SiC area;
- the Si–C signal shows a maximum between ~30 and 35 nm, corresponding to the 5 nm thick SiC area on top of the Si pyramid, as shown in Figure 1.12a;
- the signal corresponding to the π^* peak, characteristic of the sp^2-C atoms bonding, increases to a small steady value when the EELS signal reaches the UNCD film (this corresponds to the sp^2-C atoms bonding in the UNCD grain boundaries); and
- the signal corresponding to the σ^* peak, characteristic of the sp^3-C atoms bonding corresponding to diamond, increases to the maximum steady value of all signals, providing clear information about the UNCD film.

Figure 1.12d shows the EELS signal evolution (for the π^* (G) and σ^* (D) peaks, the Si–Si bond signal from Si, and the Si–C signal that would correspond to SiC) vs. distance along a pathway going from the Si substrate through the graphitic area between the Si pyramids, and into the UNCD film. Four EELS signal evolutions are clear:

- the Si–Si signal goes to zero as the EELS signal goes from the Si to the graphitic area;
- the Si–C signal is zero continuously since there is no SiC area, as shown in Figure 1.12a;
- the signal corresponding to the π^* peak, characteristic of the sp^2-C atoms bonding, increases to a relatively medium value compared to the σ^* peak signal of UNCD, and then decreases to a value comparable to that observed in Figure 1.12c, characteristic of the amount of the sp^2-C atoms bonding in the UNCD grain boundaries; and
- the signal corresponding to the σ^* peak, characteristic of the sp^3-C atoms bonding, reaches a similar value as that observed in Figure 1.12c, corresponding to reaching the UNCD film.

Another key characterization of the chemical bonds of BEN-BEG UNCD films was performed using the NEXAFS analysis previously used to characterize UNCD films grown using the chemical seeding process, as described in Figure 1.6. NEXAFS analysis of the UNCD film, for which SEM, HRTEM, and EELS analysis were presented from Figures 1.9–1.12, is presented in Figure 1.13.

Subsequently to the initial studies described above on BEN-BEG of UNCD films, other work was performed by an independent group investigating the growth of

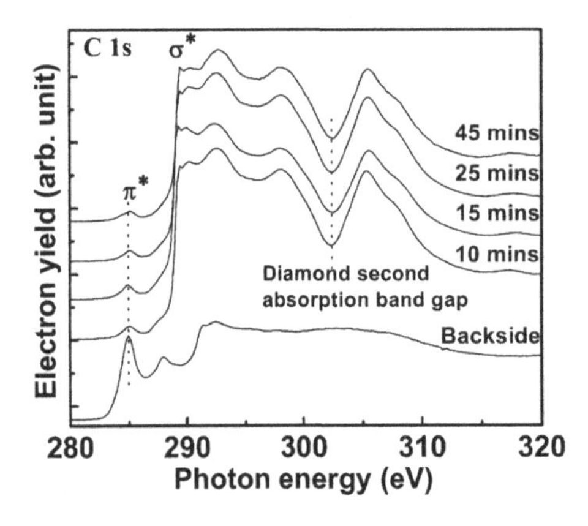

Figure 1.13 NEXAFS spectra from analysis of an MPCVD BEN-BEG UNCD film grown on a high-conductivity Si wafer. The NEXAFS analysis from the UNCD films grown from 10 to 46 min show identical spectra, indicating that the film exhibits practically the same σ^* C atoms chemical bonds intensity, characteristic of the sp^{3-}C atoms bond in diamond, and a very small π peak, characteristic of the sp^2-C atoms bond in the UNCD grain boundaries. In addition, the NEXAFS spectrum obtained on the side of the UNCD film that was at the interface with the Si substrate shows a larger π peak, in agreement with the EELS analysis presented in Figure 1.12d, which shows a graphitic structure in contact with the Si substrate surface where the UNCD film was grown by BEN-BEG (reprinted from *App. Phys. Lett.*, vol. 92, 133113, 2008 (Fig. 1) in [81] with permission from AIP Publisher).

UNCD films by the BEN-BEG process to produce UNCD films with enhanced EFE [84]. This work involved using the MPCVD process to do BEN on Si substrates bias at approximately –400 V for ≥60 min, which was required to produce an appropriate granular structure of the diamond films to enhance the EFE properties. The specific process used in the work cited in [84] included BEG under –400 V substrate bias for 60 min after a BEN step for 10 min. HRTEM studies [84] confirmed the initial R&D described above, which showed the formation of SiC–diamond nucleation structures on top of the ion bombardment Si structured surface (see Figures 1.11a and 1.12a and [81, 83]). The EFE turn-on voltage was 3.6 V/μm, which although low was still higher than the turn-on voltage achieved with N-UNCD films, as discussed in Section 1.2.1.4. HRTEM studies revealed that the main factor enhancing the EFE properties of BEN-BEG UNCD films may be the induction of nano-graphite filaments along the thickness of the films, which may facilitate the transport of electrons [84].

A question to be considered is why it is relevant to develop an optimized BEN-BEG process to grow not only UNCD, but also MCD and NCD films on large-area substrates (≥150 mm in diameter). The answer relates to key issues related to diamond films-based products: (1) Given the present state of safety considerations related to research in academic, national laboratories and industrial environments, eliminating the need for seeding with nanoparticles would greatly reduce research laboratory and industrial safety issues; (2) having a whole plasma-based seeding process without having to go through mechanical or ultrasonication seeding processes can reduce costs for industrial processes, including elimination of the laboratory space and systems required for the mechanical or wet type of seeding; and (3) for future fabrication of UNCD- or NCD-based MEMS/NEMS devices, in the necessary clean-room environment, eliminating the need for nanoparticle-based seeding may be a necessity, since nanoparticle solution-based seeding will complicate the implementation of this process in industrial clean rooms. However, it is important to note that the BEN process works on the substrates with moderate (semiconductor) and high electrical conductivity only.

MCD and NCD [76–78, 84] and UNCD [79– 81, 83] films grown using the BEN+ growth without bias or the BEN-BEG process demonstrated that the BEN process has several advantages over mechanical polishing or ultrasonic seeding processes: (1) comparable or better seeding efficiency [78–81, 83, 84]; (2) stronger adhesion to substrates [78–81, 83, 84]; and (3) an integrated fully dry nucleation/growth process using plasma processing only [76–81, 83, 84].

1.2.1.4 MPCVD Growth and Structure and Chemical Characterization of Electrically Conductive UNCD Films with Boron Atoms Replacing C Atoms in the UNCD Diamond Grains Forming B-UNCD Films or Nitrogen Atoms Inserted in Grain Boundaries Forming N-UNCD Films

Background on High-Tech and Bio-related Applications of Electrically Conductive B-UNCD and N-UNCD Films

Relevant biomedical applications of electrically conductive UNCD films include: (1) biosensors [85], (2) coatings of electrodes for a new generation of Li-ion batteries (LIBs) with ≥10 times longer life and which are safer than current LIBs [48, 49] to enable the next generation of defibrillators/pacemakers (D/P) with longer life than current D/P, which need to be replaced in 6–8 years due to LIB degradation [86]; and (3) electrodes for neural stimulation [87–89]. For these bio-related technological applications, diamond has many superior material properties compared to metals and semiconductor materials used today, which exhibit chemical corrosion by body fluids and several biocompatibility problems [59]. However, a key issue in applying other diamond, and specifically UNCD, films for electronic devices and electrodes is to achieve good functional electrical conduction properties via doping with appropriate atoms, as described here.

Doping of UNCD Films with Boron (B) Atoms to Produce B-UNCD Films
One type of doping, which has been explored extensively, involves insertion of boron (B) atoms, replacing C atoms in the diamond lattice of the UNCD grains and also in other PCD films [37]. B atoms doping of UNCD films should be compared with B atoms doping of homoepitaxial diamond films [90], which showed that B atoms inserted in the crystal lattice of diamond, with concentration of up to $8 \times 10^{20}/cm^{-3}$, results in a large expansion of the diamond lattice (0.16%), especially above the semiconductor–metal transition, due to the sum of a geometrical component from the larger atomic size of boron compared with carbon. In this respect, and directly relevant to this book, doping of UNCD films with B atoms was explored [31, 32]. The main approach used was to grow UNCD films on an ultrasonically seeded Si substrate surface, followed by MPCVD film growth process using a plasma produced with a CH_4/H_2 gas mixture and including trimethylborate ($B(OCH_3)_3$) (TMB) to insert the B atoms in the growing UNCD films, such that after achieving a concentration of $\sim 10^{20}cm^{-3}$ B atoms in the UNCD grains lattice, the diamond material undergoes an insulator–metal transition, producing the so-called "p-type" doping, inducing electrical conductivity [31, 32]. The results reported in [32] indicated that for a given concentration of B in the gas mixture, {111} planes in the crystalline grains of the UNCD films take up more boron into the crystal lattice than {110} planes, which was initially interpreted as that B-doping may facilitate the preferential growing of (111) diamond planes. However, XRD analysis showed that the (220) peak was always more intense than the (111) peak, suggesting that for all UNCD films grown, using the conditions described above, the preferred growth direction may be along the (110) planes, which may grow faster than the other planes, inducing structures elongated along the (110) axis, forming columnar structures [32]. In addition, the data reported in [32] showed that B-doped UNCD films, produced by using 250 ppm B in the gas mixture, exhibited the largest XRD I(220)/I(111) intensity peaks ratio, correlating with the observation that the UNCD films exhibited preferential {110} facets. A relevant problem with producing B-doped UNCD using the MPCVD process described above is that the MPCVD system gets contaminated permanently and then that system can grow only B-doped UNCD or other PCD films, resulting in a constrained cost of hundreds of thousands of US dollars.

Alternatively, recent R&D work demonstrated a new process for B-doping of UNCD films, performed in a dedicated RTA system after growing UNCD films

(~630 nm thick) on SiO_2 (~300 nm thick)/Si substrates using the MPCVD process in a dedicated system, as described above [38]. Subsequently, the UNCD films were coated with a spin-on-dopant (SOD) B-containing film (~ 200 nm thick), dispersed on the UNCD films' surface via a spinning speed of 3000 rpm. Subsequently, the UNCD films with the SOD films on top were baked in an RTA atmospheric oven at 200 °C for 20 min in order to evaporate any excess solvent from the SOD film. The diffusion process was carried out on three UNCD films at different temperatures (e.g., 800 °C, 900 °C, and 1000 °C) to explore the diffusion depth of B atoms inside the UNCD films, and their effect on the film electrical conductivity. Figure 1.14 shows the characterization of the B-doped UNCD films using complementary techniques to determine key parameters of the B-doped UNCD film performance. Figure 1.14a shows a high-resolution cross-section SEM image of a UNCD film grown on a SiO_2/Si substrate and with the B-based SOD layer on the surface. Figure 1.14b shows the key chemical analysis of B-doped UNCD films produce by RTA at the three temperatures described above. The Raman spectra show the characteristic peaks observed in many UNCD films [22, 23], namely: 1550 cm^{-1} G band attributed to sp^2-C atom bonds [22, 23, 70]; the 1350 cm^{-1} D band attributed to disorder-induced sp^2-C atom bonds [22, 23, 70]; and the 1450 and 1130 cm^{-1} bands both attributed to C=C and C–H vibration modes in transpolyacethylene (TPA) molecules, respectively [41, 74]. All Raman bands described are due to chemical bonds of C atoms in the grain boundary network of the UNCD films. A tiny peak at 1332 cm^{-1}, characteristic of sp^3-C atom bonding in diamond, can also be observed for the film annealed at 1000 °C, indicating that for this annealing temperature the grain size has increased to a value corresponding to the low NCD grain size range. The Raman band at 1130 cm^{-1} is reduced significantly after the UNCD film B-doping at 900 °C, and it almost disappears for the B-doping at 1000 °C. On the other hand, the Raman band at 1450 cm^{-1} is halfway and significantly reduced for the UNCD films annealed at 900 and 1000 °C, respectively, during the B-doping process.

The changes observed in the Raman spectra shown in Figure 1.14b are correlated mainly with changes in the grain boundary structure and chemistry of the UNCD films, mostly due to loss of TPA molecules and enhancement of sp^2-C atom bonds induced by the B atoms' diffusion during annealing at temperatures ≥900 °C [38, 91]. The changes observed in the Raman spectra correlate with desorption of H atoms from grain boundaries and surfaces, as seen also for many polycrystalline and crystalline diamond films annealed at temperatures >800 °C [38, 92].

The concentrations of B atoms in the subsurface region of UNCD films, measured by secondary-ion mass spectrometry (SIMS), were between 0.6 Å (~$10^{21}/cm^3$) and 2 Å (~$10^{21}/cm^3$) for UNCD films annealed between 800 and 1000 °C (see Figure 1.14c and details in [38]). On the other hand, the diffusivities of B atoms inside the UNCD films, obtained from calculations correlated with B atoms' depth profile concentration, measured by SIMS, were determined to be ~10^{-15} cm^2/s, between 1 and 3 Å, for the first 20 nm of diffusion depth (see figure 3b in [38]), and 10^{-15} cm^2/s, between 8 and 12 Å (see figure 3b in [38]) for films annealed at 800–1000 °C [38], to induce B-doping. The explanation for the increased B atoms'

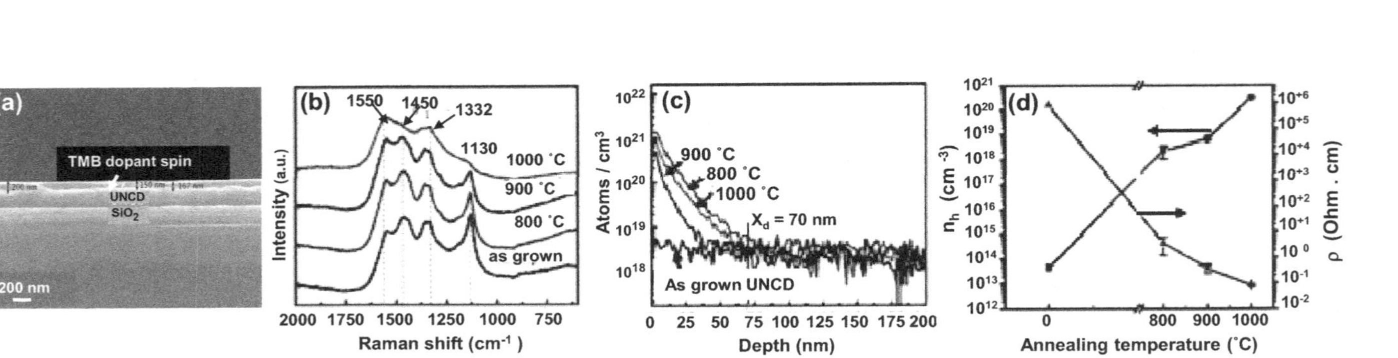

Figure 1.14 (a) Cross-section SEM image of an MPCVD-grown UNCD film on a SiO_2/Si substrate. (b) Raman spectra of B-doped UNCD films annealed at three different temperatures. (c) B atoms concentration vs. depth under the surface of B-UNCD films produced with three different annealing temperatures. (d) Data from Hall measurements of B atoms concentration and resistivity vs. annealing temperature used to produce the B-UNCD films [38] (reprinted from *MRS Commun.*, vol. 8 (3), p. 1111, 2018 (Fig. 1.14a – reprint of Fig, 1(a) in [38]; Fig. 1.14b – reprint of Fig. 2 (a) in [38]; Fig. 1.14c – reprint of Fig. 3 (a) in [38]; Fig. 1.14d – reprint of Fig. 5 (a) in [38], all with permission from Cambridge Publisher).

diffusivities in UNCD films depths $\geq$20 nm relates to the increased amount of grain boundaries in those depth ranges and correlates with prior measurements of B atoms' diffusivities ~10^{-16} cm^2/s in 2 Å under a single crystal diamond surface annealed to 1200 °C [41] vs. B atoms diffusivity of ~10^{-15} cm^2/s in 6 Å underneath MCD (10 μm grains) film surfaces annealed at 770 °C [91].

The properties of B-UNCD films revealed by Figure 1.14a–c correlate with and support the data shown in Figure 1.14d, which shows two key electrical properties of the B-UNCD films (carrier concentration and resistivity), obtained through Hall measurements [38]. The Hall measurements show that the carrier concentration in the B-UNCD films increased by several orders of magnitude as the diffusion temperature increased, correlated with decrease in resistivity by the same orders of magnitude, with carrier concentrations of ~4×10^{13} cm^{-3}, 2×10^{18} cm^{-3}, 3×10^{20} cm^{-3} for the as-deposited and doped B-UNCD films annealed at 800, 900, and 1000 °C, respectively.

A key issue to be explored in relation to the use of electrically conductive B-UNCD or other B-doped PCD films or bulk, for implantable medical devices, is whether they are biocompatible. In this respect, initial work has shown that PCDs with B atoms inserted in the lattice exhibit good biocompatibility when implanted in guinea pigs [91]. On the other hand, studies of boron nitride (BN) nanosheets of different sizes and BN nanoparticles in osteoblast-like cells ($SaOS_2$) showed that the biocompatibility of BN nanomaterials depends on their size, shape, structure, and surface chemical properties [93, 94]. Electron spin resonance measurement revealed that unsaturated B atoms located at the nanosheet edges or on the particle surface are responsible for cell death [95, 96]. Therefore, further work is necessary to determine whether B atoms on the surface of B-UNCD, B-MCD, B-NCD, and B-SCD films or bulks may also induce cell death when electrodes or devices with B atoms on the surface are implanted in the human bio-environment.

Background on Research to Produce n-Type Electrically Conductive Diamond Films for Biomedical Applications

Research focused on producing crystalline diamond or diamond-like films in electronic devices, requiring doping to achieve n-type conductivity ("electrons" as electrical carriers), have been largely unsuccessful [97–100]. Although some of the prior work [100] demonstrated that n-type doping could produce shallow donor levels close to the conduction band of diamond, the room-temperature conductivities are still too low for the application of these materials in conventional electronic devices. A significant number of studies has been performed to investigate dopant atom candidates to produce a suitable n-type diamond. The work described in the previous section revealed that B atoms can replace C atoms in the diamond lattice, acting as a p-type dopant, in a relatively straightforward process, such that the energetic state for the B atom is 0.38 eV above the energy valance band edge of diamond. On the other hand, n-type doping of diamond is more difficult. Ideal dopants for producing n-type diamond films are nitrogen (N) atoms. However, extensive research has showed that nitrogen atoms incorporate in substitutional sites in diamond, forming a deep donor level with an activation energy of 1.7 eV, which inhibits achieving suitable electrical conduction at room temperature [101] via the conventional semiconductor-type doping where the dopant atoms replace native atoms in the lattice and provide electrons to the energy conduction band. It has been reported that lithium (Li) ion implantation can successfully transform intrinsic diamond into an n-type semiconductor [102]. In addition, phosphor atoms have been explored as dopants to produce n-type diamond [103]. *P* is a benign element making up $\sim 1\%$ of the total body weight as a bone constituent in the human body [104, 105]. Recent experiments have shown that black phosphor (BP) nanosheets, especially ones with a small thickness and size, have high reactivity with oxygen and water [106] and can degrade in aqueous media. Moreover, the final degradation products are nontoxic phosphate and phosphonate [107], both of which exist in and are well tolerated by the human body [105]. Therefore, ultrasmall BPQDs with good photothermal performance and biocompatibility are potential therapeutic agents. However, their actual clinical application *in vivo* still suffers from rapid renal excretion and degradation of the optical properties during circulation in the body. In addition, some published work raises concerns about the use of P inserted in the human body [108]. So, further research is needed to confirm full sustainable biocompatibility of P atoms when integrated with other materials.

In any case, because the main technique that has been explored to produce n-type diamond is energetic ion implantation [109], the high dose used for ion implantation, such as P or Li, in diamond causes an irrecoverable graphitization of the diamond crystals. In addition, Li ions are not good for biocompatibility, which is why LIBs in defibrillator pacemakers are encapsulated in sealed, relatively biocompatible metal cases made of Ti alloys. The damage density threshold, beyond which graphitization occurs upon annealing, is found to be 10^{22} vacancies/cm^3. This value is checked against published data and is shown to be of a general nature, independent of ion species or implantation energy. The ion bombardment-induced graphitization of diamond indicates that the high conductivity was due mainly to the graphitized diamond and not to true semiconducting-type doping [102], which is a serious problem for practical applications [103] requiring the multifunctional properties of diamond for the electronic devices. The information presented above indicates that ions that have been explored to produce n-type diamond (e.g., P, Li) may not be fully suitable for producing good biocompatible material for implantable electrically conductive medical devices and electrodes for neural stimulation. That is the reason why producing high electrical conductivity N-UNCD films may provide the pathway to producing implantable diamond-based electronic devices, as described at the beginning of this section.

Growth and Characterization of N-UNCD Films for Implantable Biocompatible Electrically Conductive Electrodes and Other Electronic Devices

Growth of N-UNCD Films via the MPCVD Plus Chemical Seeding Process

The first attempt at producing N-UNCD films was done by Ding/Krauss/Auciello et al. [79], developing N-UNCD films for field emission devices [79]. N atoms grain boundary incorporated UNCD (named N-UNCD) films were grown using the MPCVD BEN-BEG process on high-conductivity Si substrates; this work will be described in the next section.

This section focuses on the R&D performed after the demonstration of the BEN-BEG synthesis of N-UNCD films. The subsequent work focused on using the conventional chemical seeding plus MPCVD growth process. The initial R&D reviewed here is described in detail in [29]. That work showed that when N_2 gas is inserted in the Ar/CH_4 gas mixture, the MPCVD-created plasma contains C_2 dimers and CN radicals due to C–N atoms' chemical reaction in the gas phase, such that the CN content increases substantially as N_2 gas is added to the Ar/CH_4 gas mixture (see figure 1a in [29]). Small additions of N_2 (1–5%) resulted in nanostructure similar to the UNCD films (3–7 nm grains and ~0.4 nm grain boundaries) (Figure 1.15a,b,d,e), while for N_2 (20%) the films transform in what was originally defined as N-UNCD films (Figure 2a in [29], patent in [30]) with grains of 7–16 nm dimensions and grain boundaries with 1–2 nm width (Figure 1.15c,d) [29, 110].

Raman spectroscopy has been used by many groups as the first easy/fast technique to characterize the atomic chemical bonds in the N-UNCD films, as described for the UNCD films in Section 1.2.1.2. However, as for the case of characterization of the

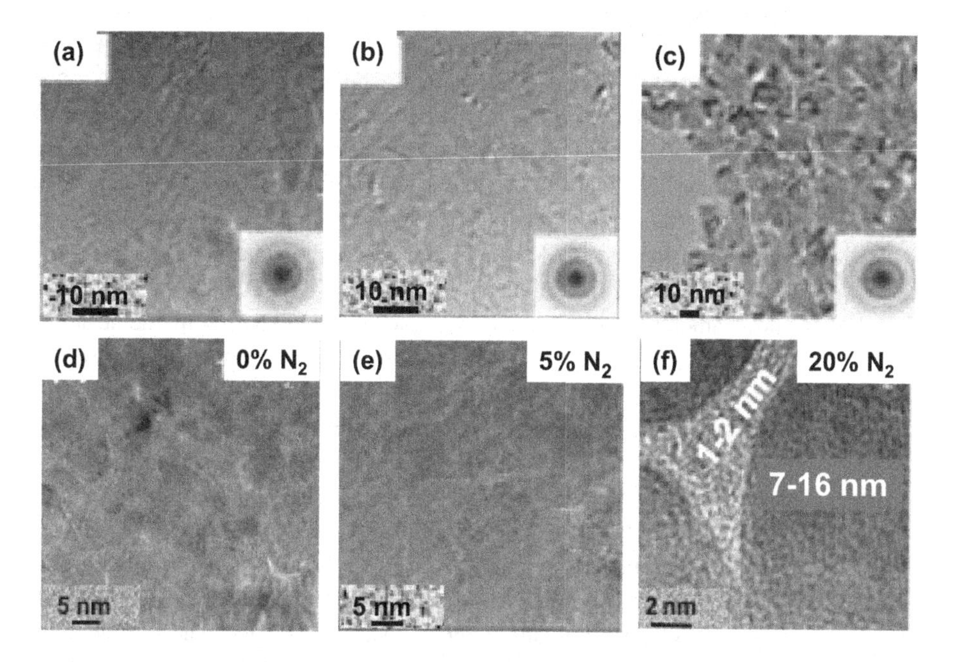

Figure 1.15 HRTEM images of N-UNCD films grown with Ar/CH_4 gas mixture with N_2 (0%) (a) low resolution, (d) high-resolution; N_2 (5%) (b) low resolution, (e) high resolution; N_2 (20%) (c) low resolution, (f) high resolution (this shows the initial characterization of N-UNCD films, revealing the enlargement of grains and grain boundaries with respect to UNCD films revealed in the original first demonstration of MPCVD-grown N-UNCD films) (reprinted from *App. Phys. Lett.*, vol. 81 (12), p. 2235, 2002 (Fig. 1) in [110] with permission from AIP Publisher).

UNCD films, although Raman spectroscopy is sensitive enough to detect the small amounts of nondiamond carbon in largely tetrahedrally C-bonded atoms such as in the UNCD and N-UNCD films, two major drawbacks, already described for the case of UNCD, exist in Raman analysis of N-UNCD films: (1) there is still a large difference in the cross-section for Raman scattering between sp^2- and sp^3-bonded C atoms [111]; and (2) Raman scattering still has a strong dependence on the diamond crystallite size [112] for N-UNCD, with larger grain size (~10 nm) than UNCD (3–5 nm). This dependence is very difficult to quantify and is especially challenging when measuring amorphous or nanocrystalline carbon films, including UNCD. The Raman spectra, obtained with a visible laser beam (He–Ne laser at a wavelength of 632 nm and spot size of 6 μm), from analysis of N-UNCD films grown with N_2 gas contents from 0 to 20% are shown in Figure 1.16a. The spectra reveal the characteristic D band peak at 1340 cm^{-1} and G band peak at 1556 cm^{-1}, both attributed to sp^2-bonded carbon, and a shallow shoulder at 1140 cm^{-1}, attributed to the nanocrystalline diamond structure [70, 111, 112] The Raman spectra show an increase of the G band peak relative to the D band peak as the N atom content in the plasma increases, with a slight shift of the peak to a higher wavenumber, indicating a slight change in the bonding configuration of the sp^2-bonded carbon at the grain boundaries

However, Raman analysis does not provide all the information needed to understand how the chemical bonding in grain boundaries of N-UNCD films produces the high electrical conductivity, as proposed by molecular dynamic simulations [113] that indicated that N atoms bonding to C atoms, with sp^2 bonding available in grain boundaries, plays a critical role in inducing the high electrical conductivity of N-UNCD films. Therefore, small-spot HRTEM-based EELS were performed to obtain information to establish correlations between the bonding structure in the diamond grains and the grain boundaries in a series of films grown with increasing amounts of nitrogen in the plasma [110]. Figure 1.16b shows a representative EELS spectrum with the HRTEM electron beam focused on a UNCD grain, indicating that it is phase pure diamond for all nitrogen doping levels, as evidenced by the near-edge electronic structure of the peak [110]. The well-defined absorption edge at 289.5 eV and the presence of the second band gap feature, revealed at ~302 eV, clearly shows that the diamond grains are sp^3-bonded carbon with a high degree of short-range ordering [110]. This effect does not change as nitrogen is added to the plasma, indicating that the electronic structure within the grains does not change. The grain boundary EELS spectra, shown in Figure 1.16c, reveals that the π^* peak, at 285 eV, increases slightly when N_2 gas increases from 0% to 20% during N-UNCD film growth, inducing N atom insertion into the N-UNCD films' grain boundaries. The σ^* peak at 289 eV, associated with sp^3 bonding in the grains, is unchanged. The relative intensities of the sp^2 vs. sp^3 peaks in the EELS spectra in Figure 1.16c indicate that the amount of sp^2 bonding increases only slightly within the grain boundaries when N_2 flows, thus N^o atoms and N^+ ions increase in the MPCVD plasma process [110].

Another key analysis on the N-UNCD films was done using the atomic scale resolution technique "atom probe analysis" to determine whether N atoms were inserted in the grain boundaries. In this study, N-UNCD films with the hypothesized

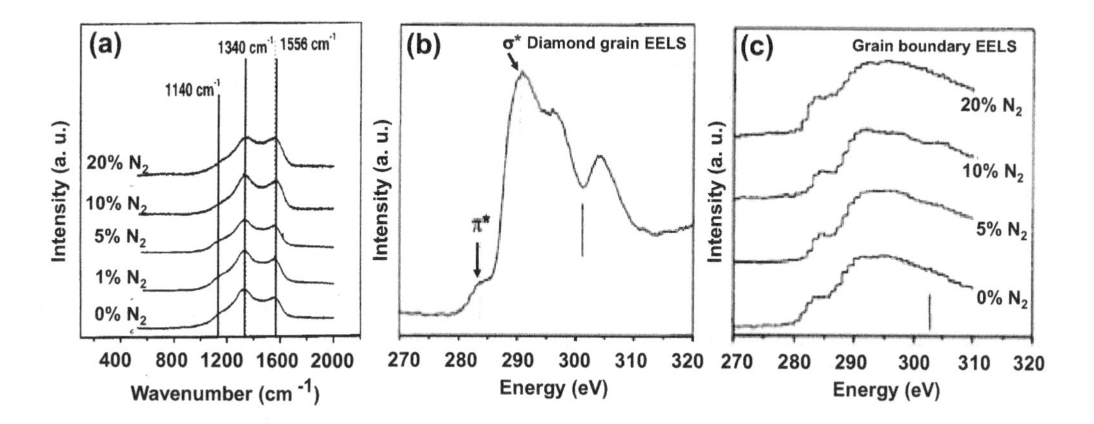

Figure 1.16 (a) Spectra from Raman analysis of N-UNCD films grown with Ar/CH_4 gas flux with N_2 gas added from 0% to 20% for the five different N-UNCD films. (b) HRTEM-EELS analysis within diamond grains of the same N-UNCD films for which Raman spectra are shown in (a). (c) HRTEM-EELS analysis in the grain boundaries of the same N-UNCD films for which Raman spectra are shown in (a) (reprinted from *App. Phys. Lett.*, vol. 81 (12), p. 2235, 2002 (Fig. 2) in [110] with permission from AIP Publisher).

N atoms inserted in the N-UNCD films' grain boundaries, as confirmed later, were grown on atomically sharp tips that were biased with a high positive potential with respect to the ground potential, while picosecond laser pulses, impinging on the specimen's microtip along the axis in a high-vacuum system, triggered electric field-induced evaporation of ions from the surface of the N-UNCD-coated tip, such that field-evaporated ions were accelerated along diverging electric field lines to project a highly magnified image of the microtip's surface onto a position-sensitive time-of-flight detector. The time of flight of the ions was used to identify their mass-to-charge-state ratio and hence their chemical identities (see figures 2, 3, and 4 in [114] for a detailed explanation of the technique). Three-dimensional tomographic atom probe tomography reconstructions of nano-diamonds provided information about the N and C atoms in the grain boundaries of N-UNCD films (see figure 6 in [114], showing maps of single-atom distributions).

The initial theory [113] indicated that when nitrogen is incorporated in the N-UNCD or other PCD films' grain boundaries, new electronic states are produced via electron release when N atoms bond chemically to sp^2 open C atom bonds in the grain boundaries. It is important to note that this mechanism of electronic conductivity change in N-UNCD films is not due to the conventional N atom substitution of C atoms in the diamond crystal lattice of the grains producing electron donor doping, such as in Si. Complementary nanostructure and chemical characterization of the N-UNCD films, performed by independent groups [111, 115] to compare with theory, provided valuable information for understanding the UNCD transformation into N-UNCD films and the electrical conductivity.

In addition to the structural and chemical bonding characterization of N-UNCD films, an important characterization is related to measuring the electrical conductivity

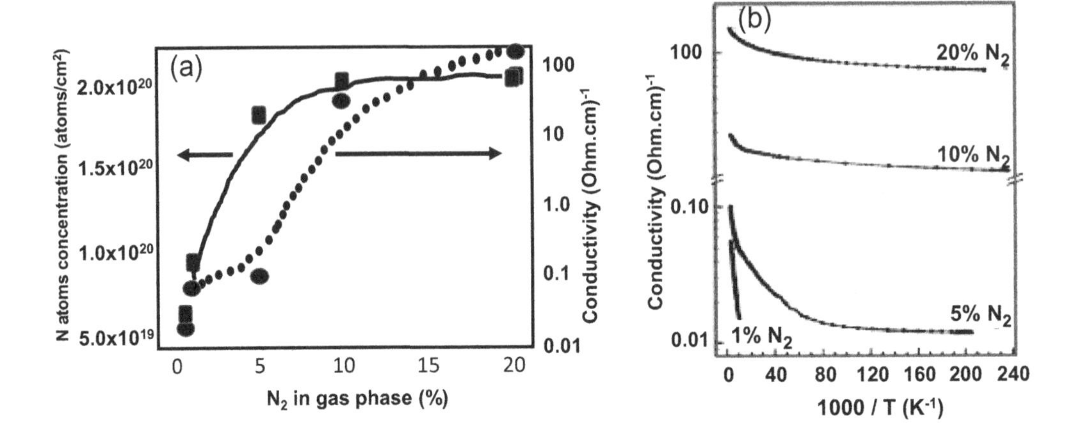

Figure 1.17 (a) Total N atoms content (left axis) and room-temperature conductivity (right axis) as a function of N_2 gas content in the MPCVD plasma. (b) Arrhenius plot of conductivity data obtained in the temperature range 300–4.2 K for a series of films synthesized using different N_2 gas concentrations in the plasma during N-UNCD film growth (reprinted from *App. Phys. Lett.*, vol. 79, p. 1441, 2001 (Fig. 3) in [29] with permission from AIP Publisher).

of the N-UNCD films. In this sense, Figure 1.17a shows SIMS data for the total nitrogen content in the films as a function of the percentage of N_2 gas added to the plasma. Along with these data is a plot of the room-temperature conductivities for the same films (see Figure 1.17b), which reveals the high electrical conductivity of the N-UNCD, critical to developing key implantable biocompatible electrically conductive electrodes and other electronic devices for neural stimulation and other medical applications.

The other key property demonstrated for the N-UNCD films developed by Auciello et al., using chemical seeding plus MPCVD growth [53] and BEN-BEG processes [79], is sustainable EFE, which is critical for a wide variety of applications. Figure 1.18 shows information on the structural form and electron emission properties demonstrated for N-UNCD film-based electron emission devices. Three important properties were demonstrated for the N-UNCD EFE films described in [53]: (1) N-UNCD films exhibit one of the lowest electric fields (~2 V/μm – Figure 1.18e) for emission onset demonstrated today, for both the conventional sharp field emitter tips (Figure 1.18a,b) and (2) flat surfaces (Figure 1.18c,d), the most recent providing a much lower cost for the electron emission manufacturing process than the more complex and costly fabrication of field emitter tips; and (3) the N-UNCD films reported in [53] were the field emitters with the longest stable emission (1000 h) demonstrated today (Figure 1.18f).

The R&D performed on the synthesis and characterization of the properties of N-UNCD films by Auciello's group has been replicated by several groups worldwide, as briefly discussed in the following.

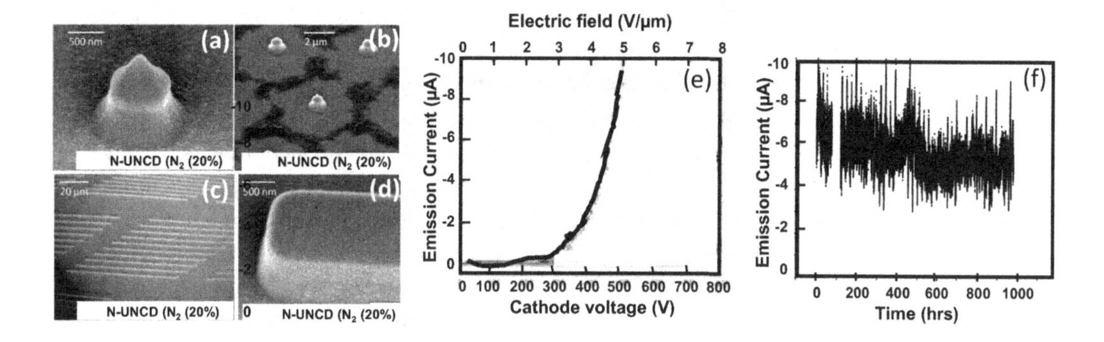

Figure 1.18 SEM images of sharp N-UNCD-coated Si field emitter tips (a) and (b) and planar UNCD-coated field emitter devices (c) and (d). (e) Emission current vs. cathode voltage and electric field emission for N-UNCD field emitter devices. (f) Electron emission current vs. emission time, showing the longest time (1000 h) field emission demonstrated today [53] (reprinted from *Proc. SPIE.* vol. 7679, p. 76791N-1, 2010 in [53] with permission from SPIE Publisher).

Electrical Conductivity

1. The temperature dependence of Hall (electron) mobility measurements on N-UNCD films, together with an enhanced electron density, was used to interpret the unusual magneto-transport features, correlated with delocalized electronic transport in N-UNCD films, which can be described as low-dimensional superlattice structures. N-UNCD films grown with 10–20% N_2 gas in the MPCVD process showed temperature dependences of conductivity as well as magnetoresistance (MR), contributions from weak localization (WL), and also from variable range hopping (VRH) transport (strong localization) to the net conductivity [116].
2. Work by another group provided evidence that WL correction may provide insight into the dominant mechanism for electrical conductivity through grain boundaries in N-UNCD films, at low temperatures ($\ll$300 K), although the details of film microstructures in connection with mechanisms including 2D WL to 3D WL interplay and electron–electron (e–e) interaction effects in both dimensions have not been elucidated yet [117]. This work provided evidence of 3D anisotropic and correlated electronic transport in N-UNCD films through conductivity and MR measurements and at magnetic fields as high as 12 Tesla. This work also provided data on Hall resistance measurements in N-UNCD films not only to establish delocalized transport in these films but also to explain the unusual magneto-transport and weak temperature variation of conductivity of the films in the 3D WL framework.
3. N-UNCD films grown by another group, using the MPCVD process with 20% N_2 gas concentration, revealed relatively higher values than those grown with 10% N_2 films, in agreement with the data shown in Figure 1.18b. The estimated values of N atom concentration in the films of about 10^{19} cm^{-3} can explain the high

conductivity of these films and suggests that in these heavily doped UNCD films transport can reach the high end of the diffusive regime, enabling band-like conduction. Although the number of N atoms in the N-UNCD films decreases as the substrate temperature is reduced, the minimum value of conductivity remains $>1 \times 10^{13}$ cm^{-2}, which is a signature for delocalized conduction, which cannot be explained by the hopping conduction mechanism [118]. The research reported in [118] indicates that the dominant effect of anisotropic WL in three dimensions is associated with a propagative Fermi surface on the conductivity correction in N-UNCD films, and is supported by magnetoresistance studies at low temperatures. Also, low-temperature electrical conductivity shows WL transport in three dimensions, combined with the effect of e–e interactions in the N-UNCD films, which is remarkably different from the conductivity in 2D WL or strong localization regimes. The corresponding dephasing time of electronic wave functions in these systems, described as $\sim T^{-p}$ with $p < 1$, follows a relatively weak temperature dependence on conductivity compared to the generally expected conductivity for bulk metals having $p \geq 1$. The temperature dependence of Hall (electron) mobility together with an enhanced electron density has been used to interpret the unusual magneto-transport features and show delocalized electronic transport in these N-UNCD films, which can be described as low-dimensional superlattice structures.

Electron Field Emission

As in the case of studies of electrical conductivity of N-UNCD films by several groups worldwide, studies on EFE of N-UNCD films were also performed by different groups worldwide, confirming all the initial studies already described. Studies included the following:

1. N-UNCD films were grown on Si substrates using the MPCVD process, but using a liquid source of N atoms in the form of triethylamine (TEA) powder dissolved in methanol spun on the surface of the substrate [119] to synthesize N-UNCD films on silicon substrates. The MPCVD process involved evaporating the N atoms from the surface of the Si substrate via heating to three different temperatures (760 °C, 830 °C, and 890 °C). The process replicated the previously demonstrated N-UNCD nanostructure of electrically conductive nanowires (see figure 3b,c in [119]). The Raman analysis showed the characteristic two well-defined peaks of the D and G bands. The electrical conductivity of the N-UNCD films grown at a substrate temperature of 830 °C reached 1174 S cm^{-1} and exhibited 3.4 V/μm, which is slightly higher than the one shown in Figure 1.18e, but a comparable EFE current density of 8.0 mA cm^{-2} at 6.2 V/μm. Summarizing, the work reported in [119] replicated the prior research already described.
2. R&D by several other groups [120–122] worldwide basically confirmed all the prior research discussed above on the synthesis and properties of MPCVD-grown N-UNCD films.

Growth of N-UNCD Films by MPCVD via the BEN-BEG Process

The first demonstration of the BEN-BEG process for growing N-UNCD films was done by Auciello et al. [79]. N-UNCD films were grown on high-conductivity Si substrates heated to 800 °C using microwave power of 600 W at a total pressure of 11 Torr. One group of films was grown with Si substrates biased at –100 V using a gas mixture of N_2 (1%) /CH_4 (1–20%) / H_2 (98–79%) [79], while a second group of films was grown with a substrate bias in the range –100 to –150 V in a gas mixture of N_2 (1%)/CH_4 (10%)/H_2 (89%) [79]. The field emission performance in terms of threshold field and emission current improved considerably as a function of increasing CH_4 concentration and negative bias voltage [79]. UV Raman analysis revealed that the field emission enhancement resulting from an increase in CH_4 concentration from 1% to 5% correlates with a decrease in the sp^3-C atoms bonding in the N-UNCD films. The dependence of field emission on negative bias voltage appears to be correlated with ion bombardment-induced damage in the film during growth. SEM images of N-UNCD films grown with –150 V bias showed smaller surface topographic features compared to films grown under 0 and –100 V bias. J vs. E_0 measurements across a length of 40 mm over the N-UNCD film showed that films grown with a substrate bias of –150 V exhibited the lowest threshold field for electron emission demonstrated at that time (~2.0 V/μm) [79].

The initial R&D described above [79] opened a new field of R&D to optimize the growth of electrically conductive N-UNCD films for high-tech and biomedical applications, the latter motivated by the fact that N atoms inserted in biocompatible UNCD films provide probably the best biologically suitable atoms in relation to interactions with the human biological environment. The big advantage of using the BEN-BEG technique to grow N-UNCD films, as also discussed for UNCD films in Section 1.2.1.3, is that the time-consuming chemical seeding process is eliminated and the overall cost of growing N-UNCD films is also substantially reduced.

1.3 Fundamentals on the Synthesis of MCD, NCD, and UNCD Thin Films via HFCVD

1.3.1 Background Information on the Need for HFCVD to Grow PCD Films

A key challenge related to diamond film technology is the ability to grow films on large areas with uniform thickness and structure for technological applications and product development at low cost. MPCVD has been used to grow uniform MCD and NCD films at small ($\leq$10 mm diameter) [123] to medium ($\leq$100 mm diameter) [124, 125] sizes (one problem the reader needs to be careful of when reading titles of papers in the literature is that most authors claim they have grown large-area diamond films even when they have grown only on 30 mm diameter [123] or 50–100 mm diameter Si substrates [124–126], which in the current Si microelectronic technologies are medium-size substrates [15, 22]). On the other hand, UNCD films have been grown using large industrial-area MPCVD systems (see Figure 1.3a) with excellent thickness

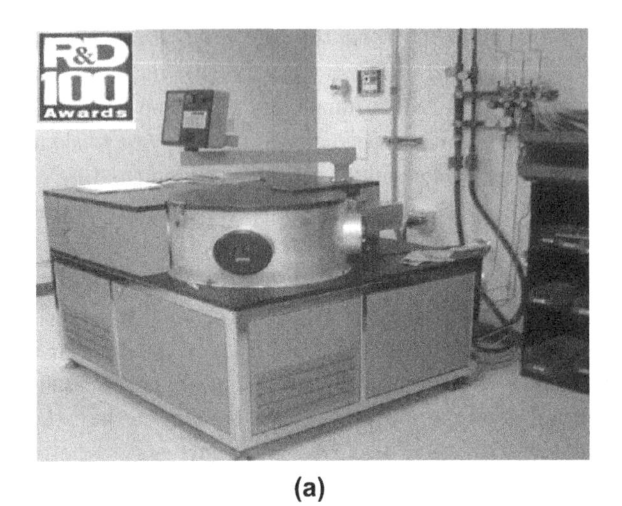

(a)

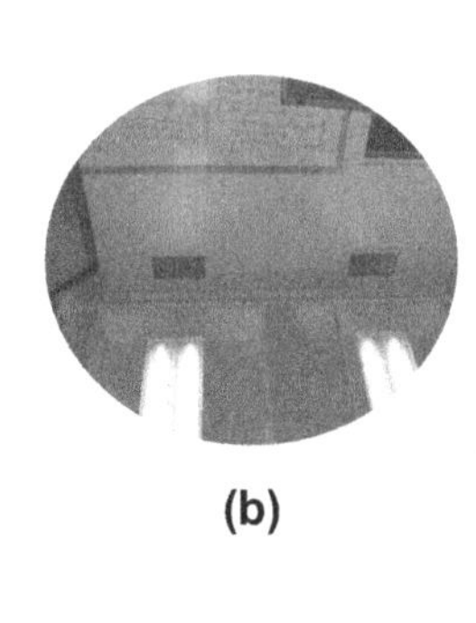

(b)

Figure 1.19 (a) Industrial HFCVD system operating at ADT, Inc. [25], producing uniform large-area UNCD films to cover, simultaneously, large number of mechanical pump seals and large-area Si wafers (b) for marketing to clients interested in developing UNCD-based MEMS technologies [25].

uniformity and nanostructure on relatively large (150–200 mm diameter) areas (see Figure 1.3b, this being the only real large-area UNCD film, grown by MPCVD, demonstrated today) [22].

Alternatively, HFCVD is capable of growing UNCD films on Si substrates with up to 300 mm in diameter and even larger area (~500×500 mm^2), and is being used by Advanced Diamond Technologies, Inc. [25] for commercial growth of UNCD films on large areas to cover large amounts of mechanical pump seals (~20–40 mm diameter) at low cost. Figure 1.19a,b shows an industrial large-area HFCVD system and a 300 mm Si wafer coated with UNCD films with excellent uniformity in thickness and nanostructure across the whole surface area of the wafer.

The HFCVD process is currently the most appropriate industrial growth process to produce large-area UNCD coatings at low cost for commercialization. That is the reason the fundamentals of growing UNCD films by HFCVD are discussed here.

1.3.2 Fundamentals and Properties of UNCD Films Grown by HFCVD

1.3.2.1 Background Information on Prior Research on HFCVD-Grown Diamond Films

The growth of UNCD films using the HFCVD technique has been reported for optimization in small substrates [127]. Most films grown by HFCVD exhibited grain size in the NCD (>10 nm) range [123–134], and no systematic analysis of the uniformity of the nanostructure and film thickness has been reported. Table 1.1 summarizes key publications on HFCVD growing of PCD films.

Table 1.1 Film growth parameters, diamond film structure and characterization techniques for films grown by HFCVD reported in the literature.

Gas mixture (%)	Filament material/ geometry	Fil/sub temp. (°C)	Filament sub distance (mm)	Seeds size (μm)	Gas pressure (Torr)	Diamond grains (μm)	Characterization technique	Reference
H_2 – 98 CH_4 – 2	W(array/ filament)	2150/800 (850)	6	0.5 on Si	20–50	0.5–1	SEM/ Raman	[123]
H_2 – 99 CH_4 –1	Ta (20/ filament)	2320/800 (850)	20	Carbide Rods	2–36	0.3–1	SEM/AFM/ Raman	[124]
H_2 – 99 CH_4 – 1	W (grid filament)	2000/450 (550)	5	20–40	7	0.04–0.07	SEM/AFM/ Raman/ XRD/IR	[125]
$H_2 + CH_4$ 1–100 Ar 0–100	Ta (coil filament)	2400/850 (900)	5	No info	100	0.05	SEM / Raman	[126]
H_2 – 99 CH_4 –1	W (4/filament)	2100/850	8	40	1–36	0.2–0.3	SEM/AFM/ XRD/Raman/	[127]
CH4 – 1 H_2 1–5 Ar-rich	Ru (helix filament)	2100–2400/ 400 (800)	5	1	20–30	0.01–0.05	SEM / Raman	[128]
CH_4 – 1 H_2 – 9 Ar – 90	W (5/filament)	2200/550 (850)	5	0.250	30	0.0045–0.0068	SEM/XRD/ Raman	[129]
CH_4 – 0.25, 0.5, 1, 1.5, 2 H_2 – 9.75, 9.5, 9, 8.5, 8 Ar – 90	W (5/filament)	2200/750	5	0.250	30	Not reported	SEM/XRD/ XPS/Raman	[131]

The information presented in Table 1.1 indicates that prior research did not involve systematic investigation of the effect of the H_2/CH_4 gas chemistry ratio and simultaneous Ar gas flow variation into the HFCVD film growth chamber to tailor the growth of diamond films using the HFCVD technique for large-area applications.

1.3.2.2 Fundamentals on Synthesis and Characterization of PCD Films By Chemical Seeding Plus HFCVD Growth Process

Based on the information presented in Section 1.3.2.1, Auciello's group performed the first series of systematic studies focused on investigating the role of Ar gas in producing HFCVD-based growth of UNCD films (3–5 nm grains) [74, 135] with high uniformity in surface roughness on 100 mm diameter Si substrates. In addition, gas chemistry was tailored to determine the conditions to grow PCD films from MCD (~1 μm) to NCD (≥10 nm grain) for applications in micro/nanoelectronics, as described later in this chapter. The studies described here involved systematic analysis of films using complementary HRTEM supported by XRD and AFM to determine the grain size, film surface chemistry, and surface roughness of the MCD, NCD, and UNCD films, respectively, across 100 mm substrates. Scanning electron microscopy and Raman analysis, and visible (532 nm analyzing laser wavelength) and far UV (244 nm analyzing laser wavelength) were used to characterize the film surface morphologies and C atom chemical bonds.

The HFCVD film growth process involved an array of W filaments ($\phi = 0.25$ mm, length ~14.2 cm) held on a molybdenum (Mo) frame in an optimized geometry, positioned 3 cm above the substrate holder (Figure 1.20a–c). The substrates were located on a metal disk and rotated during film growth to induce film thickness uniformity on up to 100 mm diameter substrates (Figure 1.20c). The filaments were heated to 2200 °C by passing AC current. The gas flows were (see Figure 1.21): (a) 200 sccm H_2, (b) 50 sccm H_2,/50 sccm Ar, (c) 30 sccm H_2/70 sccm Ar, (d) 20 sccm H_2,/80 sccm Ar, (e) 15 sccm H_2/85 sccm Ar, and (f) 10 sccm H_2/90 sccm Ar, with fixed CH_4 gas flow of 2 sccm, and a uniform pressure of 10 Torr in the film growth chamber. The geometry of the HFCVD, and specifically the filament arrays, induces a uniform distribution of C and H atoms, produced via cracking of the H_2 and CH_4 gases impacting on the hot filaments' surfaces (Figure 1.20a).

The growth of PCD films reviewed in this section was done on tungsten (W) template layers grown on SiO_2/Si substrates, since prior research by Auciello et al. demonstrated that the W surface induces the highest nucleation density for UNCD film growth [22, 136, 137]. Figure 1.21a–d shows SEM images of diamond films deposited with different Ar and H_2 gas flow ratios, keeping the CH_4 constant at 2 sccm. MCD films (grain sizes of 1–3 μm) were grown using a mixture of 3 sccm CH_4 and 200 sccm H_2 flow (Figure 1.21a). Increments of Ar concentration in the CH_4/ H_2 gas mixture revealed a remarkable reduction in grain size, from large grain size NCD films (~300 nm grain size) (Figure 1.21b), produced using a gas ratio of 2 sccm CH_4/50 sccm H_4/50 sccm Ar, to NCD films with smaller grain sizes (9–25 nm) (Figure 1.21c), grown with 2 sccm CH_4/30 sccm H_2 /70 sccm Ar. Finally, UNCD films (3–7 nm grain size) were produced with 2 sccm CH_4/10 sccm H_2 /90 sccm Ar

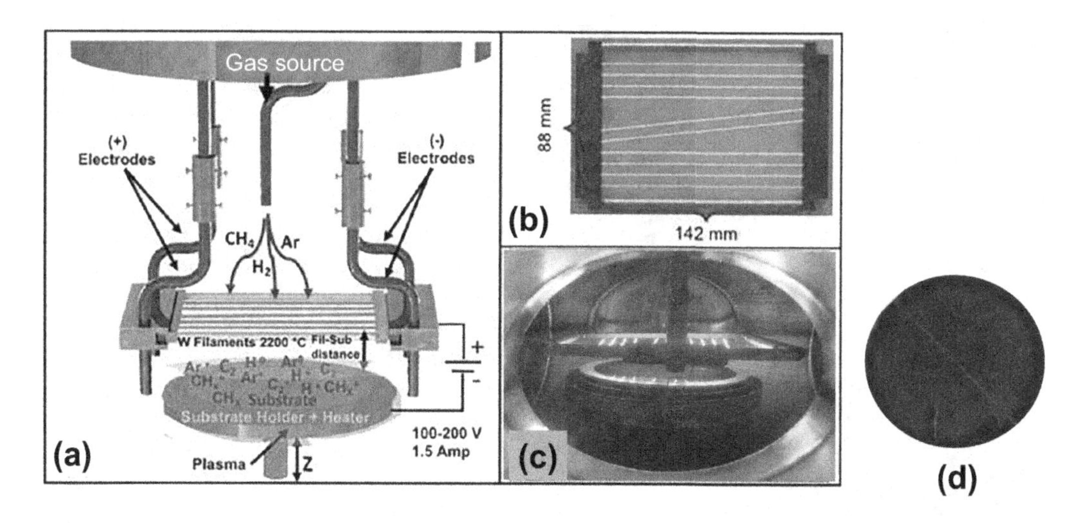

Figure 1.20 Schematic of the HFCVD geometry (a), the optimized filament array (b), and view of the HFCVD system in operation at the University of Texas-Dallas (c), growing uniform UNCD film on 100 mm diameter Si substrates (d) (reprinted from *Diam. Relat. Mater.*, vol. 78, p. 1, 2017 (Fig. 1) in [135] with permission from Elsevier Publisher).

(Figure 1.21d), which shows an electron diffraction spectrum confirming the diamond structure. The AFM analysis provided information on the surface roughness of the films, while XRD analysis provided information on the film grain size [74]. Figure 1.21e shows the variation of film surface roughness and thickness as a function of diamond grain size for the MCD, NCD, and UNCD films (the error bars represent the main distribution across the 100 mm diameter substrate). Figure 1.21e clearly shows that the gas chemistry is key to controlling the structure of PCD films, synthesized by HFCVD from MCD to UNCD [74].

The key technique to precisely measure the grain size of NCD and UNCD films was HRTEM analysis, since the grain sizes are in the very small nanoscale range. This analysis of films grown with 2 sccm CH_4/70 sccm Ar/30 sccm H_2 and 2 sccm CH_4/80 sccm Ar/20 sccm H_2 flow ratios revealed average grain sizes of 17 nm and 18 nm, respectively, classifying both of them in the NCD range (Figures 1.22a,b). On the other hand, films grown with 2 sccm CH_4/85 sccm Ar/15 sccm H_2 had an average grain size of 8 nm (Figure 1.22c), which is in the UNCD upper dimension limit [22]. Films grown with 2 sccm CH_4/90 sccm Ar/10 sccm H_2 flow ratio exhibited a grain size of 3–6 nm (Figure 1.22d), which defines the UNCD structure [21, 22]. Fast Fourier transform (FFT) diffractograms (insets in Figure 1.22a–d) and selected area electron diffraction (SAED) were performed on the NCD and UNCD films. An interplanar distance of 0.19 nm corresponding to a crystalline plane of (220) was observed. The structures of grains and grain boundaries in all films were further characterized using electron diffraction with a Fourier transformed image (see insets in Figures 1.22a–d). The grain size uniformity across 100 mm diameter films was measured by XRD. Diamond films grown with Ar/H_2 gas flow ratios (in sccm) 70/30,

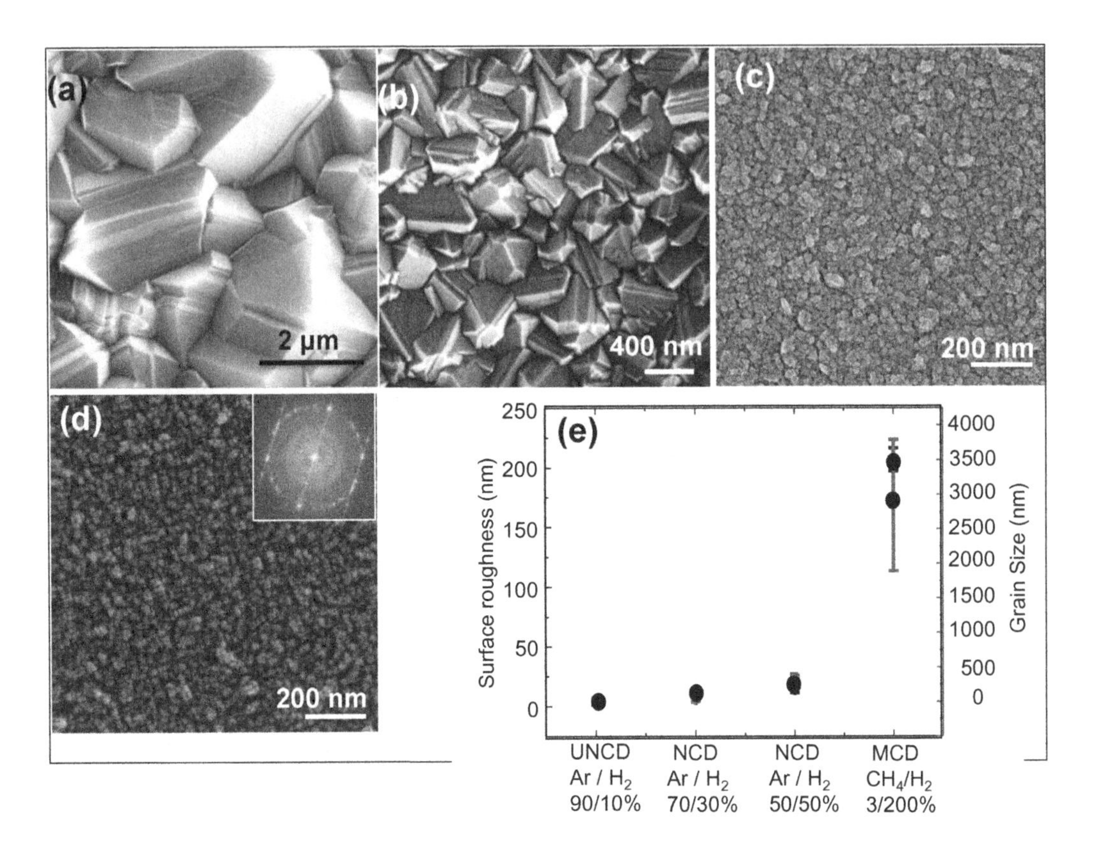

Figure 1.21 SEM images of diamond films grown by HFCVD with different gas flow ratios: (a) MCD film grown with 3 sccm CH_4/200 sccm H_2; (b) NCD film grown with 2 sccm CH_4/50 sccm H_2/50 sccm Ar; (c) NCD film grown with 2 sccm CH_4/30 sccm H_2/70 sccm Ar; and (d) UNCD film grown with 2 sccm CH_4/10 sccm H_2/90 sccm Ar. (e) Plot of the average grain size and surface roughness of the different diamond films measured across the 100 mm diameter Si substrates (reprinted from *Thin Solid Films*, vol. 603, p. 62, 2016 (Fig. 2) in [74] with permission from Elsevier Publisher).

80/20, 85/15, and 90/10, all with 2 sccm CH_4 gas flow, exhibit average grain sizes of 11, 11.8, 8.3, and 5.6 nm, respectively, as shown before in [22]. The XRD grain size was calculated using Scherrer's equation (see plot in Figure 1.22e).

Raman analysis using a 532 nm wavelength laser beam (Figure 1.23, spectra 1–4) was performed to determine the C atom chemical bonding in the HFCVD-grown MCD, NCD, and UNCD films structures. The key differences to be noted in the Raman spectra between the MCD, NCD, and UNCD films are: (1) the 1332 cm^{-1} peak is not observed for the UNCD film due to the D band signal at 1350 cm^{-1}; (2) in general, UNCD films analyzed by visible Raman (532 nm wavelength) are characterized by the appearance of sp^2-C bond peaks at 1550 cm^{-1} (G band) and at 1350 cm^{-1} (D band), such that the D band is super-imposed on the 1332 cm^{-1} peak due to the extremely small nanoscale grain size of the UNCD film with a relatively larger band gap; (3) the peak at 1550 cm^{-1} produced by the sp^2-C atom bond is stronger in UNCD

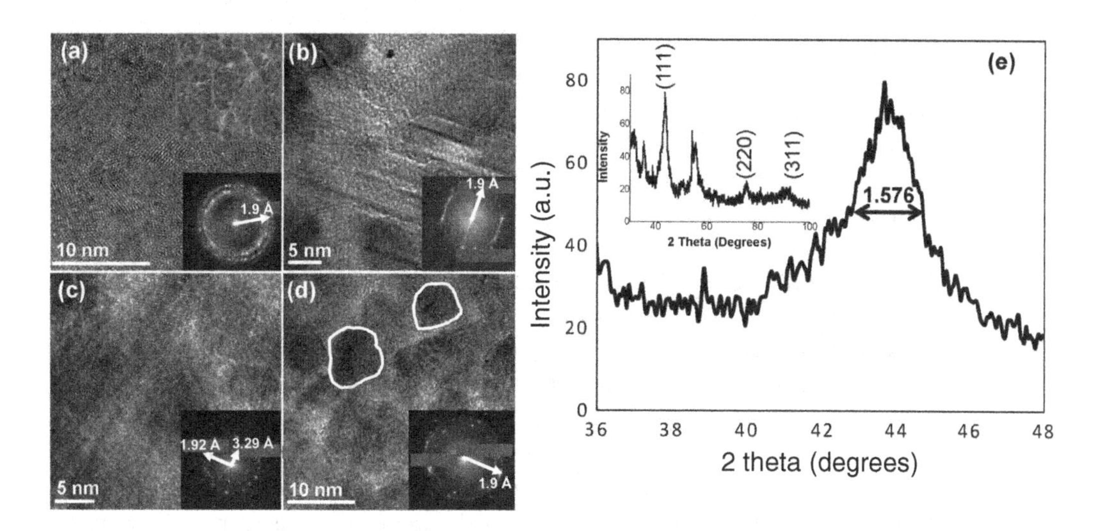

Figure 1.22 Cross-section HRTEM pictures of NCD and UNCD films grown on W-coated SiO_2/Si substrates under different gas flow ratios: (a) 70 sccm Ar/30 sccm H_2 (Fourier transformed images from diffractograms are presented in lower right insets) (NCD/17–18 nm grains); (b) 80 sccm Ar/20 sccm H_2 (NCD/17–18 nm grains); (c) 85 sccm Ar/15 sccm H_2 (UNCD/8 nm grains); and (d) 90 sccm Ar/10 sccm H_2 (UNCD/3–6 nm grains), while keeping the CH_4 gas flow constant at 2 sccm for all cases; FFT patterns as insets. (e) XRD spectra used to determine the UNCD film grain size across the 100 mm diameter substrate (reprinted from *Thin Solid Films*, vol. 603, p. 62, 2016 (Fig. 5) in [74] with permission from Elsevier Publisher).

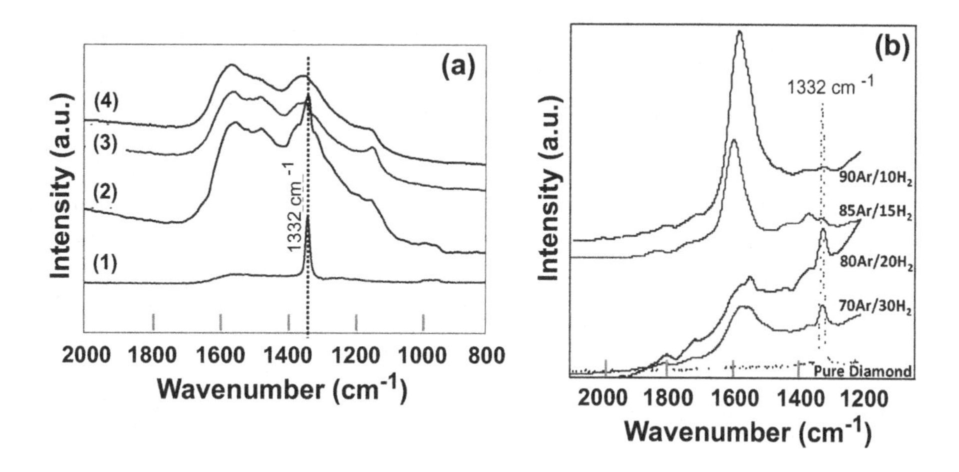

Figure 1.23 (a) Raman spectra from analysis of MCD (1), NCD (>18 nm grains (2) and ~ 8–10 nm grains (3)), and UNCD (4) films using a visible 532 nm wavelength laser beam. (b) Room-temperature Raman spectra obtained using 244 nm wavelength laser of NCD and UNCD films grown with 2 sccm flow of CH_4 gas with each of the Ar/H_2 gas flow combinations shown for each curve (notice that although very small, the diamond peak at 1332 cm^{-1} is still visible for the UNCD film (reprinted from *Thin Solid Films*, vol. 603, p. 62, 2016 (Figs. 3 and 6) in [74] with permission from Elsevier Publisher).

than in MCD and NCD due to the much larger grain boundaries network, concurrently with the substantial reduction in grain size; and (4) the peak observed at 1150 cm^{-1} corresponds to emission from trans-polyacetylene molecules in the grain boundaries.

In summary, the research on HFCVD growth of PCD films described above showed that control of the gas chemistry and flows enables producing PCD films with highly controlled structures from MCD to NCD to UNCD.

1.3.2.3 Fundamentals on Synthesis and Characterization of MCD, NCD, and UNCD Films on HfO_2 Films on Si by Chemical Seeding Plus HFCVD Growth Process

Background on the Importance of Growing MCD to UNCD Films on HfO_2 Films for Biomedical and Other Devices Based on Si

Another critical issue related to growing PCD films on Si substrates is related to the application of Si-based microchips in implantable medical devices. In this respect, Chapter 2 of this book focuses on describing the application of a hermetic/biocompatible/humor eye chemical attack-resistant UNCD coating to encapsulate an Si microchip implantable in the eye as the main component of an artificial retina to restore partial vision to people blinded by retinitis pigmentosa (see Chapter 2 and [28] for details). In relation to a similar topic, this section focuses on describing another critical issue regarding growth of PCD films on a key layer involved in Si-based CMOS devices, the main Si-based micro/nanoelectronics today [138]. The key technological issue addressed recently by Auciello's group was to develop a process to grow MCD, NCD, and UNCD films on HfO_2 layers on Si substrates, since HfO_2 is the main gate oxide used in all CMOS devices on the market today [139]. A key reason for growing MCD films on HfO_2 is that it has been proven that MCD has ~1900 W/K·m thermal transport (the closest to the highest thermal transport of single crystal diamond (~2100 W/K·m) [22], and this property may provide the means to cool down the next generation of micro/nanoelectronic devices, particularly those implanted inside the human body. The reason for growing UNCD films on HfO_2 layers is because it has been proven that UNCD is extremely biocompatible and inert to chemical attack by body fluids [28], thus it may provide a powerful encapsulating coating for Si microchips implantable inside the human body using the HfO_2 film as an insulating template layer grown around the Si microchip to then grow an insulating UNCD film.

Synthesis and Characterization of MCD, NCD, and UNCD Films on HfO_2 Films on Si

Diamond Film Growth. HfO_2 thin films are grown via atomic layer deposition (ALD) using Tetrakis (dimethylamido) hafnium ($Hf(NMe_2)_4$) as the hafnium precursor and H_2O as the oxidation precursor. HfO_2 films of 5, 10, 30, and 100 nm thicknesses were grown on SiO_2/Si substrates heated up to 100 °C, using a nitrogen purge of 20 seconds between each precursor pulse. All substrates with the HfO_2 layers on the surface were subjected to the chemical-based diamond seeding process via immersion of the substrate in a commercial solution of methanol with functionalized diamond particles (to avoid seed particles agglomeration to induce high-density pinhole-free diamond films) in an ultrasound generator, which immersed UNCD (3–5 nm) particles from the seeding solution on the substrate surface via sound wave-induced vibration. Details of the seeding process can be found in [22]. Following the seeding process, the substrates were placed inside the HFCVD chamber on a rotating substrate holder to promote film uniformity. Details of the HFCVD geometry, gas flow, and filament and substrate heating were described in Section 1.3.2.2. Briefly, the substrates were heated to ~800 °C for growing MCD and NCD films and ~650 °C for growing UNCD films. The gas flows used to grow the diamond films were: (1) for MCD and NCD films, CH_4 (3 sccm)/H_2 (200 sccm) for 8 h and 4 h, respectively; and (2) for UNCD films, CH_4 (2 sccm)/H_2 (10 sccm)/Ar (90 sccm) for 1 h, keeping the total chamber pressure at 10 Torr. The details of the HFCVD process to grow MCD to UNCD films can be found in a recent paper [74].

The next step, after the diamond film growth, was to study their structural and chemical composition properties using complementary electron microscopy techniques and Raman and XPS analyses, respectively, as described below.

Characterization of Diamond Films Structure by Electron Microscopy. Figure 1.24 shows SEM images of UNCD, NCD, and MCD films grown on HfO_2 layers on Si, demonstrating that the HfO_2 layer plays a critical role in inducing growth of dense PCD films with all structures.

Figure 1.25 shows cross-section SEM images correlated with the SEM images of the top surface shown in Figure 1.24, such that the cross-section confirms the information provided by the top surface images that UNCD, NCD, and MCD films are grown with extreme dense structures on HfO_2.

The fundamental mechanism for growth of UNCD films on HfO_2 layers was revealed by systematic detailed studies using HRTEM, Raman, and XPS analyses. Figure 1.26 shows the nanostructure of a 140 nm thick UNCD film grown uniformly on a 10 nm thick HfO_2 template layer on an Si surface, as shown in the LRTEM (low-resolution transmission electron microscopy) image of Figure 1.26a. Figure 1.26b shows a plain view HRTEM image of the UNCD film, revealing the characteristic 3–5 nm grain sizes of UNCD, with the diamond crystal structure confirmed in the diffraction pattern of the inset. Figure 1.26c shows an HRTEM cross-section image that reveals the Si substrate and HfO_2 layer, and the HfO_2/SiO_2 and UNCD/HfO_2 interfaces. The HfO_2 layer, amorphous as it is grown by ALD, crystallizes during the film growth, as indicated in the upper left inset diffraction pattern of Figure 1.26c, showing clear spots characteristic of the HfO_2 crystalline structure. The UNCD/HfO_2 interface exhibits a stripe-type structure, highlighted by the rectangle in Figure 1.26c, pointing to the upper right inset, which shows the diffraction patterns of the UNCD/HfO_2 interface. A comparison between the diffraction patterns of pure UNCD (inset in Figure 1.26b) and the UNCD/HfO_2 interface (upper right inset in Figure 1.26c) reveals the appearance of extra peaks for the diffraction at the interface, indicating the presence of a new phase at the interface. This new phase most probably appears during the first stages in the nucleation of the UNCD films, as explained later in the text. The HRTEM-EELS elemental analysis profile, shown in Figure 1.26d (scanning

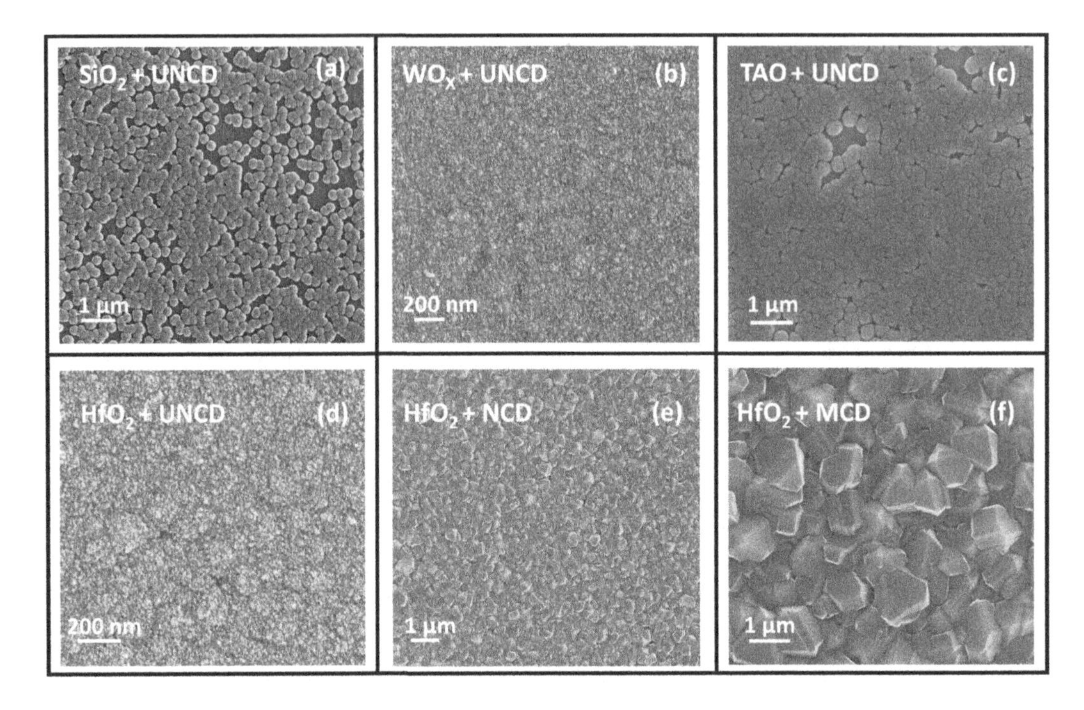

Figure 1.24 SEM images of the top surface of UNCD films grown on SiO_2 (a), WO_x (b), $TiAlO_x$ (c), and HfO_2 (d), showing that the densest UNCD film is grown on a HfO_2 layer. (e,f) Dense NCD and MCD films also grow with high-density structures on HfO_2 layers on Si (reprinted from *Diam. Relat. Mater.*, vol. 69, p. 221, 2016 (Fig. 3) in [41] with permission from Elsevier Publisher).

direction indicated with an arrow across the Si/HfO_2/UNCD layered structure shown in Figure 1.26a), confirms the elemental profile through the whole SiO_2/HfO_2/UNCD heterostructure).

Chemical Analysis of UNCD Films. Characterization of the chemical bonds in UNCD films was done using Raman and XPS analysis, as complementary studies, to confirm the results presented above from electron microscopy. Figure 1.27Aa shows visible Raman spectra from UNCD films grown on HfO_2 layers (with various thicknesses from 5 nm to 100 nm) grown by ALD on SiO_2/Si substrates. The Raman spectra were obtained with a visible 532 nm laser beam wavelength. The spectra show characteristic D band (1341.7 cm^{-1}) and G band (1588.5 cm^{-1}) peaks from the sp^2-C atom bond hybridization, identical for all UNCD films grown on the HfO_2 layers with thickness in the 5–100 nm range (see Figure 1.27Aa). Deconvolution of the Raman peaks shows that there is a contribution of three extra peaks, green (at 1489.2 cm^{-1}), violet (at 1258.3 cm^{-1}), and yellow (at 1164.3 cm^{-1}), to the D and G peaks, reported as representative of C–C atoms bonds stretching in trans-polyacetylene (TPA) molecules, broadened vibrational density of states (VDOS) of small diamond clusters, tetrahedral amorphous carbon, and TPA C–H

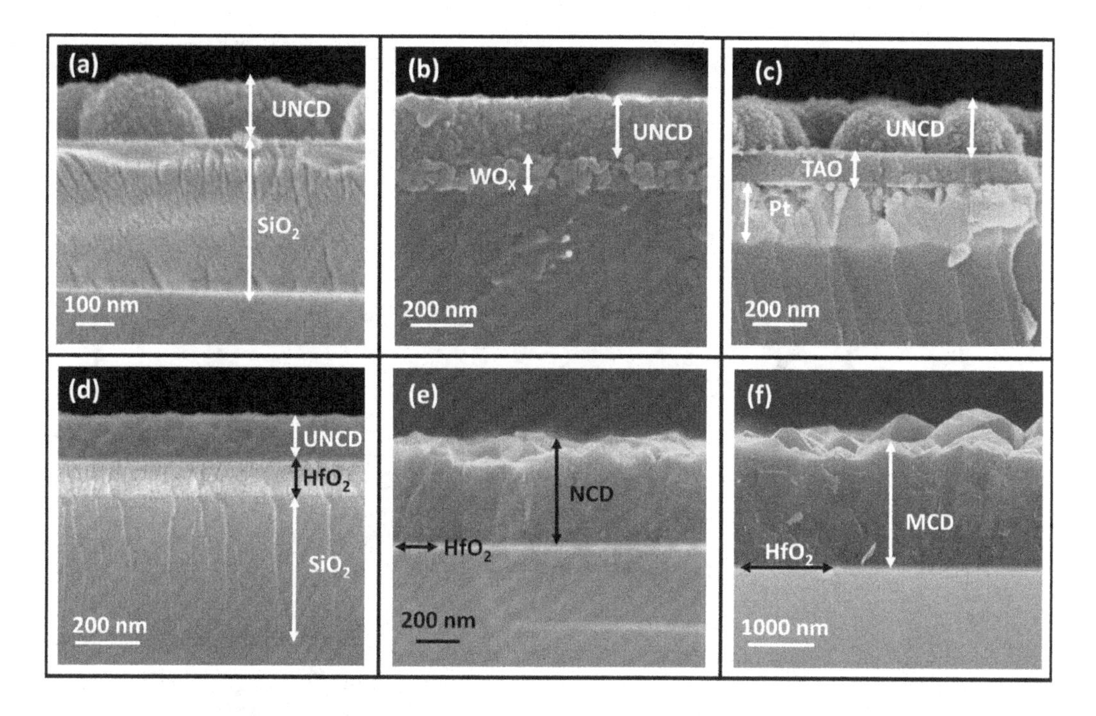

Figure 1.25 SEM images of the cross-section of UNCD films grown on SiO_2 (a), WO_x (b), $TiAlO_x$ (c), HfO_2 (d), showing that the densest UNCD film is grown on a HfO_2 layer, with a nanoscale smooth UNCD/HfO_2 interface. (e,f) Dense NCD and MCD films also grow with high-density structures and nanoscale smooth UNCD/HfO_2 interfaces on HfO_2 layers on Si (reprinted from *Diam. Relat. Mater.*, vol. 69, p. 221, 2016 (Fig. 4) in [41] with permission from Elsevier Publisher).

bending combined with C–C stretching, respectively [140, 141]. The literature [140, 141] shows that the diamond peak (1332 cm^{-1}) in UNCD films is always masked by the D mode peak of disordered carbon because visible Raman scattering from C atoms is mainly sensible to the dominating sp^2 bonds in the large network of UNCD grain boundaries, and much less to the sp^3 C bonds of diamond inside the grains. Figure 1.27Ab shows Raman spectra produced by UV (244 nm wavelength) laser beam, which provides sufficient energy to excite the σ states of both sp^2 and sp^3-C bonds of diamond, with a well-defined 1332 cm^{-1} peak from a single crystal diamond, an NCD film (10–100s nm grain size), and even a UNCD film. The disappearance of the 1150 and 1458 cm^{-1} peaks, shown in the visible (532 nm) Raman spectra, for the UV (244 nm) spectra (compare Figures 1.27Aa and Ab) indicate that those peaks are not the consequence of phonon confinement. This is due to the fact that if these peaks were zone boundary phonons, then they should be present in UV excitation as well [141]. On the other hand, the visible Raman spectrum shows that the intensity of the broad peak from the graphite D band (1340–1350 cm^{-1}) is inversely proportional to the crystal size, as indicated in

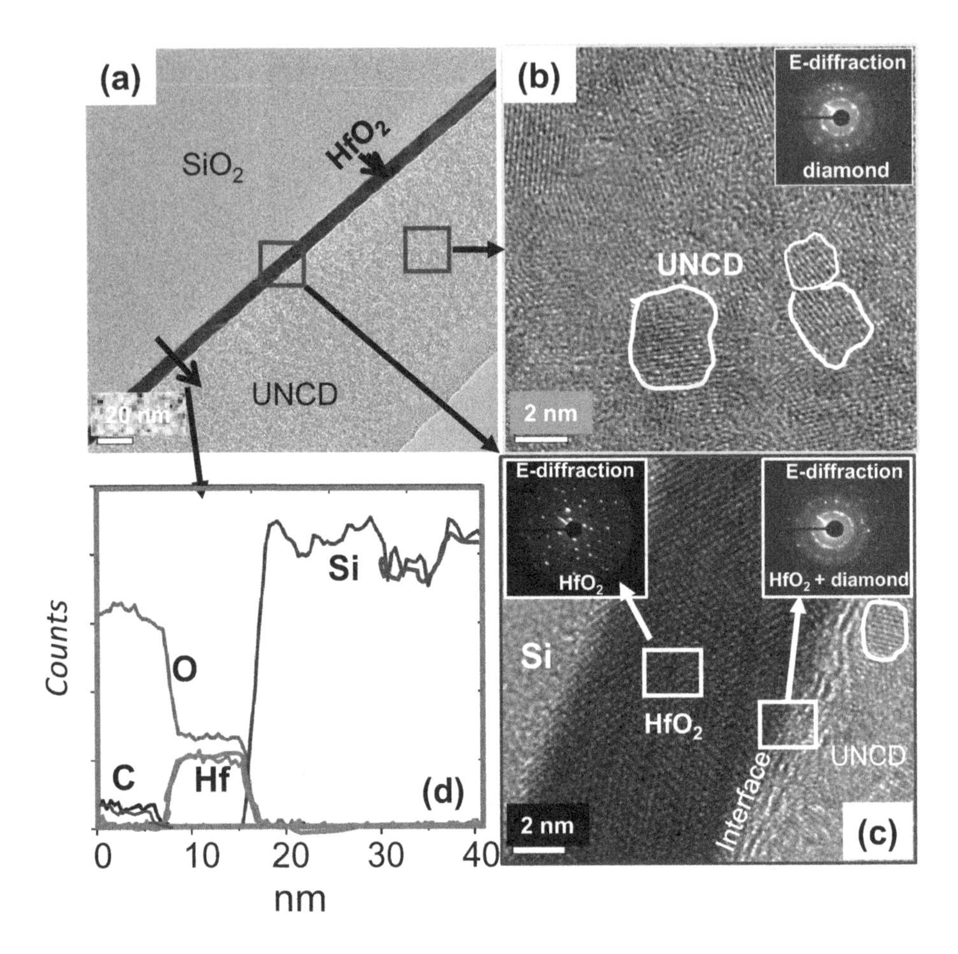

Figure 1.26 LRTEM cross-section (a) and HRTEM plain (b) and cross-section (c) images of UNCD film on UNCD/HfO_2/SiO_2 heterostructure. (d) HRTEM-EELS elemental profile analysis across the UNCD/HfO_2/SiO_2 heterostructure showing the distribution of C, O, Hf, and Si atoms correlated with the mechanism of diamond film nucleation and growth described in the main text (reprinted from *Diam. Relat. Mater.*, vol. 69, p. 221, 2016 (Fig. 1) in [41] with permission from Elsevier Publisher).

Figure 1.27Aa, although the appearance of this feature was attributed to the breakdown of the $q = 0$ selection rule. A Raman interpretation model [141] indicates that for nanocrystalline diamond, the characteristic Raman diamond peak at 1332 cm^{-1} is strongly reduced in intensity, and the D and G band peaks dominate the visible Raman spectra.

Analysis by XPS provided valuable information on the atom bonding in the initial UNCD film nucleation and growth stages. The appearance of the C1s, Hf4f, and O1s XPS peaks provided key information, complementing the HRTEM images, indicating that the UNCD films nucleate on HfO_2 due to the formation of a nanometer-thick HfC interface layer at the very initial nucleation stage. The first row of data in Figure 1.27Ba shows the XPS spectra of C1s, Hf4f, and O1s from the UNCD film

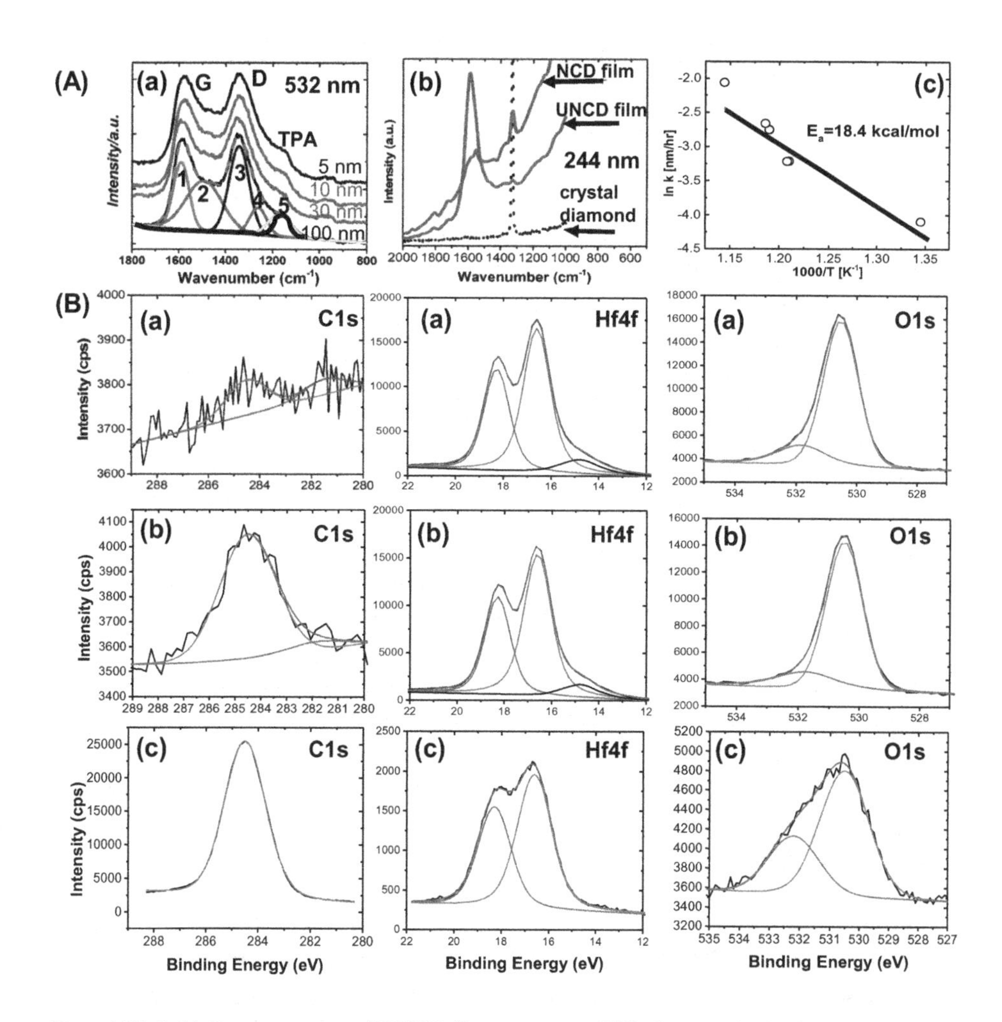

Figure 1.27 A (a) Raman spectra of UNCD films grown on HfO_2 layers, obtained with a 532 nm wavelength visible laser beam, showing the characteristic D band (1341.7 cm^{-1}) and G band (1588.5 cm^{-1}) peaks; this figure includes a deconvolution of the Raman spectrum, showing the contribution of two peaks corresponding to the G band peak (1588.5 cm^{-1}): **1** (1589 [cm^{-1}] and **2** [1500 cm^{-1}] peaks), and two peaks corresponding to the D band peak ((1341.7 cm^{-1}): 3 (1341 cm^{-1}) and 4 [1298 cm^{-1}]), peaks; the deconvoluted peak 5 at ~ 1150 cm^{-1}) correspond to the TPA peak correlated with H-rich molecules in the grain boundaries; (b) Raman spectra of a UNCD film grown on a HfO_2 layer, obtained with a 244 nm wavelength UV laser beam, showing a very small but visible 1332 cm^{-1} peak characteristic of diamond, which is exactly at the position where the strong 1332 cm^{-1} diamond peak of crystalline diamond appears (Raman spectra for a NCD film and for a crystalline diamond are shown to confirm the position of the diamond pea); (c) Arrhenius plot – growth rate (k) vs. 1/T – for UNCD films grown on HfO_2 layers at different temperatures (fitting of the experimental data points provides the activation energy for growth of UNCD films at ~18.4 kcal/mol, which is within the range of activation energies that yield high-quality diamond films).

B XPS spectra from UNCD films grown on HfO_2 layers for 5 min (a), 10 min (b), and 15 min (c). The spectra for the C1s and Hf 4f peaks reveal the presence of peaks at ~284.5 eV and ~14.7 eV, respectively, which confirm the formation of a HfC interface layer that induces nucleation of the UNCD films, the peaks for O 1s confirm the presence of O atoms contributing to form stoichiometric HfO_2 films. (Reprinted from Diam. Relat. Mater., vol. 69, p. 221, 2016 (Fig. 2) in [41] with permission from Elsevier Publisher).

grown for 5 min. Figure 1.27Bb and Bc show similar XPS data for UNCD films grown for 10 and 15 min, respectively. For UNCD films grown for 5 and 10 min, the Hf4f and O1s peaks dominate the XPS spectra, while a small but noticeable (C1s) peak appears at the binding energy of ~281.8 eV, which correlates, within experimental error, with the 281.5 eV C1s peak of HfC [142, 143]. The C1s peak of HfC does not appear in the spectrum of the UNCD film grown for 15 min, which is due to the thickness of the UNCD film covering the HfO_2 film. The C1s peak appearing at 184.5 correspond to the C atom bonds in the UNCD films.

The data from the Raman and XPS analyses shown in Figure 1.27 provide valuable information to confirm the mechanism of nucleation and growth of UNCD films on HfO_2 layers, which may play a critical role in the development of a new generation of micro/nanoelectronic devices.

In concluding this section, it is relevant to note that in addition to the development of a process to produce high-quality UNCD coatings on HfO_2 layers on Si, Figure. 1.24f and 1.25f show that high-quality MCD films can be grown on HfO_2 layers on Si substrates. This is a key result that opens the way to develop the coating technology to cool down the next generation of Si-based micro/nanoelectronics, since it has been demonstrated that MCD exhibit a thermal transport of ~1900 W/K·m [22], very close to the best thermal transport of any material perpendicular and parallel to the film surface, which is for single crystal diamond (~2100 W/K·m) [22]. However, further R&D is needed to develop a process to grow MCD at $\leq$450 °C, as has been developed for UNCD coatings [22], in order to be within the thermal budget of Si microchips to avoid destroying the Si device, as occurs when heated above 450 °C.

1.3.2.4 Fundamentals on Synthesis and Characterization of UNCD Films with Tailoring of the UNCD Surface from Super-Hydrophilic to Super-Hydrophobic for Multifunctional External and Implantable Medical Devices and Prostheses: Background for Application of Super-Hydrophobic to Super-Hydrophilic UNCD Coating surfaces

Cardiovascular and Other Blood-Interfacing Devices. Implantable medical devices that make contact with blood, such as vascular grafts, stents, and artificial heart valves, are widely used for treatment of cardiovascular diseases. Also, catheters and ports inserted in the venous system provide access to that system, to enable drug delivery in patients with cancer, including hematological malignancies, for example. Thrombus formation is a common cause of failure of these devices, where the formation of thrombi (blood-clotting aggregates) on the surface of the device due to adhesion of platelet aggregates and fibrin on supposedly hydrophilic (liquid-adsorbing) device surfaces leads to stopping of blood flow [144]. Blood consists of a complex mixture of plasma and cells. Proteins are a major constituent of plasma, which contains about 300 different proteins, with concentrations in the 35–50 mg/mL range. Rapid adsorption of proteins from plasma onto artificial surfaces is considered a main component initiating thrombus formation because the protein layer modulates subsequent reactions, attracting platelets and other molecules in the blood. The dynamics of protein adsorption are related to the chemical and physical properties of the surface and the proteins. Thus, adsorption involves interactions between charged groups at the protein surface interface and/or conformational changes in protein structure. Adsorbed proteins form a monolayer with a thickness of 2–10 nm on the surface of cardiovascular devices, such that the concentrations of proteins on the surface can be 1000-fold higher than those in the plasma [145].

A group of researchers claimed that hydrophilic surfaces may reduce adhesion of platelets and blood cells. Negative surface charge density gradients were prepared on fused silica slides using selective oxidation of a 3-mercaptopropyltrimethoxysilane (MTS) monolayer converting surface thiol groups (–SH) into negatively charged sulfonate ($-SO_3^-$) groups [146]. Gradients pre-adsorbed with fibrinogen showed an adhesion maximum in the center of the gradient region. Albumin coating of the gradients resulted in low overall platelet adhesion, with increased adhesion in regions of high negative charge density. The authors of the research described in [146] claimed that hydrophilic surfaces show less adhesion of platelets than hydrophobic surfaces. However, this claim does not have support from the surface analysis point of view, comparing true hydrophobic and hydrophilic surfaces [146].

Hydrophilic coatings based on polymers make polymer-based devices susceptible to fluids. The lubricity and water retention on the polymer surface reduce the force required to manipulate intravascular medical devices during surgical procedures. The lubricated surface helps to decrease the frictional force between devices 10- to 100-fold and helps reduce the risk of damage to blood vessel walls, enabling navigation in complex vascular pathways. Hydrophilic coatings are being implemented for balloon catheter angioplasty, neurological interventions, lesion crossing, and site-delivered drug therapies, with reduced thrombogenicity in some cases [147]. Reduction of friction between biological tissue and catheters has contributed to reduced procedure time and cost. Another benefit of hydrophilic coatings on medical devices is that they may create an interface that the human immune system does not recognize as artificial, significantly reducing the risk of problems.

Biologically compatible hydrophilic surfaces, like the demonstrated hydrophilic UNCD surface, can be used as scaffolds for efficient growth of embryonic stem cells and differentiation into other human cells for developmental biology [58] to treat different human biological conditions via replacement of natural cells by cells created in the laboratory [58].

Hydrophobic coatings can keep surgical tools and instruments that become fouled with fluids or tissue debris cleaner overall and for longer periods [147]. Because hydrophobic coatings repel fluids, the blood, urine, or tissue sheets slide off easily. In some cases, these hydrophobic coatings incorporate fluorocarbon functionality in order to improve repellency of hydrocarbons (i.e., lipids); these coatings are generally called *oleophobic*. In this sense, hydrophobic coatings demonstrate water-repellent, self-cleaning, antifouling, and/or anticorrosive effects [147]. Medical devices treated with hydrophobic coatings greatly reduce the risks of contamination and infections in patients [147]. Super-hydrophobic coatings were biologically inspired by the lotus leaf, which has an extremely high water drop contact angle of $>120°$ and low sliding angle of <10°. Micro- and nanoscale architectures on surfaces minimize water droplet adhesion. *Super-hydrophobic* surfaces refers to those exhibiting extreme water repellency. A super-hydrophobic coating has a water contact angle (WCA) greater than 120° and a sliding angle less than 10°. Many super-hydrophobic coatings become destabilized under adverse conditions and the performance is lost [147].

Other Devices Benefiting from Tailoring of Surface Wettability. In addition to the devices described above, there are other devices requiring control of material surface wettability, such as electrochemically highly stable external and implantable electrodes [148], several external and implantable biomedical devices [59], chemical and biological sensors [149, 150], microfluidics [151], and many other devices and systems.

The information presented above indicates that new paradigms in biocompatible materials with the capability for tailoring their surfaces from super-hydrophilic to super-hydrophobic are necessary. In this sense, the transformational biocompatible UNCD coating provide a new paradigm material, as described in the next section.

Fundamentals on the Synthesis and Characterization of Super-Hydrophobic to Super-Hydrophilic UNCD Coatings Using Chemical Seeding plus HFCVD Process

Background on Research on Wettability of Diamond Surfaces

Before describing the new R&D process for tailoring the wettability of UNCD film surfaces, it is relevant to provide background information on the prior R&D done on the wettability of diamond surfaces. In this sense, from the information shown in Table 1.2 and the literature cited therein, it appears that there are no reports on systematic research on the influence of grain size of PCD films on the surface chemistry termination and their effects on the wettability of the treated surfaces. Therefore, the research reviewed in this section relates to work performed by Montano et al. [152], which was focused on performing systematic studies of O_2, H_2, and CF_4-based plasma treatment of MCD, NCD, and UNCD films, and the effect of combined surface chemistry and UNCD-coated 3D microstructured surfaces, all on surface wettability and related performance.

Growth of UNCD Films by Chemical Seeding Plus HFCVD Process

Montano et al. [152] investigated wettability of MCD, NCD, and UNCD film surfaces. However, only the studies on UNCD film surface wettability are reviewed here, and the reader is referred to [152] to read about the research on MCD and NCD films.

The UNCD films were grown using the HFCVD technique. The UNCD films were grown on electrically conductive p-Si (111) wafers inserted in the HFCVD chamber onto a rotating substrate holder, located underneath an array of 10 parallel tungsten (W) filaments (see Section 1.3.2.2. and Figure 1.20 and [74]). The parameters used to grow the UNCD films are shown in Table 1.3, which correlates with the production of nanoscale dense UNCD films on Si surfaces (Figure 1.28). For the research focused on investigating the effect of 3D structures on wettability of UNCD films, these films were grown on a large array of micropatterned pillars on Si substrates (Figure 1.29) using well-established photolithography and reactive ion etching processes [152].

Table 1.2 Summary of literature on wettability studies of the surfaces of diamond films, bulk diamond, diamond powder, and diamond-like carbon (reprinted from *Carbon*, vol. 139, p. 361, 2018 in [152] with permission from Elsevier Publisher).

Sample	Treatment	Process time	Gases	Characterization	WCA	Ref.
PCD films	Plasoxidation and fluorination	SF_6: 10 min O_2: 10 min	SF_6 , O_2 plasmas	SEM, Raman, XPS	Not reported	[153]
UNCD/amorphous carbon (a-C) by MPCVD	Plasoxidation and fluorination	O_2: 10 min CHF3: 10 min	O_2, CHF_3	XPS, TOF-SIMS, AFM, and WCA	O_2: 33 ± 1° CHF_3: 91 ± 1	[154]
NCD by MPCVD	Air oxidation + hydrogenation in microwave plasma	O_2: 15 min H_2: 15 min	O_2 , H_2	WCA	O_2: ~5° H_2: ~110°	[155]
Diamond	Glow discharge plasma	CF_4: 5, 10, 15, 30, and 60 min	CF_4	AFM, Raman spectroscopy, XPS	Not reported	[156]
Type 2A Diamond (100)	Direct treatment with molecular and atomic beams in UHV	Not reported	5% fluorine in argon gas mixture	XPS and LEED	Not reported	[157]
Boron-doped epitaxial diamond layers	Plasma fluorination and oxidation	CF_4: 2 min O_2: 2 min	CF_4 , O_2	AFM and XPS	Not reported	[158]
Diamond powder commercial	Plasma fluorination	CF_4: 10 min	CF_4	FTIR DRS and XPS	Not reported	[159]
Diamond-like carbon (DLC)	Plasma fluorination	C_4F_8: 30 s CHF_3: 30 s	C_4F_8 , CHF_3	Ellipsometry, AFM, and WCA	Both gases: ~90–100°	[160]
UNCD	Plasma fluorination	CHF_3 /SF_6: 5–900 s	CHF_3, SF_6	WCA, XPS, and AFM	CHF_3: 90–100° SF_6: 110°	[161]
NCD commercial	Plasma oxidation and fluorination	O_2: 30 min C_3F_8: 30 min	O_2 , C_3F_8	XPS and WCA	O_2: 34.11° C_3F_8: 101.16°	[162]

Table 1.3 Deposition conditions used to grow UNCD films by HFCVD (reprinted from *Carbon*, vol. 139, p. 361, 2018 in [152] with permission from Elsevier Publisher).

Polycrystalline type of film	Gas mixture $H_2/CH_4/Ar$ (sccm)	Growth time (h)	Distance between filaments and sample (mm)	Film thickness (nm)	Sample code
UNCD	10–2–90	2	20	118	10
UNCD	25–2–75	2	20	135	25

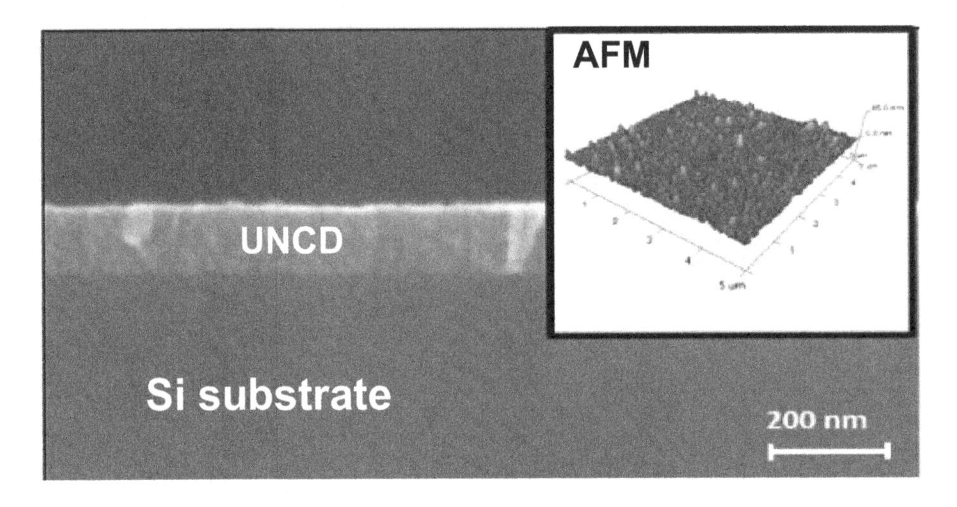

Figure 1.28 Cross-section SEM image of a UNCD film grown using HFCVD on the surface of an Si wafer; the top right insert shows an AFM image of the surface of the UNCD film, revealing an rms roughness of about 3–7 nm (reprinted from *Carbon*, vol. 139, p. 361, 2018 (Fig. 2) in [152] with permission from Elsevier Publisher).

Because of the nanoscale size (3–5 nm) of UNCD, these films can be grown with excellent thickness uniformity (Figure 1.28) and density on high aspect ratio 3D structures like pillars (Figure 1.29c,d). The parameters used to grow UNCD films on the 3D Si pillars are shown in the first row of Table 1.3. The Si pillar dimensions were: diameter, 20 μm; height, 5 μm; and distance between pillars, 20 μm.

Treatment of PCD Films Surfaces with O_2, H_2, and CF_4 Plasmas to Tailor Wettability and Characterization

UNCD Surface Treatment. UNCD films were inserted in a reactive ion etching (RIE) chamber to perform CF_4 and O_2 plasma-based treatment of the UNCD films' surfaces. The CF_4 and O_2 plasma treatments were done on a TRION Sirus T2-Tabletop RIE, using the following conditions:

1. ***O_2 plasma treatment:*** power: 50 W; O_2 flow rate: 8 sccm; process pressure: 100 mTorr; process time: 30 seconds.

2. ***CF_4 plasma treatment*:** power: 100 W; CF_4 flow rate: 12 sccm; process pressure: 50 mTorr; process time: 60 seconds.
3. ***Hydrogen plasma treatment*:** the microwave plasma treatment was performed in the MPCVD system used to grow UNCD films (see Figure 1.3) using flow of Ar (50 sccm) and H_2 (50 sccm) gases and a microwave process power of 2500 W to generate the plasma. The substrate surface temperature was monitored with a pyrometer aimed at the surface, reading 592 °C. The H-plasma process time was 10 min, followed by cooling down in 200 sccm of H_2 gas flow until the UNCD film surface reached a temperature of 100 °C.

Characterization of Key Parameters of the Treated Surfaces of UNCD Films to Produce Super-hydrophilic to Super-Hydrophobic Surfaces

Surface Chemistry. For UNCD films, the Raman analysis, with visible laser wavelength (532 nm), revealed the characteristic spectrum dominated by Raman scattering from the sp^2-bonded C atoms in the grain boundaries (see Figure 1.27A(a); 5 nm curve in Section 1.3.2.3). Raman analysis of the UNCD film using UV laser of 244 nm wavelength revealed the 1332 cm^{-1} diamond peak, although very small (see Figure 1.27A(b)/ UNCD film in Section 1.3.2.3).

Thickness and Surface Morphology of Flat and 3D Structured UNCD films. The thicknesses of the UNCD films were measured using cross-section SEM imaging (Figure 1.28), and the surface roughnesses were measured using AFM (top right insert in Figure 1.28). The AFM measurements revealed the characteristic nanoscale surface roughness of UNCD films in the range of 3–7 nm.

UNCD films were grown on a large array of micropatterned pillars on Si substrates, such that the top surface exhibited the characteristic UNCD smooth structure (Figure 1.29a). Figure 1.29b shows a low-magnification SEM image of the large array of Si pillars coated with a very conformal UNCD film. Figure 1.29c,e shows high-magnification cross-section SEM images of an Si pillar coated with UNCD film, revealing the extreme uniformity of the UNCD film thickness achieved on the high aspect ratio 3D pillar structures [152]. Finally, Figure 1.29d shows a summary of the static WCA measurements on UNCD film surfaces with different chemical treatments on flat and 3D structures, revealing the orders of magnitude change in the surface nature of UNCD films from super-hydrophilic, when treated with an O_2 plasma, to super-hydrophobic when treated with a CF_4 plasma and grown on a microstructured Si large pillar array.

The importance of controlling the wettability properties of a large range of PCD films through a relatively easy and short-duration process is that it has a direct impact on a wide range of applications to biomedical devices and protheses coated with UNCD films.

Water Contact Angle Measurements on UNCD Film Surfaces. Figure 1.30 shows the systematic measurement of the static WCA from all UNCD film flat surfaces and their XPS survey analysis, as well as the microtextured UNCD surfaces grown on the Si pillar arrays. The pictures in Figures 1.30a–c show the results from WCA measurement of flat UNCD surfaces and their change in surface energy [152]:

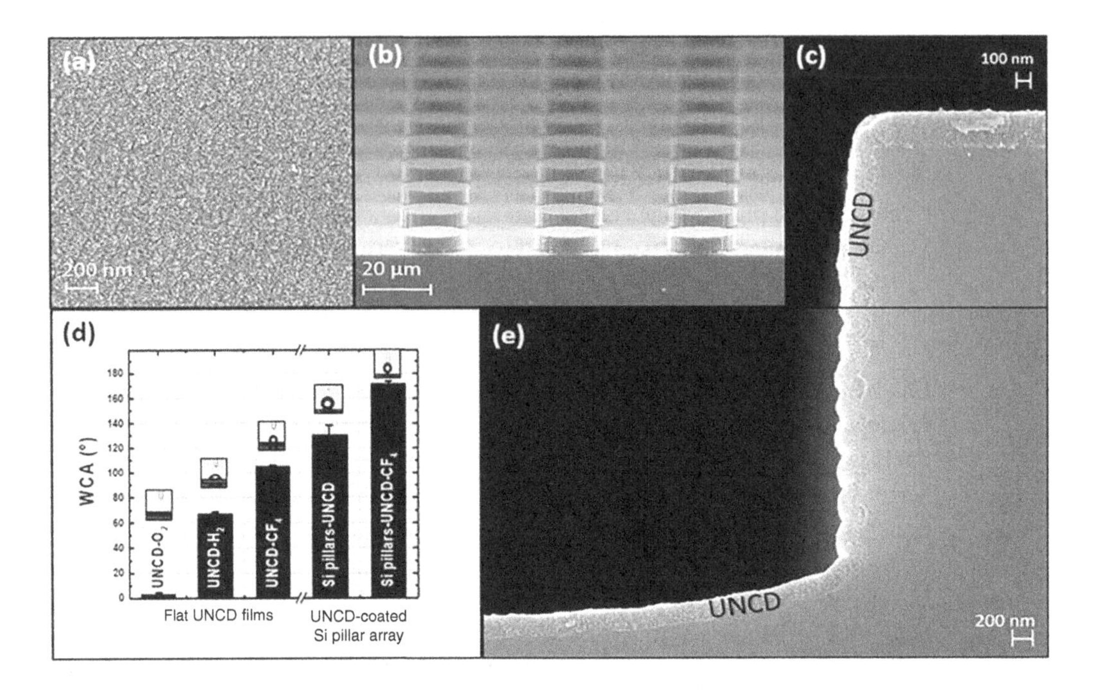

Figure 1.29 (a) SEM top view of a flat UNCD surface on top of an Si pillar. (b) Low-magnification image of UNCD-coated Si pillars array. (c,e) Part of a high-magnification image of a UNCD-coated Si pillar, showing the excellent dense/conformal coating achieved on the 3D structured Si surface. (d) WCA vs. chemical surface treatment of UNCD coatings grown on flat Si substrates and on a 3D structured Si pillar array, showing the controllable wettability properties by different means, tailoring from super-hydrophilic surface with WCA ~ 0 to super-hydrophobic surface (UNCD-coated Si pillar treated with CF_4 plasma) with WCA 178° (including the measurement error) (reprinted from *Carbon*, vol. 139, p. 361, 2018 (Fig. 3) in [152] with permission from Elsevier Publisher).

1. Super-hydrophilic (total wetting of the surface) with high surface energy, after 30 s O_2 plasma treatment (Figure 1.30a), inducing O atoms and/or O–H molecules attaching to the surface, and reacting chemically with water molecules.
2. Medium hydrophobic (with UNCD-H terminated surface by MPCVD H-plasma processing (Figure 1.30b).
3. Hydrophobic, after 60 s CF_4 plasma treatment (Figure 1.30c) with low surface energy.
4. Highly hydrophobic, pillar textured Si surface coated with a UNCD film with the surface terminated by H atoms via H-plasma treatment with an MPCVD H-plasma process, such that the H atoms, chemically bonding to C open bonds on the UNCD surface, inhibit chemical reactions of the C atoms with water (H_2O) molecules (Figure 1.30d) (WCA = 159°) [152].
5. Pillar array textured Si surface coated with UNCD films, plus exposure to a CF_4 plasma inducing a super-hydrophobic surface, induced by the combination of texture plus UNCD surface termination with F atoms, inducing the highest water

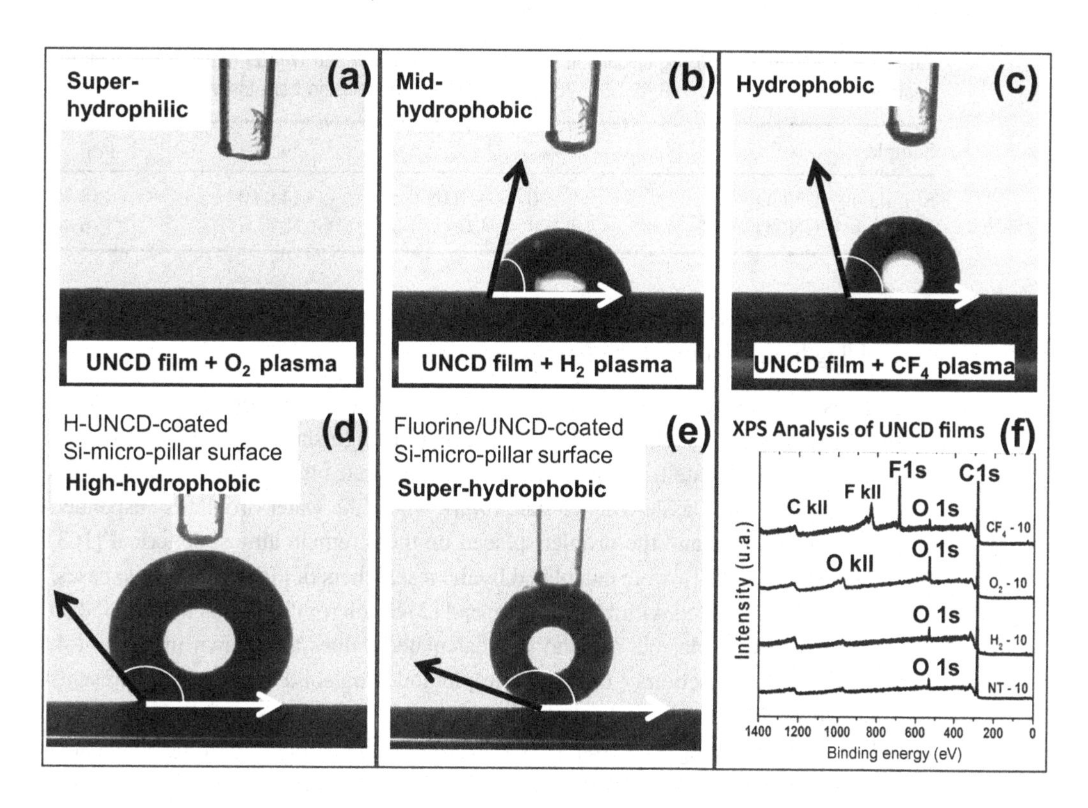

Figure 1.30 (a) Flat UNCD film surface treated with an O_2 plasma for 30 s. (b) Flat UNCD surface terminated with H atoms. (c) Flat UNCD surface treated with a CF_4 plasma for 60 s. (d) UNCD-coated Si pillar array with the UNCD surface terminated with H atoms. (e) Fluorinated UNCD-coated Si pillar array treated with a CF_4 plasma for 50 s. (f) Survey XPS spectra of as-grown UNCD film (bottom curve), H-surface terminated UNCD film (second curve from bottom), O-surface terminated UNCD film (third curve from bottom), and finally F-surface terminated UNCD film (top curve) [152] (reprinted from *Carbon*, vol. 139, p. 361, 2018 (Fig. 4) in [152] with permission from Elsevier Publisher).

drop contact angle demonstrated today (~178° water drop contact angle to the UNCD surface) [152].

XPS analysis of the surface of the treated UNCD films, shown in Figure 1.30f, showed the following results:

1. The oxygen plasma treatment resulted in replacement of H atoms on the surface of the as-grown UNCD films by O and OH groups. This agrees with results reported by other groups [154, 162].
2. In the case of CF_4 plasma treatment, a single XPS narrow peak appeared for the fluorinated UNCD surface, which is consistent with the formation of a carbon-monofluoride species on the surface, as also confirmed by other authors [158]. So, the presence of the prominent F1s peak confirms the surface fluorination of the UNCD film, which was not shown in the survey XPS spectra for the films treated

Table 1.4 Values of θ_W and θ_{CB} calculated using the Cassie–Baxter and Wenzel formulas, respectively (reprinted from *Carbon*, vol. 139, p. 361, 2018, in [152] with permission from Elsevier Publisher).

Sample	θ_{meas}	θ_W	θ_{CB}
Si pillars + UNCD	130.43 + 8.01	141.10	158.39
Si pillars + UNCD + CF_4	178. ± 1.0	NA	176.40

with O_2 or not treated at all. The bonding energy of the C–F bond is very high, thus enabling high stability of the fluorinated surface.

The results reported in [152] indicate that when a water drop interacts with a textured surface, such as the Si micropillars, two distinguished wetting states can be formed: the Wenzel state (θ_W), where the liquid penetrates the structures, wetting the whole surface; or the Cassie–Baxter state (θ_{CB}), where the water droplet is suspended on top of the textures and the droplets placed on them remain almost spherical [163, 164]. Hence, θ_W and θ_{CB} were calculated by the researchers of [152] for the two cases: (1) Si micropillars coated with UNCD film; and (2) Si micropillars coated with UNCD film exposed to CF_4 plasma. θ_W and θ_{CB} calculated values are shown in Table 1.4. The values of f (the fraction of the liquid droplet touching solid and not air) represents the fraction of the solid–liquid interface in the entire composite surface beneath the liquid, obtained by using the Cassie–Baxter formula, and r (the ratio of the actual area of the solid–liquid interface to the normally projected area), obtained by using the Wenzel formula, was used to calculate θ_W and θ_{CB}, enabling correlation with experimental measurements. The calculations indicate that the Wenzel model correlates better with the wettability state of the UNCD-coated Si pillar array without fluorine treatment, while the Cassie–Baxter model correlates better with the state of the fluorinated UNCD-coated Si pillar array (Table 1.4) .

The results described above revealed the key role that CF_4 plasma treatment of UNCD film surface has on enhancing the hydrophobic intrinsic behavior of a material such as UNCD, by controlling the surface chemistry, combined with the microstructure of the substrate, resulting in a synergistic effect that leads to the formation of Cassie–Baxter droplets, where the liquid sits partially on the air trapped underneath. The experimental observation is that water does not adsorb at all on the fluorinated UNCD micro-/nanostructured surface, which provides a new material paradigm for application to a broad range of blood-interfacing implantable medical devices.

1.3.2.5 Fundamentals on Synthesis and Characterization of PCD Films By the HFCVD-Based BEN-BEG Process

Background on HFCVD BEN-BEG of Diamond Films

As discussed in Section 1.2.1.3, the BEN-BEG processes to grow UNCD, NCD, and MCD films without any chemical pre-seeding process were first developed for the MPCVD technique, and that work involved a systematic characterization of the process using complementary HRTEM, SEM, Raman, and XPS studies, which revealed details of the film nanostructure and interfaces, critical to understanding the

BEN-BEG process (see Section 1.2.13 and references therein for details on the MPCVD BEN-BEG process). Further information on the MPCVD BEN-BEG process, relevant to the discussion on the HFCVD BEN-BEG process, should be considered. Careful reading of published research on BEN to grow NCD and MCD films indicates that either ultrasonic scratching of the substrate surface in a solution containing diamond powder was done before carrying out the BEN-BEG process [165], or polishing of the substrate surface with diamond powder was done (with micro-size particles), followed by oxygen annealing at high temperature, supposedly to eliminate diamond particles embedded on the substrate surface, but without presenting clear evidence that the cleaning happened, so most probably the particles were on the surface acting as chemical seeding [166]. On the other hand, other work involved studies of the MPCVD BEN-BEG growth process synthesizing MCD and NCD films, without any chemical pre-seeding process, but unfortunately the studies involved only Raman analysis and low-resolution SEM imaging of the film surface [167], neither of which can provide key insights into the nanostructure of the films and interfaces at the nanoscale to better understand the underlying mechanisms of the BEN-BEG process, as done for the research reported on MPCVD BEN-BEG of UNCD films, including detailed HRTEM studies [81, 83].

More recently, research was performed on the BEN-BEG process by different groups using the HFCVD technique to grow UNCD to MCD films. Table 10.5 shows a summary of the results from research related to BEN-BEG of diamond films using the HFCVD process. In this sense, Table 10.5 shows that most of the research performed was focused on HFCVD BEN-BEG of diamond films on Si^{++} substrates, mostly on small areas, except for [168], where the authors attempted to do BEN-BEG on up to 100 mm diameter Si substrate but were not able to cover the substrate with uniform film all the way to the edge.

Fundamentals and Technological Development of HFCVD BEN-BEG of UNCD Films

Based on the information presented above, Auciello's group started to perform systematic studies to investigate the fundamentals and develop and optimize the growth of uniform UNCD films on large-area substrates, using the HFCVD BEN-BEG process [135]. This work focused initially on growing uniform UNCD films on 100 mm diameter substrate (as an intermediate stage to scale to $\geq$200 mm diameter wafers) Si wafers. The Si wafers were coated with 40 nm thick tungsten (W) layers to provide high electrical conductivity surfaces and high-density UNCD film growth, as demonstrated for MPCVD growth of UNCD films [22, 137] grown on the WC nucleation layer forming W surfaces [22, 137]. The research cited in [135] showed that the HFCVD BEN-BEG process can be applied to other substrates with different electrical conductivity, including semiconductors. The importance of understanding the UNCD HFCVD BEN-BEG growth process on large-area substrates other than Si, specifically on carbide surface layers, forming materials such as W, Ti, Mo, and more, is because it opens new technological applications of the UNCD coating technology, with substantial reduction in the cost of fabrication of UNCD-coated products based

Table 1.5 Information on parameters used for HFCVD BEN-BEG synthesis of NCD to UNCD films by independent groups worldwide (reprinted from *Diam. Relat. Mater.*, vol. 78, p. 1, 2017 in [135] with permission from Elsevier Publisher).

System configuration	Gas mixture(%)	Substrate	Pressure (Torr)	Filament temp. (°C)	Substrate temp. (°C)	Bias volt (V)/ curr (A) or V/sub curr. dens.	BEN-BEG (h)	Growth time (h)	Substrate diameter	Ref.
Filament to substrate bias	H_2 – 1; CH_4 –1–5; Ar – 97–94	–	10–50	<2800	400–800	(–) 150–200/0.05 (+) 30–40/0.05	3	3	–	[169]
Grid behind filaments (+) Substrate (–)	H_2 – 98; CH_4 – 2	Si (100)	–	2100	600	240/0.005	0.5	4–20	–	[170]
Grid behind filaments (+) Substrate (–)	H_2 – 96.5; CH_4 – 3.5	Si (100)	30	2100	800–850	28/0.4 grid – 150 (0.3 mA/cm^2)	0.5	12		[171]
Grid behind filament (+) Substrate (–)	H_2 – 96.5; CH_4 – 0.7 (add N for growth)	Si (100)	15	2000	700–840	+60/–200 (0.5 mA/cm^2)	2–4	3	100 mm	[168]
Grid behind filament (+) Grid-front (–) (+) Substrate(–)	H_2 – 94; CH_4 – 4 (change for growth)	Si (111)	15–30	–	700–850	+15–170 (0.2 mA/cm^2)	0.25	0.11	3 mm	[172]
Filament–substrate bias	H_2 – 80; CH_4 – 20	Si (100)	30	2200	~800	(–)150/ vs. growth time and grain size	2	None		[173]
Filament to substrate bias	H_2 – 1; CH_4 – 7; Ar – 92	Si (100)	12	2350	950	(–)200 /not reported	0.5	1.5–2		[174]
Filament to substrate bias	H_2 – 49; CH_4 – 2; Ar – 49	40 nm W/Si (100)	10	2300	550	(–)230–250/1.5 A	0.5–2	2	100 mm	[135]

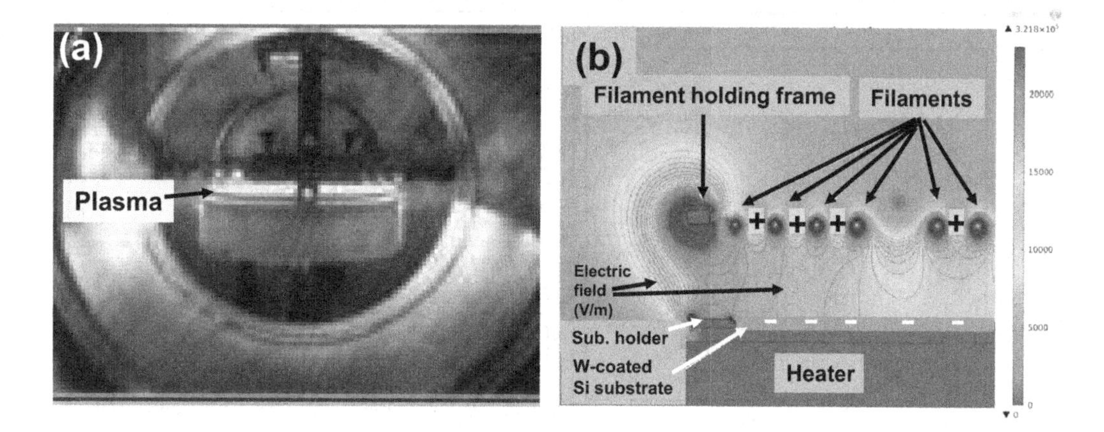

Figure 1.31 (a) The HFCVD system in operation, showing the plasma created on top of the substrate surface, inducing the UNCD film BEN-BEG process. (b) Computer simulation providing information on the electric field lines in the plasma created between the positively biased filaments and the negatively biased substrate, revealing concentration of field lines on the edges, which correlates with the film growth geometry shown below.

on elimination of the relatively time-consuming and high-cost wet chemical diamond particle seeding process currently used in commercial production of UNCD-coated industrial products [25]. In addition, HFCVD BEN-BEG of uniform UNCD films over larger areas will enable selective/directional growth of UNCD films on patterned W structured layers for fabrication of devices on large-area wafers used in the semiconductor industry [135].

The HFCVD system used for the research on HFCVD BEN-BEG of UNCD films is shown in Figure 1.20. Figure 1.31a shows a picture of the HFCVD in operation, revealing the plasma produced on top of the substrate surface, from where the positively charged C^+, CH_x^+, H^+, and Ar^+ ions are accelerated toward the negatively biased substrate, inducing sub-implantation of C atoms under the surface, inducing the UNCD film BEN-BEG process.

A key parameter in the HFCVD BEN-BEG UNCD film growth process is the incubation time when C-based diamond nanocrystals nucleate on the substrate surface during the BEN process, followed by the growth of the diamond nanoscale grains, which coalesce to produce a continuous UNCD film during the stage defined as BEG. The BEN-BEG time in the HFCVD-based growth of UNCD films was changed to be in the range 0.5–2.5 h. adding 2 h of film growth time without bias for each film. The filaments were heated to 2300 °C during film growth. A mixture of gases with a ratio of H_2 (49%), CH_4 (2%), and Ar (49%) was flown into the HFCVD chamber, previously evacuated to ~3×10^{-6} Torr. The substrate heater temperature was set up at 550 °C. Further experimental details can be found in prior papers [41, 74]. The BEN-BEG process was performed applying a voltage of ~ –220 V on the substrate to keep a constant electrical current of 1.5 A through the plasma. SEM and HRTEM analyses

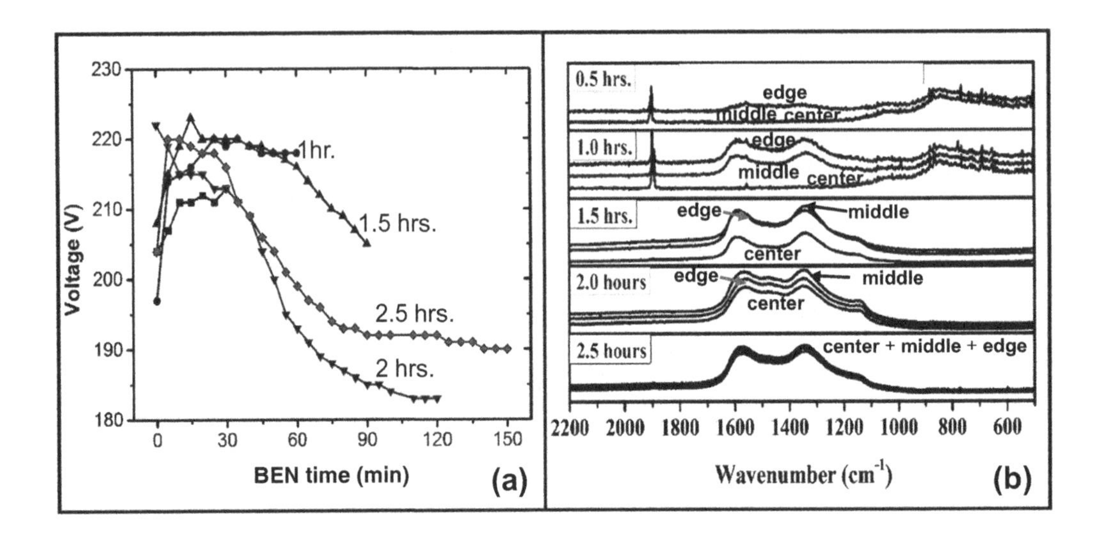

Figure 1.32 (a) Voltage between filaments and substrate vs. BEN times at a constant 1.5 A plasma current and –220 V applied to the substrate with respect to the positive biased filaments. (b) Raman spectra obtained from analysis on the center, middle, and edge areas of the UNCD films, grown on a W (40 nm thick) layer on an Si substrate surface via BEN-BEG times in the range 0.5–2.5 h, plus an additional 2 h growth time without bias. The broad peaks at ~1340 cm^{-1} (D band) and ~1588 cm^{-1} (G band) are characteristic of the sp^2-C atom bond hybridization in the grain boundaries of the UNCD films [41, 74, 135] (for UNCD films, the diamond Raman peak at 1332 cm^{-1} is buried inside the broad peak at 1340 cm^{-1}, and can only be seen when using deep UV (244 nm wavelength) laser Raman, as shown in Figure 1.27A(b); the sharp peak at ~1900 cm^{-1} corresponds to the W layer underneath the UNCD films, which is relatively thin and allows seeing Raman peaks from the underlying W layer (reprinted from *Diam. Relat. Mater.*, vol. 78, p. 1, 2017 (Fig. 2) in [135] with permission from Elsevier Publisher).

were used to characterize the surface morphology and nanostructure of the UNCD films, respectively. XRD was used to characterize the crystallinity and grain sizes of the UNCD films. Raman spectroscopy was performed, using a 532 nm wavelength laser beam, to characterize the chemical bonds in the UNCD films. The surface chemistry of the UNCD films was characterized by XPS. Details of all the characterization systems and parameters used for the analyses can be found in [135].

Figure 1.32a shows the bias voltage (V_b) between the filaments and the substrate vs. the BEN time in the range 1–2.5 h. Figure 1.32b shows Raman spectra for UNCD films grown for several BEN-BEG times in the range 0.5–2.5 h. The plasma was stabilized after several minutes of starting the BEN process. The reduction in the voltage between the filament and the substrate vs. the BEN time, observed in Figure 1.32a, can be attributed to the formation of the UNCD grains with sp^3-C atom bonding characteristic of diamond, and grain boundaries with sp^2-C atoms bonding, from which electron emission has been demonstrated to arise due to carriers originating from the formation of the sp^2-bonded carbon [22]. Therefore, the abrupt decrease and subsequent stabilization of the bias voltage V_b can be attributed to the rapid growth of nanodiamond grains, characteristic of the UNCD films and subsequent full coverage of the substrate with a dense UNCD film, respectively, as observed previously for the MPCVD BEN-BEG process [81, 83]. The results of BEN-BEG of UNCD films described in [135], as well as other BEN results shown in the literature for HFCVD BEN-BEG of NCD and UNCD films [167], without any chemical diamond seeding, are in principle supported by modeling developed to explain BEN of diamond on Si substrates [175–177], which indicate that there may be a bias-induced nucleation of carbon-based layers to subsequently induce diamond grain growth, produced by C^+ and CH_x^+ ion subplantation near the substrate surface. The modeling indicates that bias in the range –100 to 250 V would provide ion energies sufficient to pass through the surface atomic layer and enter a subsurface interstitial position at a depth in the range 5–10 nm under the surface.

Figure 1.32b shows Raman spectra for all five UNCD films for which the voltage vs. BEN time curves are shown in Figure 1.32a. Raman analysis were done on the edge, middle, and center areas of the films across the 100 mm diameter substrate. The Raman spectra obtained at ~5 mm from the edge of the UNCD film grown for 0.5 h show the characteristic D and G mode peaks, although small, present in every UNCD film grown for the last many years [22]. On the other hand, the Raman signal intensity in the middle and center of the film are very similar, but very small, indicating good uniformity in the initial stages of nucleation. As the BEN-BEG time increases, the D and G mode signals increase, more prominently at the edges, until the 2.5 h growth time, for which all Raman spectra from center, middle, and edge overlap, showing excellent uniformity of the UNCD film across the 100 mm diameter substrate. The Raman data indicates that the nucleation starts on the edges, where the electric fields concentrate first on the electrically conductive substrate. Once the insulating UNCD film nucleates on the edges, the electric field transfers to the less nucleated area between the center and the edge, and finally it transfers to the center to complete the nucleation, resulting in a uniform film growth across the 100 mm diameter wafer, as shown by the Raman spectra for the 2.5 h growth.

The proposed nucleation and initial UNCD film growth mechanism for the HFCVD BEN-BEG process, based on Figures 1.31b and 1.32b, is supported by three key complementary characterization techniques, as described below.

XRD Analysis. XRD analysis on all films (Figure 1.33) confirms the information from the Raman analysis, revealing that the UNCD films tend to grow at and near the edge faster than at the center and middle of the substrate. Figure 1.33a shows that for BEG UNCD films grown for 0.5 and 1.0 h, the main peaks are located at positions 1 (31.5°), 2 (35.5°), 3 (39.7°), and 5 (48.5°), which correspond to WC (001), WC (100), W_2C (101), and WC (101), respectively [178]. Diamond (111) appears at $2\theta = 44°$, represented by peak 4 in Figure 1.33a, which starts showing strong intensity after 2.0 h BEN-BEG of UNCD films. Diamond (220) at 75° is convoluted with the WC (111) and W_2C (221) at 73° and 76° (peaks 7 and 8 in Figure 1.33a). All diamond directions were confirmed by HRTEM diffraction (Figure 1.35). For 2.0 and 2.5 h BEN-BEG of

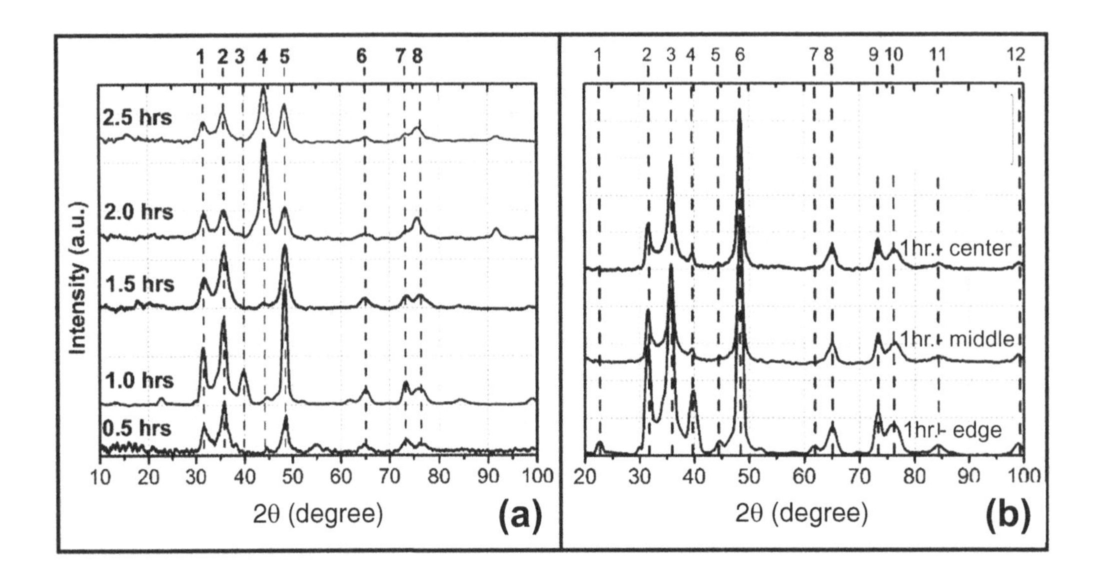

Figure 1.33 (a) XRD spectra from analysis of BEN-BEG UNCD films grown for 0.5–2.5 h plus additional growth time of 2 h without bias. : b) XRD spectra from analysis on center, middle, and edge areas of a UNCD film grown for 1 h. by BEN-BEG (the XRD peaks correspond mainly to the WC nucleation layer, correlating with the XRD spectra shown in (a) for films grown for up to 1.5 hrs., with the small diamond (111) peak (5) at 2θ = 44°, just starting to appear on the edge area, where the film nucleate faster (see discussion of the peaks identification in the text) (reprinted from *Diam. Relat. Mater.*, vol. 78, p. 1, 2017 (Fig. 3) in [135] with permission from Elsevier Publisher).

UNCD films, the diamond (311) can be seen in HRTEM, but it cannot be corroborated with XRD since it is a non-allowed diffraction due to the FCC nature of the diamond structure. Figure 1.33 (b) shows the XRD spectra from analysis on the center, middle, and edge areasof a UNCD film grown for 1 h. by BEN-BEG. The XRD peaks correspond mainly to the WC nucleation layer, correlating with the XRD spectra shown in Figure 1.33 (a) for films grown up to 1.5 hrs, with the small diamond (111) peak (5), at $2\theta = 44°$, just starting to appear on the edge area, where the film nucleates faster. However, there is no intense diamond (111) peak because the film needs to be grown for ~ 2 hrs to exhibit the dense structure. These results support the need for using complementary analytical techniques to obtain reliable information.

SEM Analysis. SEM imaging of the top UNCD film surface on the center, middle, and edge areas (Figure 1.34) confirm the data from Raman and XRD analysis, showing that the UNCD film growth from edge to center, combined with the longer BEN-BEG time (2.5 h), produces the denser UNCD films. The cross-section SEM images show good film thickness uniformity across the 100 mm diameter substrates.

HRTEM Studies. HRTEM images (top view) of BEN-BEG UNCD films grown for times in the range 0.5–2.5 h, plus 2 h growth without bias, are shown in Figure 1.35. All films showed grain sizes of 2–5 nm, characteristic of the UNCD nanostructured films for many applications [21–23, 40, 41, 46, 48–50, 58, 59, 74].

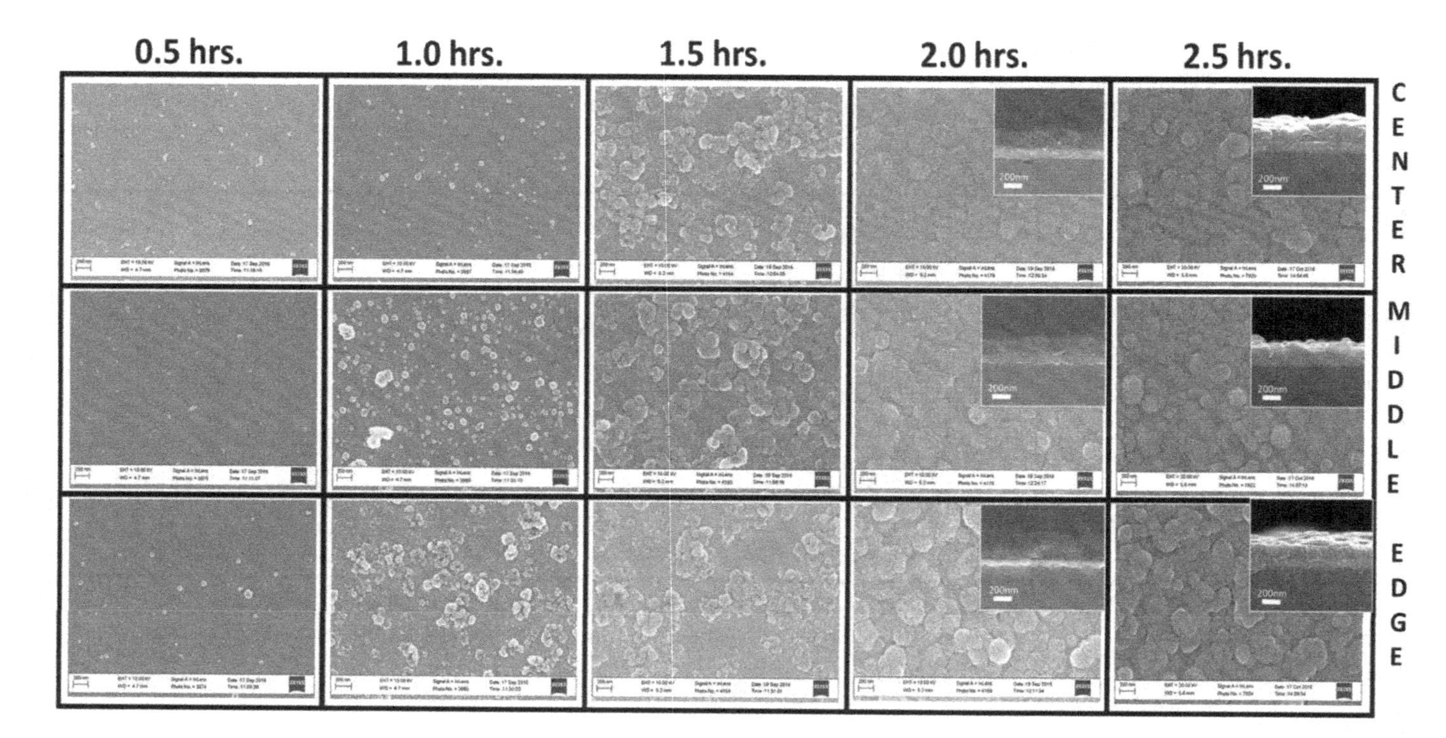

Figure 1.34 SEM top view images of HFCVD BEN-BEG UNCD films grown for times in the range 0.5–2.5 h, with additional growth time of 2 h without bias. The inset figures (top right corner) for the center, middle, and edge areas of the films grown for 2 and 2.5 h show cross-section SEM images of the full dense films grown across the 100 mm diameter substrates (reprinted from *Diam. Relat. Mater.*, vol. 78, p. 1, 2017 (Fig. 4) in [135] with permission from Elsevier Publisher).

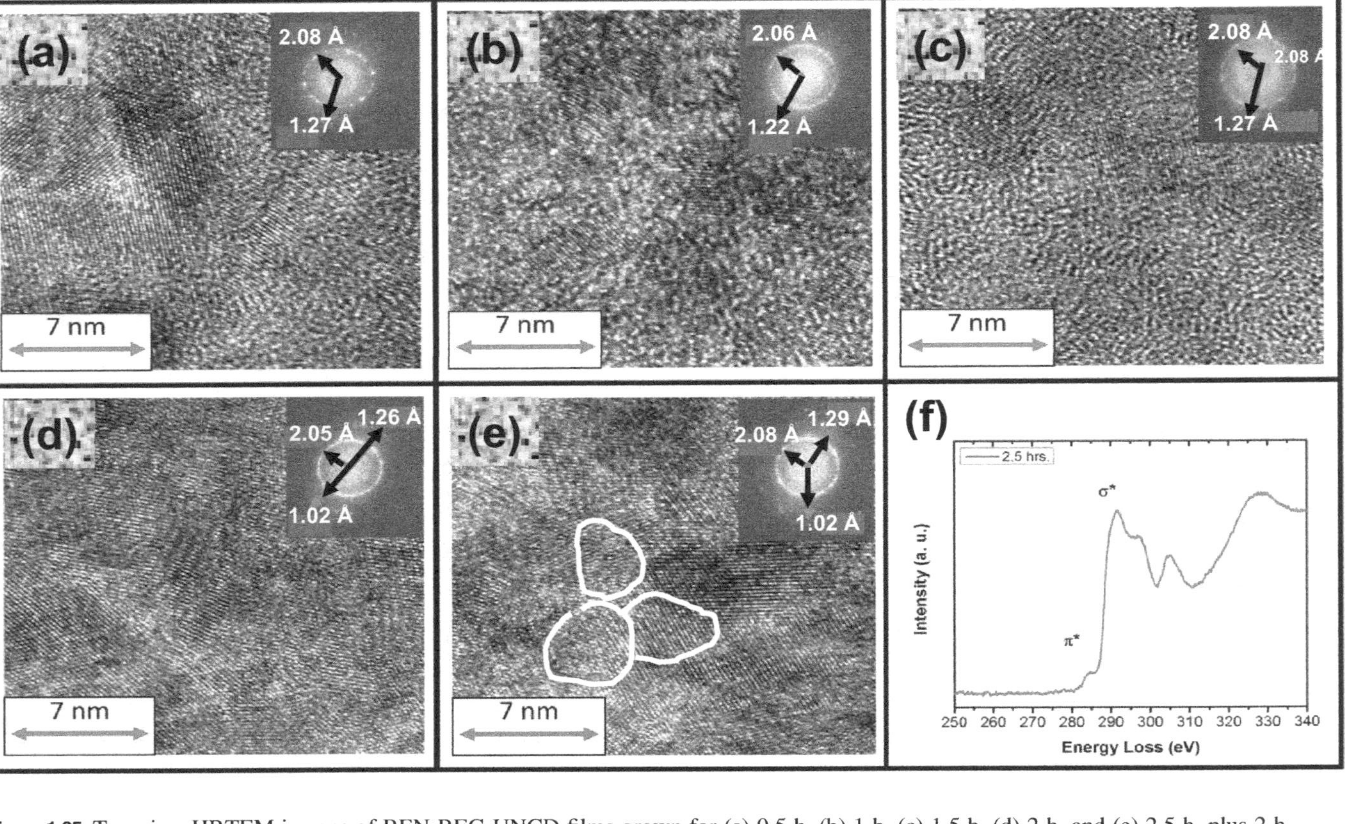

Figure 1.35 Top view HRTEM images of BEN-BEG UNCD films grown for (a) 0.5 h, (b) 1 h, (c) 1.5 h, (d) 2 h, and (e) 2.5 h, plus 2 h growth without bias. (f) Spectrum from EELS analysis corresponding to all BEN-BEG UNCD films, for which HRTEM images are shown in (a)–(e). The inset top right figures show electron diffraction patterns from the UNCD films (the arrows in the diffraction patterns in (a)–(e), pointing to 2.08, 206, and 2.05 Å, correspond to diamond (111) orientation; the arrows pointing to 1.27, 1.29, and 1.22 Å correspond to diamond (220) orientation, and the arrows pointing at 1.02 Å correspond to diamond (311) orientation (reprinted from *Diam. Relat. Mater.*, vol. 78, p. 1, 2017 (Fig. 5) in [135] with permission from Elsevier Publisher).

EELS Analysis. EELS analysis was performed also for all BEN-BEG UNCD films grown in the range 0.5–2.5 h, plus 2 h growth without bias (see a typical curve, similar for all films, in Figure 1.35f). The curve shown in this figure reveals the characteristic spectrum intensity vs. energy loss (step up at about 290 eV and peaks at about 303 and 310 eV) of UNCD obtained in EELS analysis of numerous UNCD films grown with the conventional chemical seeding process using both the MPCVD and the HFCVD processes by many groups worldwide.

EDAX Analysis. Energy-dispersive X-ray spectroscopy (EDAX) analysis was performed during the HRTEM studies, through the cross-sections of all the UNCD films, from the surface into the Si substrate. Figure 1.36 shows the EDAX cross-section analysis profiles from the surface of the UNCD films to the diamond/W/Si interfaces for all BEN-BEG UNCD films grown for 0.5–2.5 h, plus 2 h growth without bias. For all UNCD films, it can be seen that Si diffuses into the W layer and vice versa, induced by the high-temperature (600–700 °C) growth process, as observed by other groups [179]. In addition, as the time of exposure of the substrate to the BEN-BEG process increases, the electric field plus ion bombardment may induce further interdiffusion of W into the UNCD films, which can explain the observation of W particles into the diamond network for 2.0 and 2.5 h BEN-BEG process. A key feature shown by the EDAX analysis is that Si atoms diffused into the W layer do not reach the surface of the W layer, correlating with the fact that WC regions formed at the W layer surface appear to be the dominant regions contributing to nucleation of the UNCD grains.

Studies of UNCD/W Layer Interfaces via Cross-Section HRTEM. A key study enabled understanding of the nucleation process for HFCVD BEN-BEG of UNCD films on a W layer (~40 nm thick) on an Si substrate. Figure 1.37 shows cross-section HRTEM images of the UNCD–W layer interface at the UNCD film nucleation sites for the BEN-BEG UNCD films grown (a) 0.5 h, (b) 1 h, (c) 1.5 h, (d) 2 h, and (e) 2.5 h, plus 2 h growth without bias. All cross-section images were taken at diamond nucleation sites identified via electron diffraction. Determination of the diamond lattice spacing in the UNCD films was done via combination of experimental measurements and modeling using the GMS3 software, applied to different crystal structures containing W and C atoms, such as W, WC, W_2C, graphite, and diamond, detected at the UNCD–W interface. The cross-section HRTEM images, shown in Figure 1.37, revealed that crystalline W nuclei are incorporated in the diamond structure during the growth process. Although EDAX cross-section analysis shown in Figure 1.36 indicates that Si atoms can diffuse close to the WC interface, SiW nuclei or Si were not detected via XPS analysis at the WC–C interface. The prevalent diamond orientation in the UNCD grains, as indicated by the XRD analysis shown in Figure 1.33a is (111), because of minimization of the surface energy [180]. The interfaces investigated (Figure 1.37) showed the appearance of WC with different orientations (Figure 1.37a) in all the interfaces studied, suggesting that WC plays a key role in the nucleation of diamond crystals. Furthermore, the diamond (111) and the WC (101) have very similar structural spacing, allowing for heteroepitaxial growth. Other crystal structures like graphite and pure W could also be present in the interface, which subsequently starts inducing the nucleation of WC regions on the

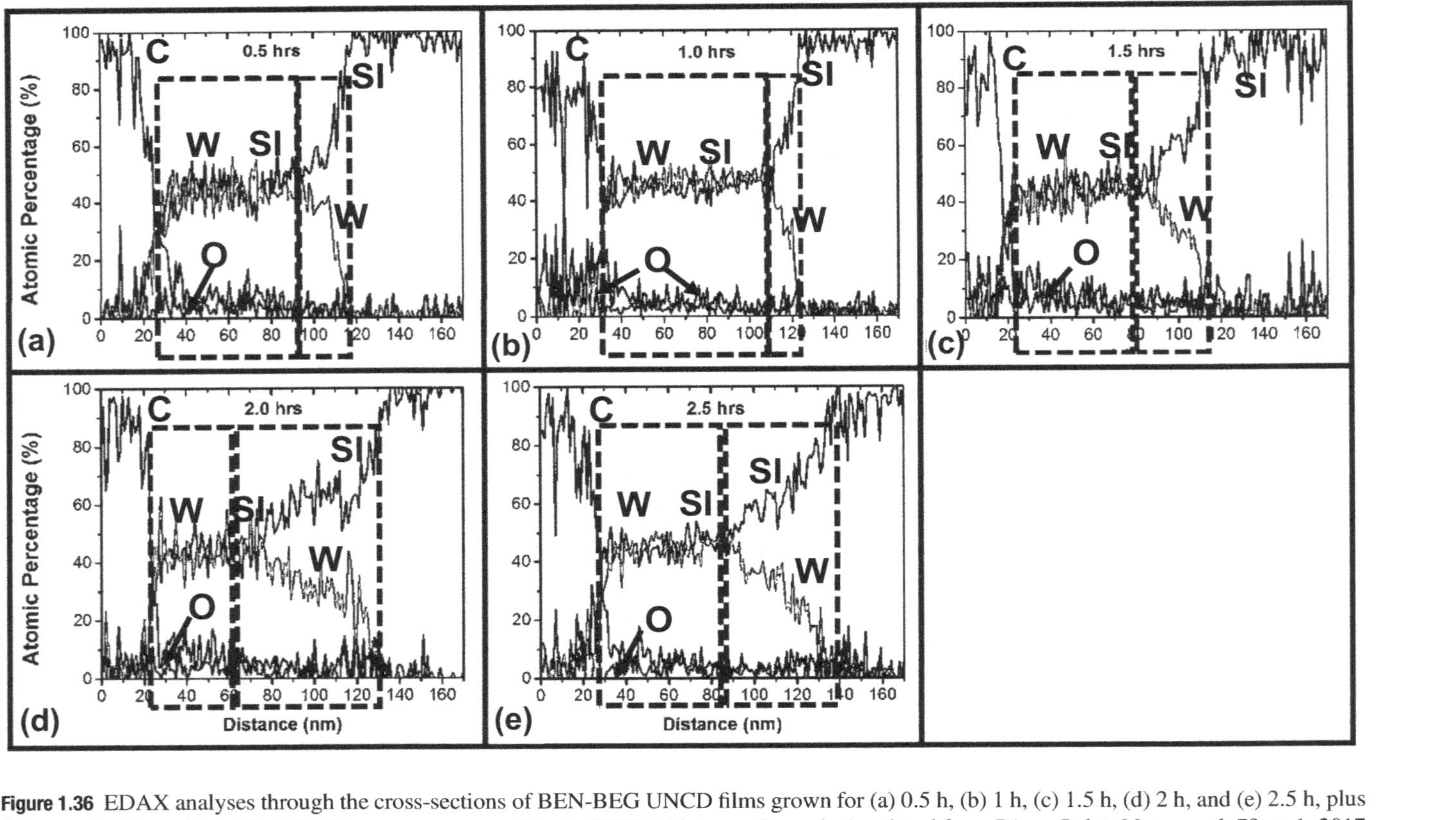

Figure 1.36 EDAX analyses through the cross-sections of BEN-BEG UNCD films grown for (a) 0.5 h, (b) 1 h, (c) 1.5 h, (d) 2 h, and (e) 2.5 h, plus 2 h growth without bias. The EDAX spectra correspond to the HRTEM images shown in (reprinted from *Diam. Relat. Mater.*, vol. 78, p. 1, 2017 (Fig. 6) in [135] with permission from Elsevier Publisher)

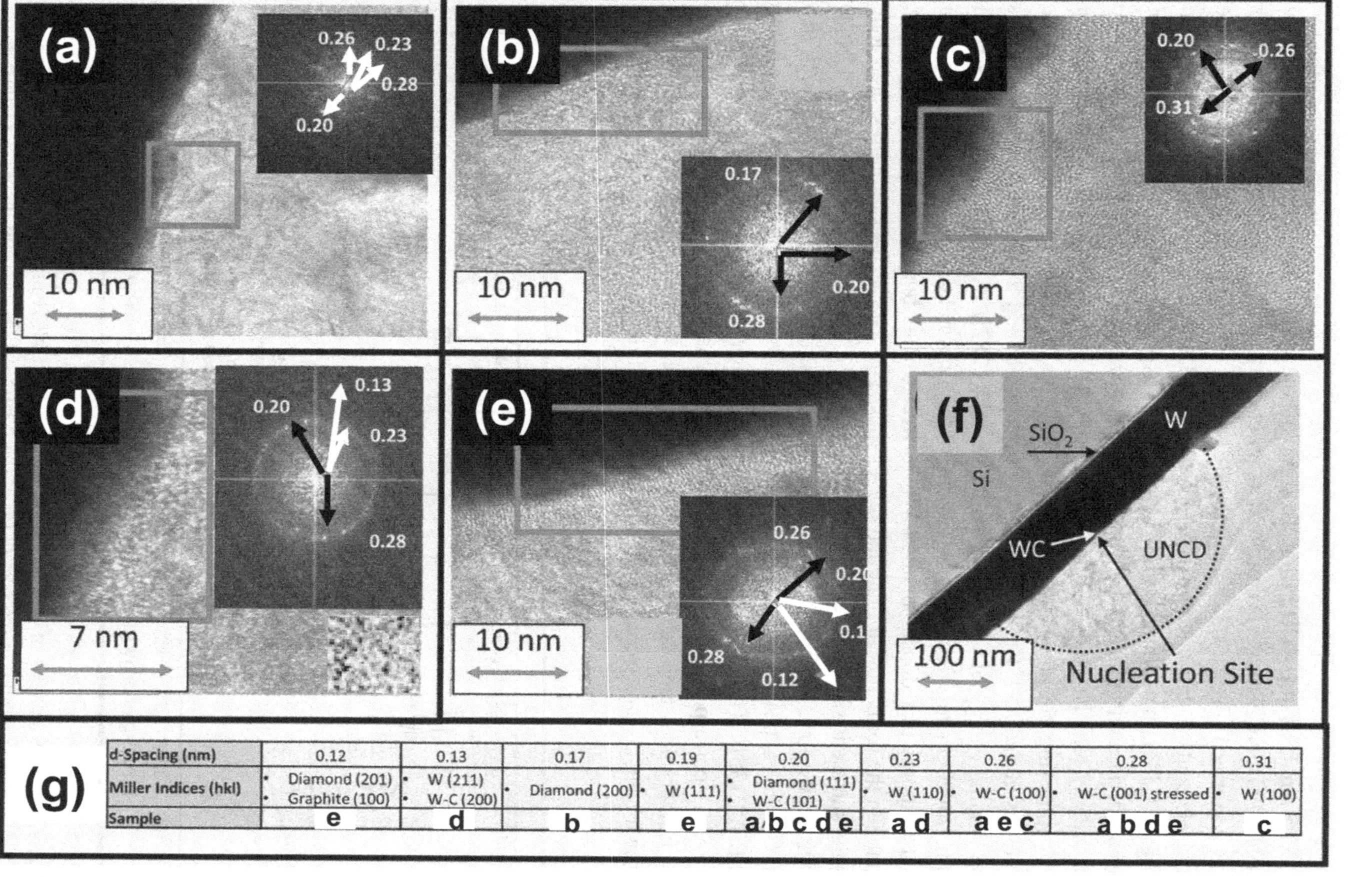

d-Spacing (nm)	0.12	0.13	0.17	0.19	0.20	0.23	0.26	0.28	0.31
Miller Indices (hkl)	• Diamond (201) • Graphite (100)	• W (211) • W-C (200)	• Diamond (200)	• W (111)	• Diamond (111) • W-C (101)	• W (110)	• W-C (100)	• W-C (001) stressed	• W (100)
Sample	e	d	b	e	a b c d e	a d	a e c	a b d e	c

Figure 1.37 Cross-section HRTEM images of BEN-BEG UNCD–W (40 nm thick) interfaces on Si substrates, grown for (a) 0.5 h, (b) 1 h, (c) 1.5 h, (d) 2 h, and (e) 2.5 h, plus 2 h growth without bias. (f) cross-section LRTEM image of UNCD film grown on W (40 nm thick) layer on Si, showing the nucleation of UNCD film on WC and the extended UNCD grown layer for which HRTEM is shown in (c). (g) table showing identification of W, WC, and diamond phases via electron diffraction from the areas in the squares, where the d-spacing values are correlated with the numbers indicated by the arrows in each inset showing the e-diffraction patterns (reprinted from Diam. Relat. Mater., vol. 78, p. 1, 2017 (Fig. 8) in [135] with permission from Elsevier Publisher).

W layer grown on the Si surface (confirmed by the HRTEM [135]) that incubates the growth of the UNCD grains.

XPS Analysis of UNCD Films. To complete the systematic analysis of the HFCVD BEN-BEG UNCD films via complementary characterization techniques, XPS analysis was performed (Figure 1.38) on all the BEN-BEG UNCD films described above. The only elements found at the surface of the UNCD films were carbon and tungsten (Figure 1.38a), in addition to O, which is due to exposure of the UNCD surface to the atmospheric environment when transferring the sample from the HFCVD system to the XPS system. No Si was found, and the oxygen in the surface disappeared after the Ar^+ sputtering cleaning with 1 KeV Ar^+ ions for 30 s. Figure 1.38a shows that the C atoms in the 0.5 h BEN-BEG UNCD films are bound to W atoms, which changes the binding energy of C atoms from 284.5 eV to 282.7 eV [181]. As the BEN-BEG film growth process develops, the amount of C atoms and CH_x radicals increases, such that C atoms that do not bind to W are free to start binding between each other, enabling the formation of diamond nuclei sites. The XPS W peaks are present as long as the surface is still being converted from W to W_2C and WC. For BEN-BEG UNCD films grown for 2.0 h and 2.5 h, the layer of crystalline diamond is dense enough that no C atoms appear bound to W and no W atoms are visible (green flat curve in figure S8B in [135]). The XPS C peak bond energy (285 eV) for UNCD films grown for 1.5–2.5 h corresponds to the C peak bond energy characteristic in the diamond phase. To further understand the Si–W and WC–C interfaces, the 0.5 h HFCVD BEN-BEG-grown UNCD films were analyzed as a function of depth from the film surface, using

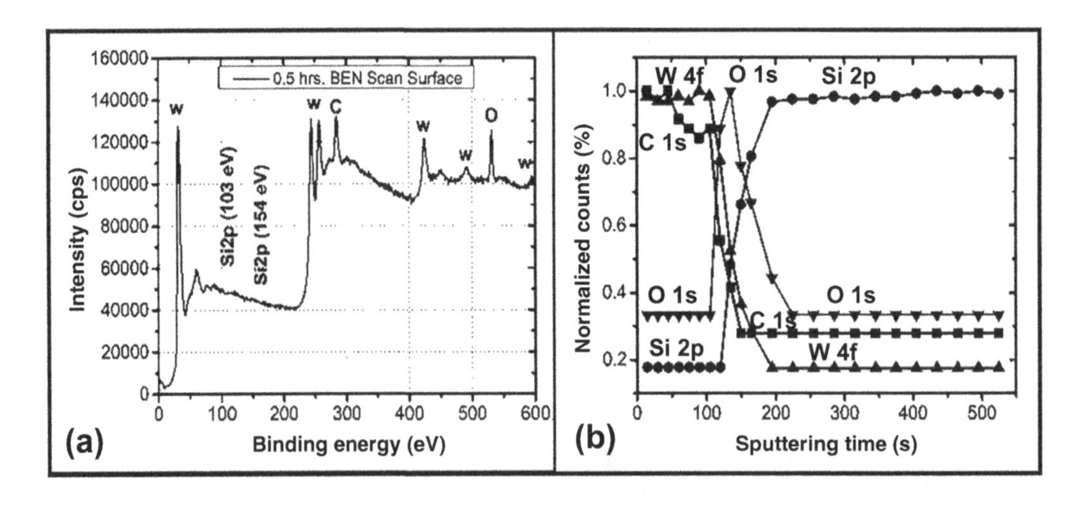

Figure 1.38 XPS analysis of HFCVD BEN-BEG-grown UNCD films, showing the presence of C, O, and W atom binding energy (a) at 0.5 h BEN-BEG time, and the XPS analysis profile spectra (b) when inducing physical etching of the film, via *in situ* Ar^+ ion beam bombardment-induced sputtering in the XPS system (reprinted from *Diam. Relat. Mater.*, vol. 78, p. 1, 2017 (Fig. 7) in [135] with permission from Elsevier Publisher).

XPS profiling, involving Ar^+ ion bombardment to produce physical etching of the film. The depth profiling showed no evidence of Si at the WC–C interface (Figure 1.38b). The W–WC interface could not be determined, but it was clear that there was a very thin layer of SiO_2 between the Si–W/WC interface, as shown by the presence of a sharp O peak at the Si–W interface (Figure 1.38b). The data shows that, initially, the CH_x^+ ($x = 1, 2, 3$) radical ions and C^+ ions produced in the plasma are accelerated toward the negatively biased substrate, producing sub-plantation at the W layer surface, inducing preferential formation of WC instead of sp^3-C bond formation characteristic of diamond. Once the W layer surface is transformed to WC, then formation of sp^3-C bonds occurs, corresponding to the diamond phase [182].

Complementary Experiment and Modeling of HFCVD BEN-BEG of UNCD Films. The mechanism proposed for explaining the growth of UNCD film by HFCVD BEN/BEG on W-covered Si wafers can be seen in Figure 1.39. The initial step, according to this figure, is the formation of a plasma between the filaments and the substrate and substrate holder at a bias voltage of ~200 V, applied between the positively biased filaments and the negatively biased substrate. Figure 1.39a shows that the plasma covers the entire wafer surface. However, Figure 1.39b shows that there is an electric field–enhanced concentration at the relatively sharp substrate holder edges, inducing higher amounts of CH_x^+ and C^+ ions striking the edge of the wafer, initially inducing sub-plantation of the ions on the W layer to induce formation of the W_2C/WC nucleation sites as a precursor for the formation of diamond nanocrystals. Figure 1.39c represents the actual nucleation and growth of the diamond nanocrystals, such that as the W surface layer gets transformed to W_2C/WC, the nucleation and growth of the diamond cluster propagates toward the center. The WC layer then spreads across the substrate surface from edge to center with different crystal orientations until enough nuclei of W (001) and (101) are grown to induce formation of diamond (111), based on the fact that the WC and diamond crystals have similar lattice parameters and low activation energy of 1.39 eV [183]. The optimum time determined for the HFCVD BEN-BEG of UNCD film on a W surface is ~2.5 h, based on the observation that UNCD films produced for this growth time are very uniform, as shown in Figure 1.39d, revealing a fully coated 100 mm diameter W-coated Si substrate with a ~200 nm thick UNCD film.

The HFCVD BEN-BEG growth process to grow UNCD films on Si and other substrates opened the path to new technological applications of UNCD films grown by this unique process, as described in the next section.

Selective Growth of UNCD Films on W-Patterned Layers on a Si Substrate. The electric fields generated between the filaments and the W-coated Si substrate during growth of UNCD films by the HFCVD BEN-BEG process can be used to induce selective growth of diamond films on patterned metallic layers grown on Si or other semiconducting or insulating substrates to produce UNCD film-based patterned devices. This process was demonstrated by producing BEN-BEG of UNCD films on W film–patterned lines on Si substrates. Figure 1.40a shows a BEN-BEG UNCD film grown on patterned W lines resembling a diamond grown

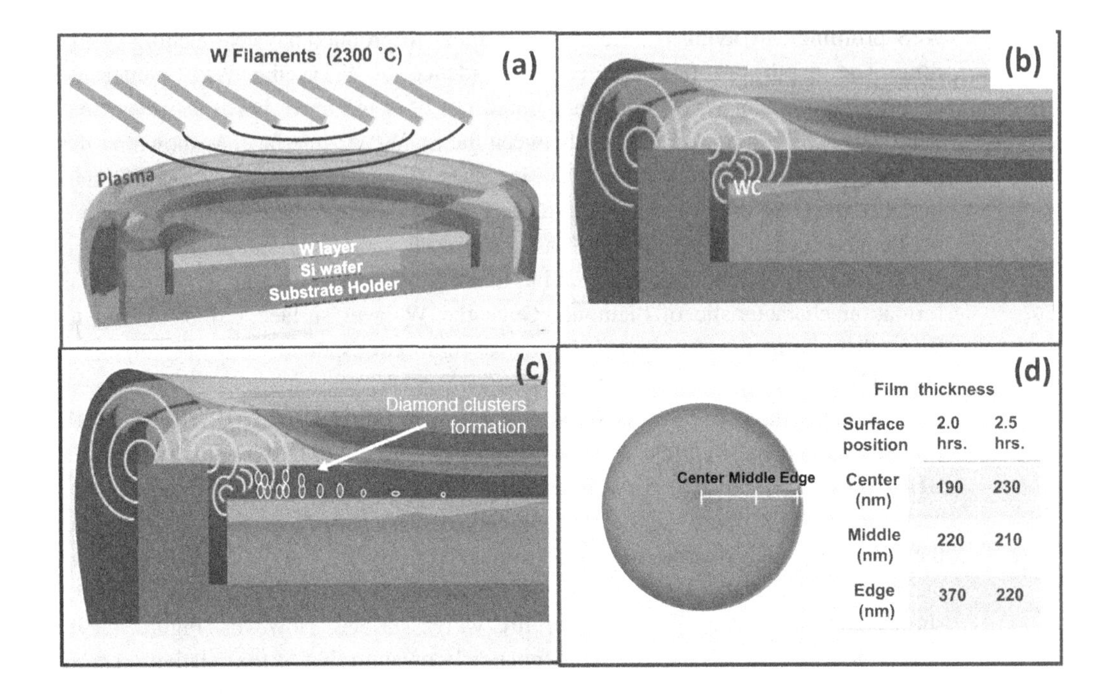

Figure 1.39 (a) Overall schematic of electric field lines across filaments over the W-coated 100 mm diameter Si substrate. (b) Schematic showing the BEN-BEG process inducing WC formation on the surface of the W layer grown on the Si substrate and Si–W reaction areas at the Si–W layer interface. (c) Model representation of plasma–electric field lines inducing diamond nuclei formation from the substrate edge toward the center, due mainly to the initial electric field concentration at the W-coated Si substrate edges during the BEN-BEG growth process. (d) Picture of uniform BEN-BEG UNCD film across 100 mm W-coated Si substrate after 2.5 h of BEN-BEG followed by 2 h of growth without bias (reprinted from *Diam. Relat. Mater.*, vol. 78, p. 1, 2017 (Fig. 10) in [135] with permission from Elsevier Publisher).

on an SiO_2/Si substrate. Figure 1.40b shows the plasma concentrated on the W lines during the HFCVD BEN-BEG of UNCD film selectively grown on the electrically conductive W lines grown on an insulating SiO_2/Si substrate. This experiment demonstrates the feasibility of using the HFCVD BEN-BEG technique for growing UNCD and other diamond films selectively on W-patterned layers on Si-based substrates.

Conclusions on the HFCVD BEN-BEG Process

HFCVD BEN-BEG of UNCD films on W-coated 100 mm diameter Si-based substrates has shown key outcomes:

1. The HFCVD BEN-BEG process produces UNCD films with excellent uniformity on up to 100 mm diameter substrates, using optimized growth times in the range

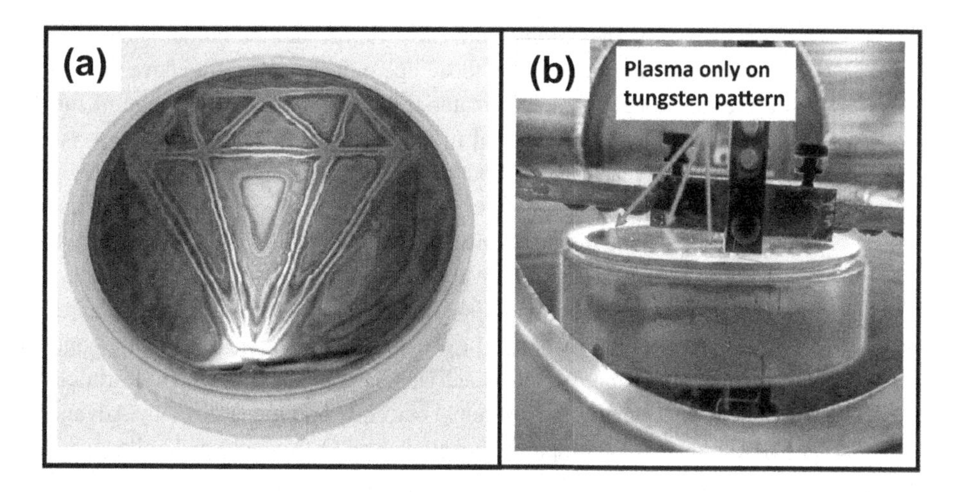

Figure 1.40 (a) Optical picture of UNCD film grown selectively on electrically conductive W-patterned lines with a diamond shape, grown on an SiO_2/Si substrate exposed to 2.0 h of HFCVD BEN-BEG with additional growth time of 2 h without bias. (b) Picture taken during the HFCVD BEN-BEG UNCD film growth process, showing the plasma concentrated on the electrically conductive W lines, connected to the electrically conductive substrate holder with thin wire (compare this picture with the one shown in Figure 1.31a, which shows the bright plasma uniformly distributed across the whole W-coated Si substrate) (reprinted from *Diam. Relat. Mater.*, vol. 78, p. 1, 2017 (Fig. 12) in [135] with permission from Elsevier Publisher).

2.0–2.5 h; however, if the BEN-BEG process is sustained beyond 2.5 h the UNCD film starts to be etched away.

2. The data show that 2.0 h of HFCVD BEN-BEG time induces nucleation of diamond nanocrystals with orientation (001) and (101), which stimulate the formation of diamond grains with (111), (220), and (311) orientations.
3. The concentration of electric fields on the edge area of the substrates has a key influence on the BEN-BEG growth mode, starting from the edge and progressing toward the center area.
4. The HFCVD BEN-BEG process can provide a pathway for producing selectively grown UNCD and other diamond films on patterned electrically conductive W lines on large-area Si substrates for fabrication of diamond-based electronics in the future.

1.4 Conclusions

There are key conclusions related to the UNCD coating technology reviewed in this chapter, with a view to applications in a new generation of medical devices and prostheses:

1. The fundamental and applied science focused on understanding the bases for the synthesis and properties of the novel material named UNCD, in thin-film (coating)

form, since its invention in the early 1990s, and the numerous applications demonstrated until now and those upcoming in the future have proven that UNCD represents a new paradigm material that exhibits a broad range of functionalities applicable to a large range of multifunctional devices.
2. The past 20 years of R&D on UNCD coatings have been extremely productive in advancing the science and technological applications, and many groups worldwide are now working on the science and technology of UNCD coatings, which provides a pathway for substantial advances in the future.
3. UNCD coatings are already in commercial industrial products (e.g., UNCD-coated mechanical pump seals and bearings, AFM tips, and electrically conductive/ corrosion-resistant boron-doped UNCD-coated metal electrodes in electrolysis-based water purification systems) currently commercialized by Advanced Diamond Technologies (a company co-founded by O. Auciello and colleagues in 2003, profitable in 2014, and partially sold to a large company in 2019).
4. The next frontier is the application of UNCD coatings as a unique biocompatible material made of C atoms (the element of life in all human DNA, cells, and molecules). UNCD-coated dental implants are in an advanced state of clinical trials (20 patients have already received UNCD-coated dental implants since 2018, which are demonstrating far superior performance compared with current metal-based implants). The clinical trials are being conducted by Original Biomedical Implants (OBI-USA and OBI-México), two companies founded by Auciello and colleagues, in collaboration with Dr. Gilberto Lopez (a world-class craniofacial surgeon) in Querétaro-México (see Chapter 5). Other UNCD coating-based prostheses to be developed include hips, knees, and many more.
5. The two techniques to grow UNCD coatings reviewed in this introductory chapter, microwave plasma chemical vapor deposition (MPCVD) and hot filament chemical vapor deposition (HFCVD), are the two key techniques that can be used to grow the UNCD coatings for different applications. Each of these techniques needs to be explored to produce the UNCD coatings for particular applications in order to determine which one of them is the most appropriate for application to particular devices or systems.

In conclusion, UNCD coating provides a new paradigm material with superior biocompatibility for new generations of medical devices and prostheses.

Acknowledgments

O. Auciello acknowledges different sources that supported the science and technological applications of UNCD coatings during the last 20 years, namely: Distinguished Endowed Chair Professor grant from University of Texas-Dallas; SENACYT-Panamá, Department of Energy-Basic Energy Sciences grants, DARPA grants, ONR grants, National Science Foundation grants; and industrial funding (Rubio-Pharma-México, UHV-Nanoranch, Samsung, INTEL, Lam, Lockheed-Martin). He also acknowledges the contributions of the three main scientists that jointly with him performed the R&D to develop the UNCD coating technology using the MPCVD method, namely: D. M. Gruen, A. R. Krauss, and J. A. Carlisle at Argonne National Laboratory. He also acknowledges the contributions to the R&D related to the HFCVD technique, namely: M. J. Yacamán, M. J. Arellano-Jimenez, D. Berman-Mendoza, R. Garcia, K. Kang, J. Montes-Gutierrez, P. Tirado, and A. G. Montaño-Figueroa.

References

[1] B. V. Spitsyn, L. L. Bouilov, and B. V. Derjaguin, "Vapor growth of diamond on diamond and other surfaces," *J. Cryst. Growth*, vol. 52, p. 219, 1981.

[2] S. Matsumoto, Y. Sato, M. Tsutsumi, and N. Setaka, "Growth of diamond particles from methane-hydrogen gas," *J. Mater. Sci.*, vol. 17, p. 3106, 1982.

[3] S. Matsumoto, "Development of CVD diamond synthesis techniques," in *Proc. 1st Symposium on Diamond and Diamond-like Films, Electrochem. Soc. Proc.*, New York, vol. 50, p. 89, 1989.

[4] Y. Hirose and N. Kondo, "Program and book of abstracts," *Japan Appl. Phys. Spring Meeting Proceedings*, p. 34, 1988.

[5] M. Kamo, Y. Sato, S. Matsumoto, and N. Setaka, "Diamond synthesis from gas phase in microwave plasma," *J. Cryst Growth*, vol. 62(3), p. 642, 1983.

[6] S. Matsumoto, "Chemical vapor deposition of diamond in RF glow discharge," *J. Mater. Sci. Lett.*, vol. 4(5), p. 600, 1985.

[7] M. Matsumoto, M. Hino, and T. Kobayashi, "Synthesis of diamond films in a RF induction thermal plasma," *Appl. Phys. Lett.*, vol. 51, p. 737, 1987.

[8] K. Suzuki, S. Sawabe, H. Yasuda, and T. Inzuka, "Growth of diamond thin films by DC plasma chemical vapor deposition," *Appl. Phys. Lett*, vol. 50 (12), 728, 1987.

[9] K. Kurihara, K. Sasaki, M. Kawaradi, and N. Koshino, "High-rate synthesis of diamond by DC plasma jet chemical vapor deposition," *Appl. Phys. Lett.*, vol. 52, p. 437, 1988.

[10] P. K. Backmann and R. Messier, "Emerging technology of diamond thin films," *C&EN*, vol. 67(20), p. 24. 1989.

[11] S. J. Harris and D. G. Goodwin, "Growth on the reconstructed diamond (100) surface," *J. Phys. Chem.*, vol. 97, p. 23, 1993.

[12] B. Dischler and C. Wild, *Low-Pressure Synthetic Diamond: Manufacturing and Applications*, Heidelberg: Springer, 1998.

[13] P. K. Bachmann, H. J. Hagemann, H. Lade, et al., "Thermal properties of C/H, C/H/O, C/H/N and C/H/X grown polycrystalline CVD diamond," *Diam. Relat. Mater.*, vol. 4: p. 820, 1995.

[14] T. Sharda and S. Bhattacharyya, "Advances in nanocrystalline diamond," in *Encyclopedia of Nanoscience and Nanotechnology*, vol. 2, H.S. Nalwa, Ed. Stevenson Ranch, CA: American Scientific Publishers, p. 337, 2004.

[15] J. E. Butler and A. V. Sumant, "The CVD of nanodiamond materials," *Chem. Vap. Deposition*, vol.14, p. 145, 2008.

[16] D. M. Gruen, S. Liu, A. R. Krauss, L. Luo, and X. Pan, "Fullerenes as precursors for diamond film growth without hydrogen or oxygen additions," *Appl. Phys. Lett.*, vol. 64, p. 1502, 1994.

[17] T. G. McCauley, T. D. Corrigan, A.R. Krauss, O.Auciello et al., "Electron emission properties of Si field emitter arrays coated with nanocrystalline diamond from fullerene precursors," *Proc. MRS, Symposium. "Electron Emission from Highly Covalent Materials,"* 1998, vol. 498, p. 227.
[18] D. M. Gruen, Nanocrystalline diamond films, *Annu. Rev. Mater. Sci.* vol. 29: p. 211, 1999.
[19] R. E. Smalley, "Discovering the fullerenes," *Rev. Mod. Phys*. vol. 69, p. 723, 1997.
[20] S. Jiao, A. V. Sumant, M. A. Kirk, et al. "Microstructure of ultrananocrystalline diamond films grown by microwave Ar–CH_4 plasma chemical vapor deposition with or without added H_2," *J. Appl. Phys.*, vol. 90, p. 118, 2001.
[21] N. Naguib, J. Birrell, J. Elam, J. A. Carlisle, and O. Auciello, "A method to grow carbon thin films consisting entirely of diamond grains 3–5 nm in size and high-energy grain boundaries," US Patent #7,128,8893, 7,556,982, 2006.
[22] O. Auciello and A. V. Sumant, "Status review of the science and technology of ultrananocrystalline diamond (UNCD^{TM}) films and application to multifunctional devices," *Diam. Relat. Mater.*, vol. 19, p. 699, 2010.
[23] O. A. Shenderova and D. M. Gruen, Eds., *Ultrananocrystalline Diamond: Synthesis, Properties and Applications*, 2nd ed. Oxford: Elsevier, 2012.
[24] D. Zhou, D. M. Gruen, L. C. Qin, T. G. McCauley, and A. R. Krauss, "Control of diamond film microstructure by Ar additions to CH_4/H_2 microwave plasmas," *J. Appl. Phys*. vol. 84, p. 1981, 1998.
[25] Advanced Diamond Technologies, Inc. Homepage. www.thindiamond.com.
[26] Original Biomedical Implants, Inc. Homepage. www.originalbiomedicalimplants.com.
[27] A. R. Konicek, D. S. Grierson, P. U. P. A. Gilbert, W. G. Sawyer, A. V. Sumant, and O. Auciello, "Origin of ultralow friction and wear in ultrananocrystalline diamond," *Phys Rev Lett.*, vol. 100 (23), p. 235502, 2008.
[28] X. Xiao, J. Wang, J. A. Carlisle, et al., "*In Vitro* and *in vivo* evaluation of ultrananocrystalline diamond for coating of implantable retinal microchips," *J. Biomed. Mater.*, vol. 77B, p. 273, 2006.
[29] S. Bhattacharyya, O. Auciello, J. Birrell, et al., "Synthesis and characterization of nitrogen doped ultrananocrystalline diamond thin films," *Appl. Phys. Lett.*, vol. 79, p. 1441, 2001.
[30] D. M. Gruen, A. R. Krauss, O. Auciello, and J. A. Carlisle, "N-type doping of NCD films with nitrogen and electrodes made therefrom," US patent #6,793,849 B1, 2004.
[31] H. Zeng, U. P. Arumugam, S. Siddiqui, and J. A. Carlisle, "Low temperature boron doped diamond," *Appl. Phys. Lett.*, vol. 103, p. 223108, 2013.
[32] W.-X. Yuan, Q. X. Wu, Z. K. Luo, and H. S. Wu, "Effects of boron doping on the properties of ultrananocrystalline diamond films," *J Electr. Mater.*, vol. 43 (4), p. 1302, 2014
[33] K. Okano, H. Naruki, Y. Akiba, et al., "Characterization of boron-doped diamond film," *Japan. J. Appl. Phys.*, vol. 28 (1), p. 1066, 1989.
[34] G. M. Swain and R. Ramesham, "The electrochemical activity of boron-doped polycrystalline diamond thin film electrodes," *Anal. Chem.*, vol. 65 (4), p. 345, 1993.
[35] R. Ramesham, "Selective growth and characterization of doped polycrystalline diamond thin films," *Thin Solid Films*, vol. 229, p. 44, 1993.
[36] S. W. Vernon, M. Swope, J. E. Butler, T. Feygelson, and G. M. Swain, "The structural and electrochemical properties of boron-doped nanocrystalline diamond thin-film

electrodes grown from Ar-rich and H_2-rich source gases," *Diam. Relat. Mater.*, vol. 18 (4), p. 669, 2009.

[37] V. V. S. S. Srikanth, P. S. Kumar, and V. B. Kumar, "A brief review on the in-situ synthesis of boron-doped diamond thin films," *Int J Electrochem*, vol. 2012, Article ID 218393, 2012.

[38] P. Tirado, J. J. Acantar-Peña, E. de Obaldia, et al., "Boron doping of ultrananocrystalline diamond films by thermal diffusion process," *MRS Commun.*, vol. 8 (3), p. 1111, 2018.

[39] X. Xiao, J. Birrell, J. E. Gerbi, O. Auciello, and J. A. Carlisle, "Low temperature growth of ultrananocrystalline diamond," *J. Appl. Phys.*, vol. 96, p. 2232, 2004.

[40] J. A. Carlisle, D. M. Gruen, O. Auciello, and X. Xiao, "A method to grow pure nanocrystalline diamond films at low temperatures and high deposition rates," US Patent #7,556,982, 2009.

[41] J. J. Alcantar-Peña, G. Lee, E. M. A. Fuentes-Fernandez, et al., "Science and technology of diamond films grown on HfO_2 interface layer for transformational technologies," *Diam. Relat. Mater.*, vol. 69. p. 221, 2016.

[42] S. Sudarsan, J. Hiller, B. Kabius, and O. Auciello, "Piezoelectric/ultrananocrystalline diamond heterostructures for high-performance multifunctional micro/nanoelectromechanical systems," *Appl. Phys. Lett.*, vol. 90, p. 134101, 2007.

[43] M. Zalazar, P. Gurman, J. Park, et al., "Integration of piezoelectric aluminum nitride and ultrananocrystalline diamond films for implantable biomedical microelectromechanical devices," *Appl. Phys. Lett.*, vol. 102, p. 104101, 2013.

[44] O. Auciello, J. Birrell, J. A. Carlisle, et al., "Materials science and fabrication processes for a new MEMS technology based on ultrananocrystalline diamond thin films," *J. Phys Condens. Matter*, vol. 16, p. R539, 2004.

[45] O. Auciello, S. Pacheco, A. V. Sumant, et al., "Are diamonds a MEMS best friend?," *IEEE Microwave Mag*, vol. 8, p. 61, 2008.

[46] O. Auciello, "Science and technology of ultrananocrystalline diamond (UNCD^{TM}) film-based MEMS and NEMS devices and systems," in *Science and Technology of UNCD Films*, O. A. Shenderova and D. M. Gruen, Eds. New York: Elsevier, p. 383, 2013.

[47] Y.-W. Cheng, C-K. Lin, Y.-C. Chu, et al., "Electrically conductive ultrananocrystalline diamond-coated natural graphite-copper anode for new long-life lithium-ion battery," *Adv. Mater.*, vol. 26 (1–5), p. 3724, 2014.

[48] Y. Tzeng, O Auciello, C-P. Liu, C-K. Lin, and Y-W Cheng, "Nanocrystalline-diamond/carbon and nanocrystalline-diamond/silicon composite electrodes for Li-based batteries," US Patent #9,196,905, 2015.

[49] Y. C. Link, K. J. Sankaran, Y. C. Chen, et al., "Enhancing electron field emission properties of UNCD films through nitrogen incorporation at high substrate temperature," *Diam. Relat. Mater.*, vol. 20 (2), p. 191, 2011.

[50] A. R. Krauss, M. Q. Ding, O. Auciello, et al., "Electron field emission for ultrananocrystalline diamond films," *J. Appl. Phys.*, vol. 89, p. 2958, 2001.

[51] M. Hajra, M. Ding, O. Auciello, et al. "Effect of gases on the field emission properties of ultrananocrystalline diamond-coated silicon field emitter arrays," *J. Appl. Phys.*, vol. 94 (6), p. 4079, 2003.

[52] K. Panda, J. J. Hyeok, J. Y. Park, et al., "Nanoscale investigation of enhanced electron field emission for silver ion implanted/post-annealed ultrananocrystalline diamond films," *Sci. Rep.*, vol. 7, article number 16325, 2017.

[53] S. A. Getty, O. Auciello, A. V. Sumantet al. "Characterization of nitrogen-incorporated ultrananocrystalline diamond as a robust cold cathode material," in *Micro-and Nanotechnology Sensors, Systems, and Applications-II*, T. George, S. Islam, and A. Dutta, Eds. Bellingham, WA: SPIE, p. 76791N-1, 2010.

[54] J. M. Garguilo, F. A. M. Koeck, R. J. Nemanich, et al., "Thermionic field emission from nanocrystalline diamond-coated silicon tip arrays," *Phys. Rev.*, vol. B 72, p. 165404, 2005.

[55] J. Wang, M. A. Firestone, O. Auciello, and J. A. Carlisle, "Surface functionalization of ultrananocrystalline diamond films by electrochemical reduction of aryl diazonium salts," *Langmuir*, vol. 20, p. 11450, 2004.

[56] W. Yang, O. Auciello, J. E. Butler, et al., "Direct electrical detection of hybridization at DNA-modified nanocrystalline diamond thin films," *J. Electrochem. Soc.*, 2007.

[57] P. Bajaj, D. Akin, A. Gupta, et al. "Ultrananocrystalline diamond film as an optimal cell interface for biomedical applications," *Biomed. Microdevices*, vol. 9 (6), p. 787, 2007.

[58] B. Shi, Q. Jin, L. Chen, and O. Auciello, "Fundamentals of ultrananocrystalline diamond (UNCD) thin films as biomaterials for developmental biology: embryonic fibroblasts growth on the surface of (UNCD) films," *Diam. Relat. Mater.*, vol. 18 (2), p. 596, 2008.

[59] O. Auciello , "Novel biocompatible ultrananocrystalline diamond coating technology for a new generation of medical implants, devices, and scaffolds for developmental biology," *Biomater. Med. Appl. J.*, vol. 1 (1), 1000103, 2017.

[60] J. E. Butler and H. Windischmann, "Developments in CVD-diamond synthesis during the past decade," *MRS Bull.*, vol. 23 (9), p. 22, 1998.

[61] M. A. Prelas, G. Popovici, and K.L. Biglow (Eds.). *Handbook of Industrial Diamonds and Diamond Films*. Chichester: Wiley, 2009.

[62] E. Kohn, P. Gluche, and M. Adamschik, "Diamond MEMS: a new technology," *Diam. Relat. Mater.*, vol. 8, p. 934, 1999.

[63] S. Rotter, *Proc. Applied Diamond Conference/Frontier Carbon Technologies-ADC/FCT '99*, M. Yoshikawa, Y. Koga, Y. Tzeng, C. P. Klages and K. Miyoshi, Eds. Tokyo: MYU, K.K, p. 25, 1999.

[64] A. V. Sumant, D. S. Grierson, A. R. Konicek, et al., "Surface composition, bonding, and morphology in the nucleation and growth of ultra-thin, high quality nanocrystalline diamond films," *Diam. Relat. Mater.*, vol. 16, p. 718, 2007.

[65] P. W. May, J. N. Harvey, J. A. Smith, and Y. A. Mankelevich, "Re-evaluation of the mechanism for ultrananocrystalline diamond deposition from $Ar/CH_4/H_2$ gas mixtures," *J. Appl. Phys.*, vol. 99, p. 104907, 2006.

[66] P. W. May, N. L. Allan, M. N. R. Ashfold, J. C. Richley, and Y. A. Mankelevich, "Simplified Monte Carlo simulations of chemical vapour deposition diamond growth," *J. Phys.: Condens. Matter.*, vol. 21, p. 364203, 2009.

[67] A. V. Sumant, O. Auciello, H.-C. Yuan, et al., "Large area low temperature ultrananocrystalline diamond (UNCD) films and integration with CMOS devices for monolithically integrated diamond MEMS/NEMS-CMOS systems," *Proc. SPIE*, vol. 7318, p. 17, 2009.

[68] Z. Xu, Z. He, Y. Song, et al., "Topic review: application of Raman spectroscopy characterization in micro/nano-machining," *Micromachines (Basel)*, vol. 9(7), p. 361, 2018.

[69] W. H. Weber and R. Merlin, Eds., *Raman Scattering in Materials Science*, Berlin: Springer, 2000.

[70] J. Birrell, J. E. Gerbi, O. Auciello, et al. "Interpretation of the Raman spectra of ultrananocrystalline diamond," *Diam. Relat. Mater.*, vol. 14 p. 86, 2005.
[71] J. Stohr, *NEXAFS*, New York, Springer, 1992.
[72] D. Zuiker, A. R. Krauss, D. M. Gruen, et al., "Characterization of diamond thin films by core-level photo-absorption and UV excitation Raman spectroscopy," *Mat. Res. Soc. Proc.*, vol. 437, p. 211, 1996.
[73] P. Tirado, J. Alcantar-Peña, E. de Obaldia, R. Garcia, and O. Auciello, "Effect of the gas chemistry, total pressure, and microwave power on the grain size and growth rate of polycrystalline diamond films grown by microwave plasma chemical vapor deposition technique," *Proc. IEEE-7th International Engineering, Sciences and Technology Conference*, p. 85, 2019.
[74] E. M. A. Fuentes-Fernandez, J. J. Alcantar-Peña, G. Lee, et al., "Synthesis and characterization of microcrystalline diamond to ultrananocrystalline diamond films via hot filament chemical vapor deposition for scaling to large area applications," *Thin Solid Films*, vol. 603, p. 62, 2016.
[75] J. Filik, "Raman spectroscopy: a simple non-destructive way to characterize diamond and diamond-like materials," *Spectroscopy Europe*, vol. 17 (5), p. 10, 2005.
[76] B. R. Stoner, G.-H. M. Ma, S. D. Wolter, and J. T. Glass, "Characterization of bias-enhanced nucleation of diamond on silicon by *in vacuo* surface analysis and transmission electron microscopy," *Phys. Rev. B*, vol. 45, p. 11067, 1992.
[77] S. Gerber, S. Sattel, H. Ehrhardt, et al., "Investigation of bias enhanced nucleation of diamond on silicon," *J. Appl. Phys.*, vol. 79, p. 4388, 1996.
[78] Y. C. Lee, S. J. Lin, C. T. Chia, H. F. Cheng, and I. N. Lin, "Effect of processing parameters on the nucleation behavior of nano-crystalline diamond film," *Diam. Relat. Mater.*, vol. 14, p. 296, 2005.
[79] M. Q. Ding, R. Krauss, O. Auciello, et al., "Studies of field emission from bias-grown diamond thin films," *J. Vac. Sci. Tech B*, vol. 17, p. 705, 1999.
[80] Y. C. Lee, S. J. Lin, C. Y. Lin, et al., "Pre-nucleation techniques for enhancing nucleation density and adhesion of low temperature deposited ultra-nano-crystalline diamond," *Diam. Relat. Mater.*, vol. 15, p. 2046, 2006.
[81] Y. C. Chen, X. Y. Zhong, A. R. Konicek, et al., "Synthesis and characterization of smooth ultrananocrystalline diamond films via low pressure bias-enhanced nucleation and growth," *Appl. Phys. Lett.*, vol. 92, p. 133113, 2008.
[82] O. Auciello and R. Kelly, *Ion Bombardment Modification of Surfaces: Fundamentals and Applications*, Amsterdam: Elsevier, 1984.
[83] X. Y. Zhong, Y. C. Chen, N. H. Tai, et al. "Effect of pretreatment bias on the nucleation and growth mechanisms of ultrananocrystalline diamond films via bias-enhanced nucleation and growth: an approach to interfacial chemistry analysis via chemical bonding mapping," *J. Appl. Phys.*, vol. 105, p. 034311, 2009.
[84] K-Y. Teng, C. Huang-Chin, G-C. Tzeng, et al., "Bias enhanced nucleation and growth processes for improving the electron field emission properties of diamond films," *J. Appl. Phys.*, vol. 111, p. 053701, 2012.
[85] A. Härtl, E. Schmich, J. A. Garrido, et al., "Protein modified nanocrystalline diamond thin films for biosensor applications," *Nat. Mater.*, vol. 3, p. 736, 2004.
[86] O. Auciello, "Unpublished: Auciello's wife defibrillator/pacemaker needed to be replaced in 2019, after only 6 years from implantation."

[87] S. F. Cogan, "Neural stimulation and recording electrodes," *Ann. Rev. Biomed. Eng.*, vol. 10, p. 275, 2008.
[88] Y. Lu, T. Li, X. Zhao, et al., "Electrodeposited polypyrrole/carbon nanotubes composite films electrodes for neural interfaces," *Biomaterials*, vol. 31, p. 5169, 2010.
[89] A. Hung, R. Greenberg, D. M. Zhou, J. Judy, and N. Talbot, "High-density array of micro-machined electrodes for neural stimulation," US patent #7676274 B2, 2010.
[90] F. Brunet, P. Germi, M. Pernet, et al., "The effect of boron doping on the lattice parameter of homoepitaxial diamond films," *Diam. Relat. Mater.*, vol. 7(6), p. 869, 1998.
[91] J. Seo, H. Wu, S. Mikael, et al., "Thermal diffusion boron doping of single-crystal natural diamond," *J. Appl. Phys.*, vol. 119, p. 205703, 2016.
[92] J. Cui, J. Ristein, and L. Ley, "Electron affinity of the bare and hydrogen covered single crystal diamond (111) surface," *Phys. Rev. Lett.*, vol. 81, p. 429, 1998.
[93] T. Sung, G. Popovici, M. A. Prelas, and R. G. Wilson, "Boron diffusion coefficient in diamond," *MRS Proc.*, vol. 416, p. 467, 1996.
[94] G. Popovici, T. Sung, S. Khasawinah, M. A. Prelas, and R. G. Wilson, "Forced diffusion of impurities in natural diamond and polycrystalline diamond films," *J. Appl. Phys.*, vol. 77, p. 5625, 1995.
[95] D. J. Garrett, A. L. Saunders, C. McGowan, et al., "In vivo biocompatibility of boron-doped and nitrogen included conductive-diamond for use in medical implants," *J. Biomed. Mater. Res B*, vol. 1048 (1), p, 19, 2015.
[96] S. Mateti, C. S. Wong, Z. Liu, et al., "Biocompatibility of boron nitride nanosheets," *Nano Res.*, vol. 11, p. 334, 2018.
[97] K. Okano, S. Koizumi, S. R. Silva, and G. A. J. Amaratunga, "Low threshold cold cathodes made of nitrogen-doped chemical-vapor-deposited diamond," *Nature*, vol. 398, p. 140, 1996.
[98] S. Koizumi, M. Kamo, Y. Sato, H. Ozaki, and T. Inuzuka, "Growth and characterization of phosphorous doped {111} homoepitaxial diamond thin films," *Appl. Phys. Lett.*, vol. 71, p. 1065, 1997.
[99] B. D. Yu, Y. Miyamoto, and O. Sugino, "Efficient *n*-type doping of diamond using surfactant-mediated epitaxial growth," *Appl. Phys. Lett.*, vol. 76, p. 976, 2000.
[100] J. F. Prins, "N-type semiconducting diamond by means of oxygen-ion implantation," *Phys. Rev. B*, vol. 61, p. 719, 2000.
[101] T. Collins and E. C. Lightowlers, "Electrical properties," in *The Properties of Diamond*, J. E. Field, Ed. London: Academic Press, pp. 79, 1979.
[102] S. Prawer, C. Uzan-Saguy, G. Braunstein, and R. Kalish, "Can n-type doping of diamond be achieved by Li or Na ion implantation?," *Appl. Phys. Lett.*, vol. 63, p. 2502, 1993.
[103] S. Koizumi, M. Kamo, Y. Sato, et al., "Growth and characterization of phosphorus doped *n*-type diamond thin films," *Diam. Relat. Mater.*, vol. 7 (2–5), p. 540, 1998.
[104] D. L. Childers, J. Corman, M. Edwards, and J. J. Elser, "Sustainability challenges of phosphorus and food: solutions from closing the human phosphorus cycle," *Bio Science*, vol. 61, p. 117, 2011.
[105] J. Shao, H. Xie, H. Huang, et al., "Biodegradable black phosphorus-based nanospheres for in vivo photothermal cancer therapy," *Nat. Commun.*, 2016. DOI: 10.1038/ncomms12967.
[106] J. O. Island, G.A. Steele, H. S. van der Zant, and A. Castellanos-Gomez. "Environmental instability of few-layer black phosphorus," *2D Mater.*, vol. 2, p. 011002, 2015.
[107] X. Ling, H. Wang, S. Huang, F. Xia, and M. S. Dresselhaus, "The renaissance of black phosphorus," *Proc. Natl Acad. Sci. USA*, vol. 112, p. 4523, 2015.

[108] I. Pravst, "Risking public health by approving some health claims? The case of phosphorus," *Food Policy*, vol. 36, p. 726, 2011.

[109] M. S. Dresselhaus and R. Kalish, *Ion Implantation in Diamond, Graphite and Related Materials*. Berlin: Springer, 1992.

[110] J. Birrell, J. A. Carlisle, O. Auciello, D. M. Gruen, and J. M. Gibson, "Morphology and electronic structure in nitrogen-doped ultrananocrystalline diamond," *Appl. Phys. Lett.*, 81, p. 2235, 2002.

[111] F. L. Coffman, R. Cao, P. A. Pianetta, et al. "Near edge X-ray absorption of carbon materials for determining bond hybridization in mixed sp^2/s^3 bonded materials," *Appl. Phys. Lett.*, vol. 69, p. 568, 1996.

[112] R. J. Nemanich, J. T. Glass, and G. Lucovsky, "Raman scattering characterization of carbon bonding in diamond and diamond-like films," *J. Vac. Sci. Technol. A*, vol. 6, p. 1783, 1988.

[113] P. Zapol, M. Sternberg, L. A. Curtiss, T. Frauenhein, and D. M. Gruen, "Tight binding molecular dynamics simulation of impurities in ultranancrystalline diamond grain boundaries," *Phys. Rev. B*, vol. 65, p. 045403, 2001.

[114] P. R. Heck, F. J. Staderman, D. Isheim, et al., "Atom-probe analyses of nanodiamonds from Allende," *Meteorit. Planet. Sci.*, vol. 49(3), p. 453, 2014.

[115] A. Gicquel, F. Silva, and K. Hassouni, "Diamond growth mechanism in various environments," *J. Electrochem. Soc.*, vol. 147, p. 2218, 2000.

[116] T. C. Choy, A. M. Stoneham, M. Ortuno, and A. M. Somoza, "Negative magnetoresistance in ultrananocrystalline diamond: strong or weak localization?," *Appl. Phys. Lett.*, vol. 92, p. 012120, 2008.

[117] S. Bhattacharyya, "Two-dimensional transport in disordered carbon and nanocrystalline diamond films," *Phys. Rev. B*, vol. 77, p. 233407, 2008.

[118] K. V. Shah, D. Churochkin, Z. Chiguvare, and S. Bhattacharyya, "Anisotropic weakly localized transport in nitrogen-doped ultrananocrystalline diamond films," *Phys. Rev. B*, vol. 82, p. 184206, 2010.

[119] W. Yuan, L. Fang, Z. Feng, et al., "Highly conductive nitrogen-doped ultrananocrystalline diamond films with enhanced field emission properties: triethylamine as a new nitrogen source," *J. Mater. Chem. C*, vol. 4, p. 4778, 2016.

[120] K. Jothiramalingan, S. Haenen, and K, Haenen, "Nitrogen incorporated ultrananocrystalline films for field electron emission applications," in *Diamond: Novel Applications of Diamond*, N. Yang, Ed. Siegen: Springer, p. 123, 2015.

[121] K. J. Sankaran, J. Kurian, H. C. Chen, et al., "Origin of a needle-like granular structure for ultrananocrystalline diamond films grown in a N_2/CH_4 plasma," *J. Phys. D Appl. Phys.*, vol. 45, p. 365303, 2012.

[122] C. S. Wang, G. H. Tong, H. C. Chen, W. C. Shih, and I. N. Lin, "Effect of N_2 addition in Ar plasma on the development of microstructure of ultra-nanocrystalline diamond films," *Diam. Relat. Mater.*, vol. 19, p. 147, 2010.

[123] B.-R. Huang, C-T. Chia, M-C. Chang, and C-L. Cheng, "Bias effects on large area polycrystalline diamond films synthesized by the bias enhanced growth technique," *Diam. Relat. Mater.*, vol. 12, p. 26, 2003.

[124] R. Alves, A. Amorim, J. Eichenberger Neto, et al., "Filmes de diamante CVD em grandes áreas obtidos por crescimentos sucessivos em etapas," *Matéria Rio J.*, vol. 13 (3), p. 569, 2008.

[125] J. Weng, F. Liu, L. W. Xiong, J. H. Wang, and Q. Sun, "Deposition of large area uniform diamond films by microwave plasma CVD," *Vacuum*, vol. 147, p. 134, 2018.

[126] Q. Suna and J. Wang, "Study on the large area diamond film deposition in a self built overmoded microwave power chemical vapor deposition device," *Chem. Eng. Trans.*, vol. 62, p. 1129, 2017.
[127] S. T. Lee, Y. W. Lam, Z. Lin, Y. Chen, and Q. Chen, "Pressure effect on diamond nucleation in a hot-filament CVD system," *Phys. Rev. B*, vol. 55, p. 15937, 1997.
[128] S. Schwarz, S. M. Rosiwal, M. Frank, D. Breidt, and R. F. Singer, "Dependence of the growth rate, quality, and morphology of diamond coatings on the pressure during the CVD-process in an industrial hot-filament plant," *Diam. Relat. Mater.*, vol. 11, 589, 2002.
[129] T. Hao, H. Zhang, C. Shi, and G. Han, "Nano-crystalline diamond films synthesized at low temperature and low pressure by hot filament chemical vapor deposition," *Surf. Coat. Tech.*, vol. 201, p. 801. 2006.
[130] P. W. May, J. A. Smith, and Y. A. Mankelevich, "Deposition of NCD films using hot filament CVD and Ar/ CH_4/H_2 gas mixtures," *Diam. Relat. Mater.*, vol. 15, p. 345, 2006.
[131] X. Liang, L. Wang, H. Zhu, and D. Yang, "Effect of pressure on nanocrystalline diamond films deposition by hot filament CVD technique from CH_4/H_2 gas mixture," *Surf. Coat. Tech.*, vol. 202, p. 261, 2007.
[132] K. Uppireddi, B. R. Weiner, and G. Morell, "Synthesis of nanocrystalline diamond films by DC plasma-assisted argon-rich hot filament chemical vapor deposition," *Diam. Relat. Mater.*, vol. 17, p. 55, 2008.
[133] D. C. Barbosa, F. A. Almeida, R. F. Silva, et al., "Influence of substrate temperature on formation of ultrananocrystalline diamond films deposited by HFCVD argon-rich gas mixture," *Diam. Relat. Mater.*, vol. 18, p. 1283, 2009.
[134] D. C. Barbosa, P. Hammer, V. J. Trava-Airoldi, and E. J. Corat, "The valuable role of renucleation rate in ultrananocrystalline diamond growth," *Diam. Relat. Mater.*, vol. 23, p. 112, 2012.
[135] J. J. Alcantar-Peña, E. de Obaldia J. Montes-Gutierrez, et al., "Fundamentals towards large area synthesis of multifunctional ultrananocrystalline diamond films via large area hot filament chemical vapor deposition bias enhanced nucleation/bias enhanced growth for fabrication of broad range of multifunctional devices," *Diam. Relat. Mater.*, vol. 78, p. 1, 2017.
[136] N. Naguib, J. Birrell, J. Elam, J. A. Carlisle, and O. Auciello, "Use of tungsten interlayer to enhance the initial nucleation and conformality of ultrananocrystalline diamond (UNCD) thin films," US Patent #20070257265 A1.
[137] N. N. Naguib, J. W. Elam, J. Birrell, et al. "Enhanced nucleation, smoothness and conformality of ultrananocrystalline diamond (UNCD) ultrathin films via tungsten interlayers," *Chem. Phys. Lett.*, vol. 430, p. 345, 2006.
[138] H. Fussstetter, H. Richter, and M. Umeno, "Sub-quarter-micron silicon issues in the 200/ 300 mm conversion era," *Microelectron. Eng.*, vol. 56 (1–2), p. 1, 2001.
[139] R. Wallace and O. Auciello, "Science and technology of high-dielectric constant (K) thin films for next generation CMOS," in *Thin films and Heterostructures for Oxide Electronics*, S.B. Ogale, Ed. New York: Springer, p. 79, 2005.
[140] F. Klauser, D. Steinmüller-Nethl, R. Kaindl, E. Bertel, and N. Memmel, "Raman studies of nano- and ultrananocrystalline diamond films grown by hot-filament CVD," *Chem. Vap. Depos.*, vol. 16 (4–6), p. 127, 2010.
[141] A. C. Ferrari and J. Robertson, "Interpretation of Raman spectra of disordered and amorphous carbon," *Phys. Rev. B*, vol. 61 (20), p. 14095. 2000.

[142] P. Špatenka, H. Shur, G. Erker, and M. Rump, "Formation of hafnium carbide thin films by plasma enhanced chemical vapor deposition from bis(η-cyclopentadienyl) dimethyl hafnium as precursor," *Appl. Phys. Mater. Sci. Process.*, vol. 60 (3), p. 285, 1995.
[143] L. Ramqvist, K. Hamrin, G. Johansson, A. Fahlman, and C. Nordling, "Charge transfer in transition metal carbides and related compounds studied by ESCA," *J. Phys. Chem. Solids*, vol. 30 (7), 1835, 1969.
[144] I. H. Jaffer, J. C. Fredenburgh, J. Hirsh, and J. I. Weitz, "Medical device-induced thrombosis: what causes it and how can we prevent it?," *J. Thrombosis Haemostasis*, vol. 13 (Suppl. 1), p. S72, 2015.
[145] C. J. Wilson, R. E. Clegg, D. I. Leavesley, and M. J. Pearcy, "Mediation of biomaterial–cell interactions by adsorbed proteins: a review," *Tissue Eng.*, vol. 11, p. 1, 2005.
[146] L. E. Corum and V. Hlady, "Screening platelet–surface interactions using negative surface charge gradients," *Biomaterials*, vol. 31(12), p. 3148, 2010.
[147] E. Hughes, "Advances in hydrophilic and hydrophobic coatings for medical devices," *Medical Design Briefs Magazine*, 2017.
[148] M. W. Varney, D. M. Aslam, A. Janoudi, H.-Y. Chan, and D. H. Wang, "Polycrystalline-diamond MEMS biosensors including neural microelectrode-arrays," *Biosensors*, vol. 1, p. 118, 2011.
[149] W. P. Kang, Y. Gurbuz, J. L. Davidson, and D. V. Kerns, "A new hydrogen sensor using a polycrystalline diamond-based Schottky diode," *J. Electrochem. Soc.*, vol. 141, p. 2231, 1994.
[150] Y. Gurbuz, W. P. Kang, J. L. Davidson, D. L. Kinser, and D. V. Kerns, "Diamond microelectronics gas sensors," in *8th International Conference on Transducers*, p. 745, 1995.
[151] S. Guillauden, X. Zhu, and D. A. Aslam, "Fabrication of 2 μm wide polycrystalline diamond channels using silicon molds for micro-fluidic applications," *Diam. Relat. Mater.*, vol. 12, p. 65, 2003.
[152] A. Gabriela Montano-Figueroa, J. J. Alcantar-Peña, P. Tirado, et al., "Tailoring of polycrystalline diamond surfaces from hydrophilic to superhydrophobic via synergistic chemical plus microstructuring processes," *Carbon*, vol. 139, p. 361, 2018.
[153] K. Teli, M. Hori, and T. Goto, "Co-deposition on diamond film surface during reactive ion etching in SF_6 and O_2 plasmas," *J. Vac. Sci. Technol. A*, vol. 18, p. 2779, 2000.
[154] C. Popov, H. Vasilchina, W. Kulisch, et al., "Wettability and protein adsorption on ultrananocrystalline diamond/amorphous carbon composite films," *Diam. Relat. Mater.*, vol. 18, p. 895, 2009.
[155] L. Osterovskaya, V. Perevertailo, V. Ralchenko, A. Saveliev, and V. Zhuraviev, "Wettability of nanocrystalline diamond films," *Diam. Relat. Mater.*, vol. 16, p. 2109, 2007.
[156] S. F. Durant, V. Baranauskas, A. C. Peterlevitz, et al., " Characterization of diamond fluorinated by glow discharge plasma treatment," *Diam. Relat. Mater.*, vol. 10, p. 490, 2001.
[157] A. Fredman and C. D. Stinespring, "Fluorination of diamond (100) by atomic and molecular beams, *Appl. Phys. Lett.*," vol 57, p. 1194, 1990.
[158] A. Denisenko, A. Romanyuk, C. Pietzka, J. Scharpf, and E. Kohn, "Surface structure and surface barrier characteristics of boron-doped diamond in electrolytes after CF_4 plasma treatment in RF-barrel reactor," *Diam. Relat. Mater.*, vol. 19, p. 423, 2010.

[159] T. Ando, J. Tanaka, M. Ishii, et al. "Diffuse reflectance Fourier-transform infrared study of the plasma-fluorination of diamond surfaces using a microwave discharge in CF_4," *J. Chem. Soc. Faraday Trans.*, vol. 89, p. 3105, 1993.
[160] M. Schvartzman, A. Mathur, J. Hone, C. Jahnes, and S. J. Wind, "Plasma fluorination of carbon-based materials for imprint and molding lithographic applications," *Appl. Phys. Lett.*, vol. 93, p. 153105, 2008.
[161] W. Kulisch, A. Voss, D. Merker, et al., "Plasma surface fluorination of ultrananocrystalline diamond films," *Surf. Coat. Technology*, vol. 302, p. 448, 2016.
[162] Y.-S. Park, H.-G. Son, D.-H Kim, et al., "Microarray of neuroblastoma cells on the selectively functionalized nanocrystalline diamond thin film surface," *Appl. Surf. Sci.*, vol. 361, p. 269, 2016.
[163] M. Nosonovsky and B. Bhushan, "Biomimetic super-hydrophobic surfaces: multiscale approach," *Nano Lett.*, vol. 7, p. 2633, 2007.
[164] J. Kim and S.-O. Choi, "Super-hydrophobicity," in *Waterproof and Water Repellent Textiles and Clothing*, J. Williams, Ed. Cambridge: Woodhead Publishing, p. 267, 2018.
[165] Q. Yang, C. Xiao, and A. Hirose, "Plasma enhanced deposition of nano-structured carbon films," *Plasma Sci. Technol.*, 7(1), p. 2660, 2005.
[166] B. R. Stoner and J. T. Glass, "Textured diamond growth on (100) p-SiC via microwave plasma chemical vapor deposition," *Appl. Phys. Lett.*, vol. 60, p. 698, 1992.
[167] T. Soga, T. Sharda, and T. Jimbo, "Precursors for CVD growth of nanocrystalline diamond," *Phys. Solid State*, vol. 46(4), p. 720, 2004.
[168] K. Janischowsky, W. Ebert, and E. Kohn, "Bias enhanced nucleation of diamond on silicon (100) in a HFCVD system," *Diam. Relat. Mater.*, vol. 12(3–7), p. 336, 2003.
[169] K. Uppireddi, B. R. Weiner, and G. Morell, "Synthesis of nanocrystalline diamond films by DC plasma-assisted argon-rich hot filament chemical vapor deposition," *Diam. Relat. Mater.*, vol. 17(1), p. 55. 2008.
[170] T. D. Makris, R. Giorgi, N. Lisi, L. Pilloni, and E. Salernitano, "Bias enhanced nucleation of diamond on Si (100) in a vertical straight hot filament CVD," *Diam. Relat. Mater.*, vol. 14(3–7), p. 318, 2005.
[171] X. T. Zhou, H. L. Lai, H.Y Peng, et al. "Heteroepitaxial nucleation of diamond on Si (100) via double bias assisted hot filament chemical vapor deposition," *Diam. Relat. Mater.*, vol. 9(2), p. 134, 2000.
[172] S. Pecoraro, J. C. Arnault, and J. Werckmann, "BEN-HFCVD diamond nucleation on Si (111) investigated by HRTEM and nano-diffraction," *Diam. Relat. Mater.*, vol. 14(2), p. 137, 2005.
[173] Y. Li, J. Li, Q. Wang, Y. Yang, and C. Gu, "Controllable growth of nanocrystalline diamond films by hot filament chemical vapor deposition method," *J Nanosci. Nanotechnol.*, vol. 9(2), p. 1062, 2009.
[174] S. G. Ansari, T. L. Anh, H-K. Seo, et al., "Growth kinetics of diamond film with bias enhanced nucleation and $H_2/CH_4/Ar$ mixture in a hot-filament chemical vapor deposition system," *J Cryst. Growth*, vol. 265(3–4), p. 563, 2004.
[175] J. Robertson, J. Gerber, S. Sattel, et al., "Mechanism of bias-enhanced nucleation of diamond on Si," *Appl. Phys. Lett.*, vol. 66, p. 3287, 1995.
[176] J. Gerber, J. Robertson, S. Sattel, and H. Ehrhardt, "Role of surface diffusion processes during bias-enhanced nucleation of diamond on Si," *Diam. Relat. Mater.*, vol. 5(3–5), p. 261, 1996.

[177] M. M. García, I. Jiménez, O. Sánchez, C. Gómez-Aleixandre, and L. Vázquez, "Model of the bias-enhanced nucleation of diamond on silicon based on atomic force microscopy and x-ray-absorption studies," *Phys. Rev. B*, vol. 61, p. 383, 2000.
[178] T. Dash, B. B. Nayak, M. Abhangi, et al., "Preparation and neutronic studies of tungsten carbide composite," *Sci. Technol.*, vol. 65(2), p. 241, 2014.
[179] A. de Luca, A. Poravoce, M. Texier, et al., "Tungsten diffusion in silicon," *J. Appl. Phys.*, vol. 115(1), p. 013501, 2014.
[180] H. Liu and D. S. Dandy, "Studies on nucleation process in diamond CVD: an overview of recent developments," *Diam. Relat. Mater.*, vol. 4(10), p. 1173, 1995.
[181] Z. Liu, P. Li, F. Zhai, et al., "Amorphous carbon modified nano-sized tungsten carbide as a gas diffusion electrode catalyst for the oxygen reduction reaction," *RSC Adv.*, vol. 5 (87), p. 70743, 2015.
[182] S. Liu, S. E. Xie, J. Sun, C. Ning, and Y. Jiang, "A study on nano-nucleation and interface of diamond film prepared by hot filament assisted with radio frequency plasma," *Mater. Lett.*, vol. 57(11), p. 1662, 2003.
[183] F. A. M. Koeck and R. J. Nemanich, "Substrate–diamond interface considerations for enhanced thermionic electron emission from nitrogen doped diamond films," *J. Appl. Phys.*, vol. 112(11), p. 113707, 2012.

2 Ultrananocrystalline Diamond (UNCD™) Film as a Hermetic Biocompatible/Bioinert Coating for Encapsulation of an Eye-Implantable Microchip to Restore Partial Vision to Blind People

Orlando Auciello

2.1 Background Information on Vision, Conditions that Induce Blindness in Humans Born with Vision, and Concepts for Artificial Retina Devices to Restore Partial Vision

2.1.1 Human Image (Vision) Formation

Human vision is produced by photons in light emanating from the outside world, which penetrate the eye through the cornea and the lens [1] (Figure 2.1a), the latter focusing the light onto the retina at the back of the eye and inducing photonic-based excitation of the retina photoreceptors (Figure 2.1b). The photonic excitation is transformed into electrical charges, which excite the bipolar cells connected to the photoreceptors (Figure 2.1b). The bipolar cells amplify the electrical charges and inject them into the ganglion cells, from where the charges are transmitted through their axons, forming the optical nerve, to the part of the brain called the thalamus or lateral geniculate nucleus (LGN), deep in the center of the brain, where the images are finally formed.

2.1.2 Retinitis Pigmentosa

Retinitis pigmentosa (RP) is a biological condition induced by a progressive degeneration of the human retina's photoreceptors, which are responsible for the first action in the generation of human vision, as shown in Figure 2.1. The degeneration of the photoreceptors starts in the peripheral retina (see the white square on the retina in Figure 2.1a), progressing toward the macula, near the optic nerve, to finally induce blindness when the last photoreceptor dies. Retinitis pigmentosa affects over one million individuals worldwide per year [1]. It is a hereditary condition that has been linked to more than 3000 mutations spanning 50 genes that encode proteins critical for photo-transduction [1, 2]. A mutation in the *USH2A* gene is known to cause 10–15%

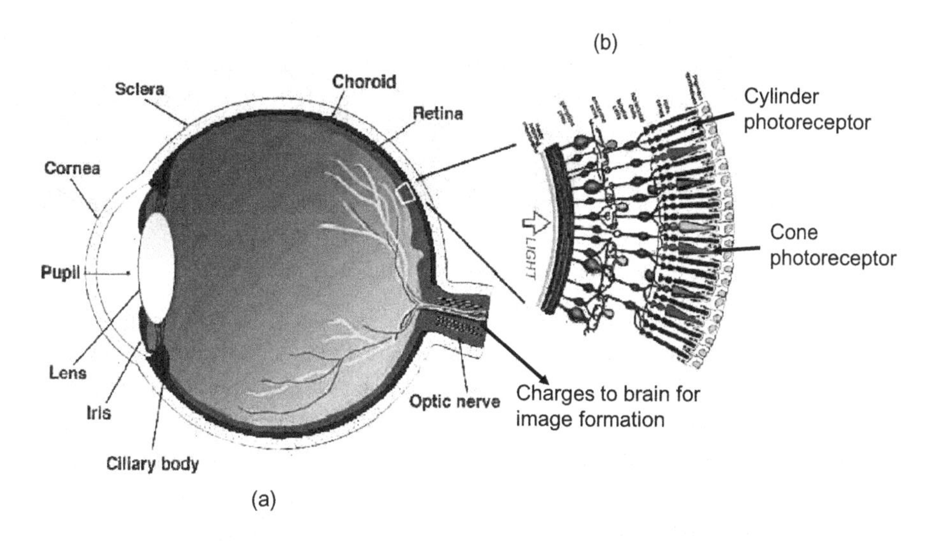

Figure 2.1 (a) Schematic of the human eye showing the main components. (b) Schematic of the retina showing the main three layers and connection to the brain, where visual images are finally formed for vision, as explained in the text.

of a syndrome form of RP known as Usher's syndrome when inherited in an autosomal recessive fashion. Although each mutation may act differently, typically, RP is manifested by the death of rod photoreceptors in the peripheral retina, responsible for night vision. In general, RP results in the reduction of peripheral and photopic (color) vision, the latter provided by the cone photoreceptors. In the early stages of RP, photopic vision is spared because many of the cone photoreceptors in the central retina are still functional, but eventually the cone photoreceptors also degenerate, resulting in severe visual impairment [3]. Actually, it has been determined that in severe end-stage RP, approximately 95% of photoreceptors, 20% of bipolar cells, and 70% of ganglion cells degenerate [4]. A significant fraction of the inner retinal neurons is not destroyed, but the absence of photoreceptors prevents useful vision.

Retinitis pigmentosa is the main biological condition that leads to blindness for people born with normal vision, and usually starts in childhood, although the beginning of the condition and how quickly it evolves to blindness varies from person to person. The person with RP loses peripheral vision at the same time or soon after their night vision declines. The condition is known as "tunnel vision." Figure 2.2 shows the progression from normal vision to "tunnel vision" to blindness generated by RP.

2.2 Vision Restoration Technologies

2.2.1 Gene-Based Retinal Degeneration Therapies

Currently, there is no cure for RP, although several treatments are being explored. Biologically based gene therapy approaches to treat RP are being investigated and are

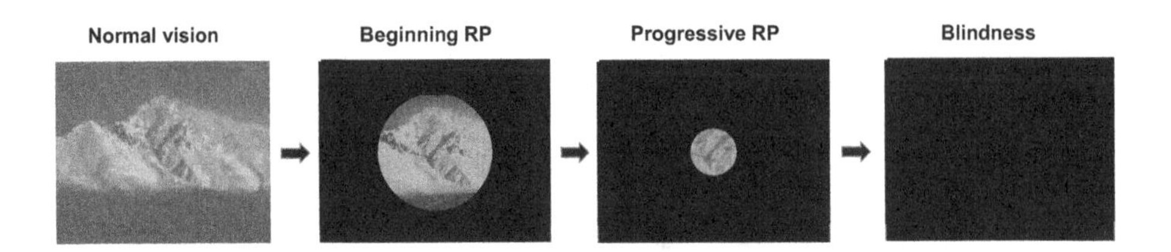

Figure 2.2 Evolution from normal vision to blindness due to RP.

under validation in animal studies [5, 6]. Prior gene therapy-based clinical trials to treat Leber's congenital amaurosis has shown encouraging results for the potential future use of gene therapy to treat RP [7, 8]. A potential impediment for gene therapies is that there are many genetic mutations that cause RP, each of which would require a unique gene therapy solution. Another RP biomolecule-based treatment under development uses ontogenetics, which focus on inducing light sensitivity in the remaining functional retinal neurons via introduction of photovoltaic-sensitive molecules or transmembrane proteins [9]. The first clinical trial involving the use of ontogenetics started recently, using channelrhodopsin-2 [10].

2.2.2 Electronic Retinal Prostheses-Based Retinal Degeneration Therapies

Electronic retinal prostheses represent another RP therapy, and one of them is the main focus of this chapter. Retinal prostheses functionality involves electrical stimulation of the surviving functional natural retinal neurons (ganglion and bipolar cells) to induce responses that are transmitted to the brain and interpreted by patients as a visual percept.

Electronic-based retinal prostheses evolved from years of R&D focused on using electronic excitation of functional vision-producing cells (e.g., ganglion and bipolar cells) to restore vision lost due to RP. The first attempts to generate artificial vision involved excitation of the cortex via application of electric current to the occipital lobe, which produced perception of light in patients [11]. This work led to the development of an 80-electrode implantable cortical visual prosthesis that proved capable of inducing visual percepts in a blind patient [12]. Several groups are now working on developing visual cortical prostheses [13]. An advantage of cortical visual prostheses over retinal prostheses is that the effectiveness of the first is largely independent of ocular health. However, major challenges for cortical prostheses are the long-term stability of the electrode–tissue interface as current electrodes are made mainly of metals that, in general, can be chemically corroded [14] by the eye humor, thus compromising the creation of visual perception.

Several types of retinal prostheses are currently under development:

1. a flexible digital viewing device, including information and navigation tools [15];
2. a microsystem visual prosthesis involving a spiral electrode around the optic nerve, connected to an implanted stimulator, which receives images from an

external camera and translates them into electrical signals that stimulate the optic nerve [15];

3. a miniature telescope implanted in the eye's posterior chamber, which works by increasing (about three times) the size of the image projected onto the retina to overcome a central blind spot [16];
4. the Tübingen MPDA Project, involving a microchip located behind the retina connected to micro-photodiode arrays (MPDA), collect incident light and transform it into electrical current that stimulates the retina ganglion cells [17];
5. an artificial silicon retina involving a microchip containing photo diodes, which detect light and convert it into electrical pulses that stimulate ganglion cells (discontinued);
6. a photovoltaic retinal prosthesis, with a subretinal photodiode array and an infrared image projection system mounted on video goggles, which transmits information from a video camera to be processed by a pocket PC and then displayed on pulsed near-infrared (850–915 nm) video goggles, projecting the images onto the retina via the eye optics to activate photodiodes in the subretinal implant, converting light into pulsed bi-phasic electric current in each pixel [18];
7. Dobelle Eye, involving a camera mounted on glasses that transmits images to a stimulator chip implanted in the brain's primary visual cortex that injects electrical charges directly to the brain cells [19];
8. an intracortical visual prosthesis with intracortical electrode arrays, similar to the Dobelle system, but with increased spatial resolution based on more electrodes per unit area [20];
9. Bionic Vision Australia, involving a microchip with 98 stimulating electrodes projected for implantation in the supra-choroidal space (a device with 1024 electrodes is under development) [21];
10. Harvard/MIT Retinal Implant, involving a subretinal stimulator chip with an electrode array beneath the retina, which receives images beamed from a camera mounted on glasses, decodes them, and stimulates ganglion cells with electrical pulses [22];
11. Argus II Retinal Prosthesis, involving a chip that in the final rendition would be implanted inside the eye on the ganglion cell layer (Figure 2.3), receive an image from a camera on glasses, and inject processed electrical charges to the ganglion cells through a large electrode array, before finally transmitted to the brain via the ganglion cell axons bundle (optical nerve) [23]. The Argus II device was developed by a team of researchers from universities, national laboratories, and Second Sight (the company currently commercializing the device) during a 10-year project funded by the US Department of Energy. The Argus II is currently the most advanced artificial retina device, and the only one currently implanted commercially in the USA and Europe to restore partial sight to blind people. It was named one of the top 25 inventions for 2013 by *TIME Magazine*.

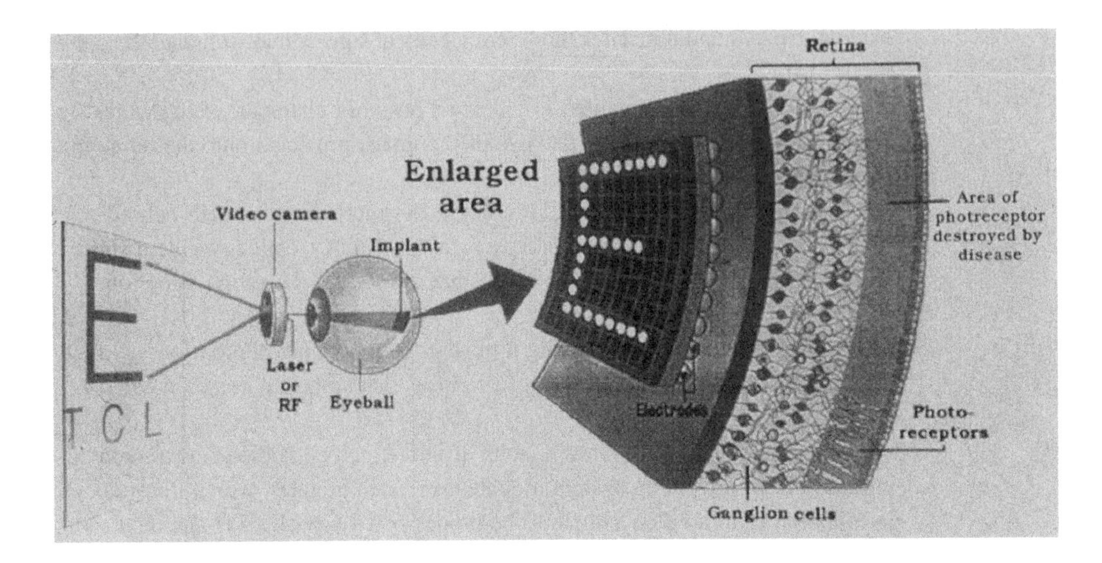

Figure 2.3 Schematic showing a CCD (charge-coupled device) camera mounted on glasses outside the eye, capturing images and sending wireless electromagnetic signals with image information to an Si-based microchip implanted inside the eye, as desired in its final future rendition, with a large array of electrodes connected to the ganglion cells; the Si microchip should be coated with a biocompatible/hermetic/humor corrosion-resistant coating. UNCD is probably the best biocompatible/eye fluid corrosion-resistant coating, as demonstrated in 10 years of animal studies (reprinted from *MRS Bull.* vol. 39, p. 621, 2014 (Fig. 3) in [24] with permission from Cambridge Publisher).

2.2.3 Argus II Retinal Prosthesis

The microchip of the Argus II device should ideally be implanted totally inside the eye with an array of electrodes connected to the ganglion cell layer of the retina. The idea is that the implanted microchip receives wireless electromagnetic signals carrying information about images from the outside world images, processes the image, and transmits the information in the form of electrical charges injected into the ganglion cells, with enough resolution that the ganglion cell transmits the charges to the brain via the optical nerve – formed by the axons of the ganglion cells – to help restore limited vision and enabled people blinded by RP to be able to read and move around without assistance, and with the ultimate objective of – in the future – being able to recognize people's faces [23]. The basic principle of the retinal implant, developed by a DOE consortium, was to translate the visual information captured by a CCD camera mounted on a pair of glasses into a pattern of electrical stimulation charges that are subsequently injected into the retina at the ganglion cell layer level for subsequent transmission to the brain to restore image formation.

Clinical trials of a prototype artificial retina demonstrated that blind people can see images induced by electronic stimulation of brain cells. The initial prototype device, used in early clinical trials, involved a low-resolution device that included an array of

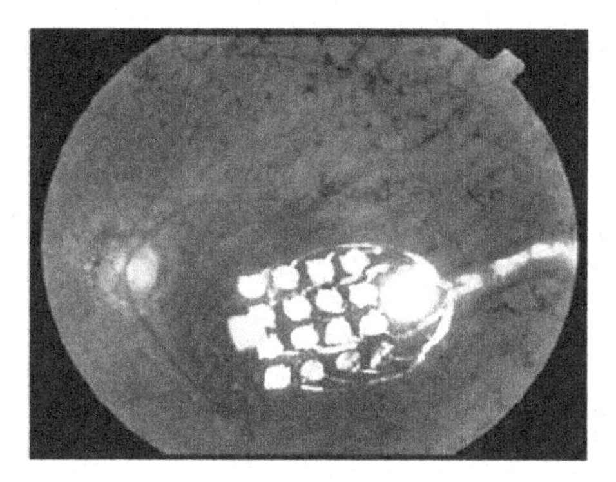

Figure 2.4 Optical picture taken through the eye of a clinical trial patient implanted with the initial 16-electrode Pt array in a polymer matrix (see the detailed description in [25]).

16 electrodes, made of Pt circular electrodes produced by a metal synthesis technique, described in [25]. The circular Pt electrodes and wires, connecting the electrodes to the microchip, are located outside the eye and encapsulated in medical-grade silicone, except for the surface of the platinum electrodes (Figure 2.4), which need to be exposed to make physical contact with the ganglion cells on the retina to efficiently transfer the electrical charges from the microchip. The electrode array developed by Second Sight (the company marketing the Argus II device) was composed of 16 Pt discs arranged in a 4 × 4 square array (Figure 2.4). The white silicone tube to the left of the electrodes is a handle that allows the surgeon to hold and manipulate the electrode array during implantation inside the eye. A detailed description of the electrode array's development and performance, briefly described here, can be found in a comprehensive review [25].

The reason the Si microchip-based artificial retina exhibited serious problems when initially implanted in rabbits' eyes in animal studies was that the Si microchip was etched chemically by the eye saline's humor. Therefore, subsequent studies required the Si microchip to be inserted in a Ti-based box outside the eye and connected to the ganglion cells through Pt wires, encapsulated in a biocompatible polymer, permanently going through the eye wall to connect to Pt electrodes in the polymer matrix, as shown in Figure 2.4. The problem with having Pt wires, even if encapsulated in a biocompatible polymer, permanently through the eye's wall is that this arrangement has the potential to induce infections. In order to improve on this device, it is desirable to develop a microchip that can be totally implanted inside the eye and receive the image from the CCD camera outside the eye, via wireless, as shown in Figure 2.3, in such a way that it can stimulate the ganglion cells in the retina at hundreds to thousands of individual locations, in the same way that a display uses thousands of pixels to create an image.

In relation to the critical problem of developing a technology that enables insertion of the Si microchip inside the eye, this chapter focuses on describing the critical topics relevant to the development of a robust, long-term artificial retina device, fully implantable inside the eye. To achieve this goal, it is critical to encapsulate the Si microchip with a biocompatible/hermetic/corrosion-resistant encapsulation to protect the microchip from the corrosive action of the eye's humor.

2.3 Scientific and Technological State-of-the-Art of Biocompatible/ Bioinert Encapsulation Technologies for Si Microchips Implantable in the Eye

2.3.1 Process and Design Considerations for Hermetic/Biocompatible/ Bioinert Encapsulants for Implantable Si Microchips (Artificial Retina)

Hermetic coatings for encapsulation of implantable biomedical devices should have a double functionality of protecting the implantable microdevice and the surrounding biological tissue in order to enable high-performance, long-lasting, failure-free electronic devices. Packaging technologies currently under development include hard-case and thin-film encapsulating coatings. The hard-case technology involves metals, ceramics, or glass as materials to protect the implantable Si microchip. The problem with this approach is that a hard case is bulky and does not facilitate miniaturization. Therefore, encapsulation of the Si microchip by a biocompatible/ corrosion-resistant coating provides a better platform, mainly because of size considerations. Therefore, this chapter focuses on describing the science and technology of a novel material named ultrananocrystalline diamond (UNCD), the main topic of this book, as a potentially strong candidate for a hermetic bioinert/biocompatible/ corrosion-resistant Si microchip-encapsulating coating. The materials synthesis, characterization, and electrochemical properties of the UNCD coating are reviewed for application as an encapsulating coating for the artificial retinal microchips that are implantable inside the human eye, and eventually in other parts of the human body.

2.3.2 Materials Previously Investigated for Biocompatible/Hermetic Encapsulating Coatings for Implantable Microchips

The materials used for producing coatings for encapsulation of implantable Si microchips are challenging because of the size, material properties, mechanical structure and rigidity, biocompatibility, resistance to chemical attack by the eye's saline humor, the need for a long lifespan, and the maximum allowable temperature. Reliability, long-term stability, and acceptable cost are essential requirements for the coating. Coating materials under evaluation for encapsulating an Si-based microchip, as the main component of the so-named artificial retina, include SiC [26], polytetrafluoroethylene and polyimide [27–29], and SiO_2 [30]. However, R&D has shown that SiO_2 coatings exhibit dissolution and decay in retinal implants inserted in animals for up to six months [30]. SiN coatings exhibit structural defects (pinholes) plus chemical reaction with the saline components, resulting in compositional changes of the coating and thus loss of insulating properties and consequently loss of protection of the microphotodiode encapsulated by the coating [31]. Polyimide and other polymers are inexpensive and flexible, but tend to absorb significant quantities of fluids, which induces electrical leakage over time. Implantable medical devices should last decades, however, and polymers are not considered hermetic for that period of time [32]. The science and technology of microchip embedded capacitors for implantable artificial retinas, related to oxide films for capacitors because of the double functionalities of the oxide films for microchip embedded capacitor application and as potential components of hermetic encapsulating coatings, have been described in detail by Auciello. All the materials mentioned above have been described in detail in the cited references, and thus they will not be described here since the main topic of this chapter is the UNCD coating.

2.4 Carbon-Based Biocompatible UNCD Coating for Si Microchip Encapsulation and Growth Techniques

The science and technology of the UNCD coatings used to encapsulate the Si-based microchip for artificial retinas is described in Chapter 1. Therefore, only a brief summary of the specific process used to grow UNCD coatings to encapsulate the Si microchip is presented here. The hermetic UNCD encapsulating coating for artificial retina microchips is produced using $Ar/CH_4/H_2$ gas mixtures. A small amount of hydrogen gas (1–20%) is added in order to decrease the electrical conductivity of the film and render it electrochemically inert. The gas mixture is flown into the chamber of microwave plasma chemical vapor deposition (MPCVD) or hot filament chemical vapor deposition (HFCVD) systems, evacuated to a relatively high vacuum (~10^{-7}–10^{-8} Torr). The description of the MPCVD and HFCVD systems can be seen in Chapter 1.

A key parameter to grow the UNCD coating for encapsulation of the Si-based microchip is the growth temperature. The temperature to which the Si microchip is heated to grow the UNCD encapsulating coating is 400 °C, which is within the thermal budget of the Si microchip. The reason why the UNCD coating technology is the only one that can be used to encapsulate an Si microchip is because the UNCD coating technology is the only patented [33] process that enables growth of a diamond film at such a relatively low temperature [34, 35]; all other techniques need heating of the substrate to 600–800 °C, which will destroy the Si microchip. In order to compare the change in properties (density and electrical impedance) for the encapsulating UNCD films as a function of film growth temperature, UNCD films were also grown at temperatures in the 600–800 °C range. Conductive double-polished Si (N-type, R 0.001~0.005 $\Omega\cdot cm^{-1}$) strips (3 × 25 × 0.55 mm) were used as substrates to grow UNCD films to perform impedance measurements. One end of the coated Si strips were covered with a Si cover slide (Figure 2.5a) to avoid growth of the insulating UNCD coating and to provide a place for electrical connection to the impedance

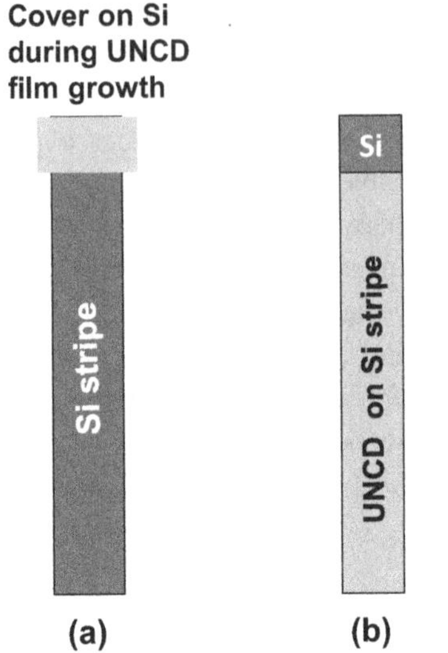

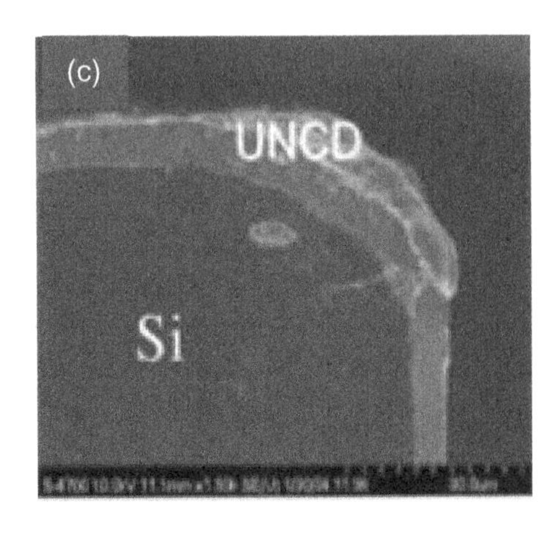

Figure 2.5 Schematic of Si strips covered by a mask (light gray) to inhibit growth of the UNCD film (a) and coated with a UNCD film (b) to perform electrical impedance measurements as described in the text. (c) Cross-section SEM image of an Si microchip encapsulated with a hermetic/biocompatible/corrosion-resistant UNCD coating for implantation in rabbits' eyes (reprinted from *MRS Bull.* vol. 39, p. 621, 2014 (Fig. 3) in [24] with permission from Cambridge Publisher).

system for measuring the leakage current as a part of electrochemical tests in the eye's humor. On the other hand, many Si microchips were coated at 400 °C for subsequent implantation in rabbits' eyes for *in-vivo* testing of the biocompatibility/bioinertness to undesirable biological actions and resistance to chemical corrosion by the eyes' humor. Cross-section scanning electron microscopy (SEM) imaging was extensively performed (see the example of the UNCD-coated Si microchip in Figure 2.5c) to determine the UNCD–Si interface structure, to ensure excellent hermeticity for optimum performance of the UNCD film as a corrosion-resistant protective encapsulating coating.

Uniformity and density of the hermetic coatings is one of the key properties to ensure the quality of the coatings. The MPCVD and HFCVD techniques, described in detail in Chapter 1, are suitable for producing extremely uniform, dense, pinhole-free UNCD films that can be used for encapsulation of the Si microchip implantable inside the eye, on the retina (see a more detailed description of the technology of UNCD coating applied to the Argus II device in [24, 36]. An important feature of both the MPCVD and HFCVD techniques is that they enable growing the UNCD films with excellent uniformity in thickness (±1–2%) and nanostructure over large areas from 200 mm to 400 mm in diameter, which enables UNCD-coated products that are already on the market by Advanced Diamond Technology (co-founded by Auciello and colleagues in 2003), currently the only company worldwide commercializing UNCD coatings.

2.5 Characterization of Hermetic/Biocompatible/Corrosion-Resistant/Bioinert Si Microchip-Encapsulating UNCD Coating

2.5.1 Characterization of Chemical, Microstructural, and Morphological Properties of UNCD Coatings

The chemical, microstructural, and morphological properties of UNCD coatings were characterized using a variety of complementary analytical techniques, as discussed here.

2.5.1.1 Raman Analysis of Polycrystalline Diamond Films

Visible Raman spectroscopy analysis enabled the characterization of chemical bonds in the UNCD films. The Raman analysis was performed using a Renishaw Raman Microscope, with a He–Ne laser at a wavelength of 632.8 nm (energy 1.96 eV). Due to the resonant Raman effect, the Raman scattering cross-section for sp^2 (graphite) bonded carbon atoms, mainly in the grain boundaries, is much larger than for sp^3 (diamond) bonded carbon atoms inside the UNCD grains when a laser with a wavelength in the visible region is used, because the energy of the incident photons is much lower than the energy of the band gap [35, 36].

Two peaks, D band (1330 cm^{-1}; the activated A_{1g} mode due to the finite crystal size) and G band (1580 cm^{-1}; E_{2g} modes of single crystal graphite), are observed in the UNCD visible Raman spectra, which represent the sp^2-bonded carbon atoms in the grain boundaries. The characteristic Raman peak of a diamond grain (1332 cm^{-1}), which usually overlaps with the D peak, is observed when sufficiently large amounts of hydrogen are added to the gas mixture to produce the plasma to grow the films, with the larger amount of hydrogen resulting in the growth of diamond grains beyond the characteristic 3–5 nm grains of UNCD, toward larger grains with sizes in the range 10–100 nm [34, 35]. In this sense, Figure 2.6a shows the Raman spectra of polycrystalline diamond films grown using plasmas containing 2%, 10%, and 20% hydrogen, keeping the flow of CH_4 gas at 1% and varying the Ar gas flow to control the total flow at 100%. The Raman spectra in Figure 2.6a shows that when the hydrogen flow is >10%, the typical UNCD structure is not retained. The strong and sharp peak of 1332 cm^{-1} appears over the D peak due to a grain size increase of the diamond film to a value where the laser light accesses appropriately the sp^3 bonds of C atoms in the diamond lattice.

In addition to the Raman analysis of diamond films grown with different hydrogen flows, Raman analyses were performed on polycrystalline diamond film at different temperatures, as shown in the visible laser beam ($\lambda = 633$ nm) Raman spectra in Figure 2.6b and UV laser beam ($\lambda = 266$ nm) Raman spectra in Figure 2.6c. The Raman analysis reveals that when the diamond film growth temperature decreases, the peak positions gradually shift from 1330 cm^{-1} and 1560 cm^{-1} (at 800 °C) to 1360 cm^{-1} and 1540 cm^{-1} (at 400 °C). The height ratio of 1560 cm^{-1} and 1330 cm^{-1} peaks reverses between 700 °C and 600 °C. Visible Raman is more sensitive to the sp^2-bonded carbon in grain boundaries as mentioned above; the changes of the peaks reflect the changes in the H vs. C atom content in grain boundaries.

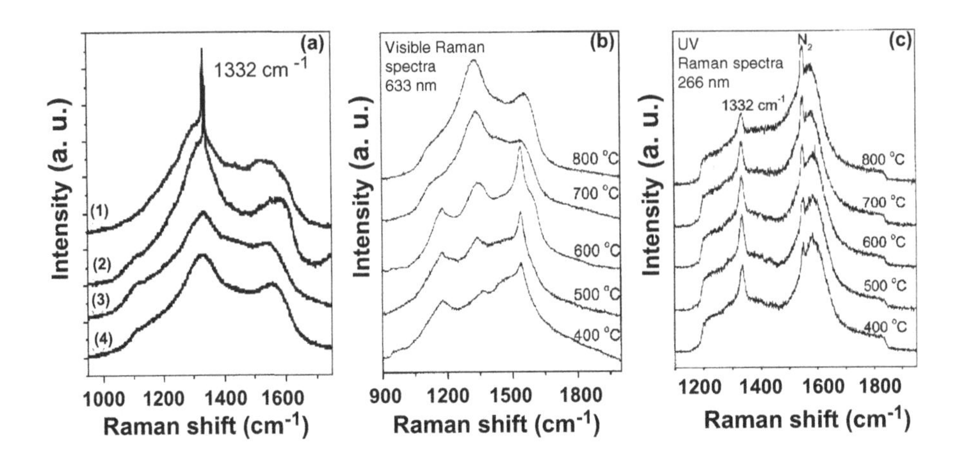

Figure 2.6 (a) Raman spectra of polycrystalline diamond films grown at 800 °C, with different H_2 percentages in the Ar/CH_4 gas mixture: (1) 20%, (2) 10%, (3) 2%, and (4) 0%. All curves have been normalized against the intensity of the peak around 1590 cm^{-1}. The Raman spectrum (4) is characteristic of the UNCD films with 3–5 nm grain size. (b) Visible laser-based Raman spectra of UNCD films grown with H_2 flow of 0% at different temperatures. (c) UV laser-based Raman spectra of UNCD films grown with H_2 flow of 0% at different temperatures. (Reprinted from *J. Biomed. Mater.*, vol. 77B, p. 273, 2006 (Fig. 6) in [36] with permission from Wiley Publisher).

The UV laser beam (4.66 eV at 266 nm) Raman spectroscopy is used on the analysis of UNCD because this laser beam energy can excite efficiently both π C atom bonds in grain boundaries and the σ C atom bond states, thus probing both sp^2- and sp^3-C atom bonds efficiently. The results show that the diamond peak at 1332 cm^{-1} increases in intensity as the deposition temperature decreases from 800 °C to the 400–700 °C range (Figure 2.6c), which can be due to higher retention of hydrogen atoms in the grain boundaries. The broad feature center at 1600 cm^{-1} is due to sp^2-bonded carbon.

The Raman analysis described above is closely correlated with the SEM and high-resolution transmission electron microscopy (HRTEM) analyses to obtain comprehensive information about the nanostructures of the diamond film to correlate them with the *in-vivo* animal tests to determine the optimum diamond film nanostructure to produce the best encapsulating coating.

2.5.1.2 SEM, LRTEM, and HRTEM Analysis of Polycrystalline Diamond Films

The microstructure information of the surface and cross-section morphologies of the polycrystalline diamond coating were investigated using a Hitachi S-4700 field emission scanning electron microscope (FE-SEM). HRTEM observations were carried out using a Philips CM30 microscope operated at 300 kV and a high-resolution JEOL 4000EXII system.

The SEM analysis of polycrystalline diamond UNCD films grown with different hydrogen flow percentages is shown in Figure 2.7. The SEM top and cross-section

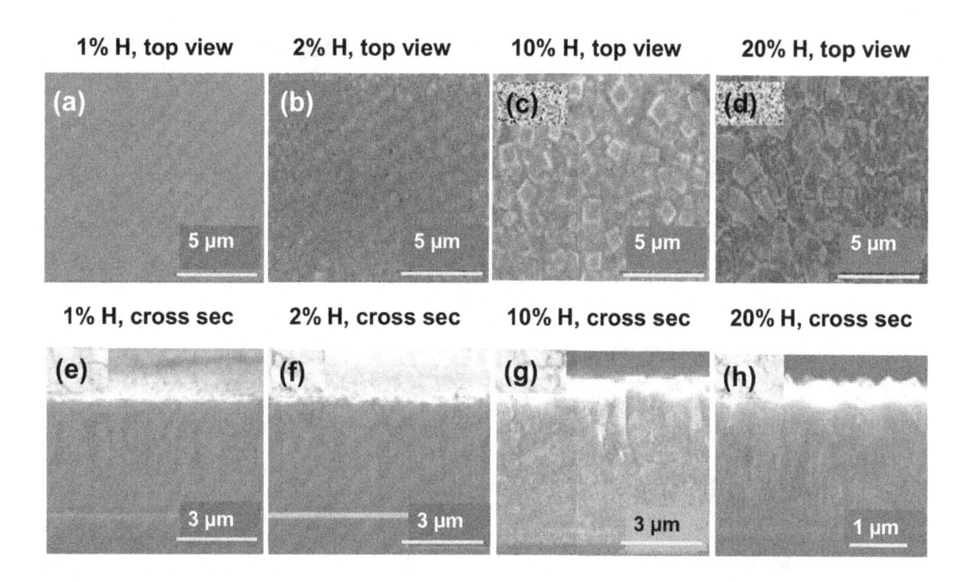

Figure 2.7 SEM images of polycrystalline diamond films grown with different hydrogen flow percentages in the Ar/CH4 (1%) gas mixture, showing the microstructure changes induced by hydrogen incorporation into the films. (a) (top view) / (e) cross-section images relates to the nanostructure characteristic of UNCD films (the best for encapsulating Si microchips) with grain sizes in the range 3–5 nm, produced only with 1% H2 flow, during the MPCVD film growth process. The UNCD structure provides the densest/pinhole-free diamond films that exhibited the best performance in the in-vivo animal studies involving implantation of UNCD-coated Si microchips for up to 3 years. (b) top view / (f) cross-section of small size grains (20–50 nm) (nanocrystalliue diamond (NCD) film. (c) top view / (g) cross-section of medium-large size nano-grains (100-500 nm) NCD film. (d) top view / (h) cross-section of micron-size grains (1–2 µm) microcrystalline diamond (MCD) film. The NCD and MCD films are not good encapsulating coatings. (reprinted from *J. Biomed. Mater.*, vol. 77B, p. 273, 2006 (Figs. 4 and 5) in [36] with permission from Wiley Publisher).

images show that the incorporation of hydrogen to the gas mixture during film growth results in dramatic changes to the diamond film microstructure. Films with H_2 flow in the range 10–20% during growth, exhibit grain size in the range 0.5–1 µm (Figure 2.7c,d) and have the associated rough surface (Figure 2.7g,h). The SEM images of Figures 2.7a–h correlate very well with the Raman spectra shown in Figure 2.6a (1 and 2), which shows the very well-defined 1332 cm^{-1} peak of microcrystalline diamond (MCD).

Low-resolution (Figure 2.8a) and intermediate resolution (Figure 2.8b) cross-section transmission electron microscope (TEM) images show sharp nanoscale resolution UNCD–Si interface, which is exceptionally good for providing the hermetic coating needed to encapsulate the Si microchip for implantation inside the eye. In addition, HRTEM images of UNCD films grown at 400 °C (Figure 2.8c) and 800 °C (Figure 2.8d) show the film growths at both temperatures exhibit the same nanostructure with grains in the range 3–5 nm, characteristic of UNCD films. This nanostructure is responsible for the exceptionally sharp UNCD–Si interface, which is critical for the UNCD coating to provide outstanding encapsulation protection to the Si microchip, as revealed in the extensive animal studies.

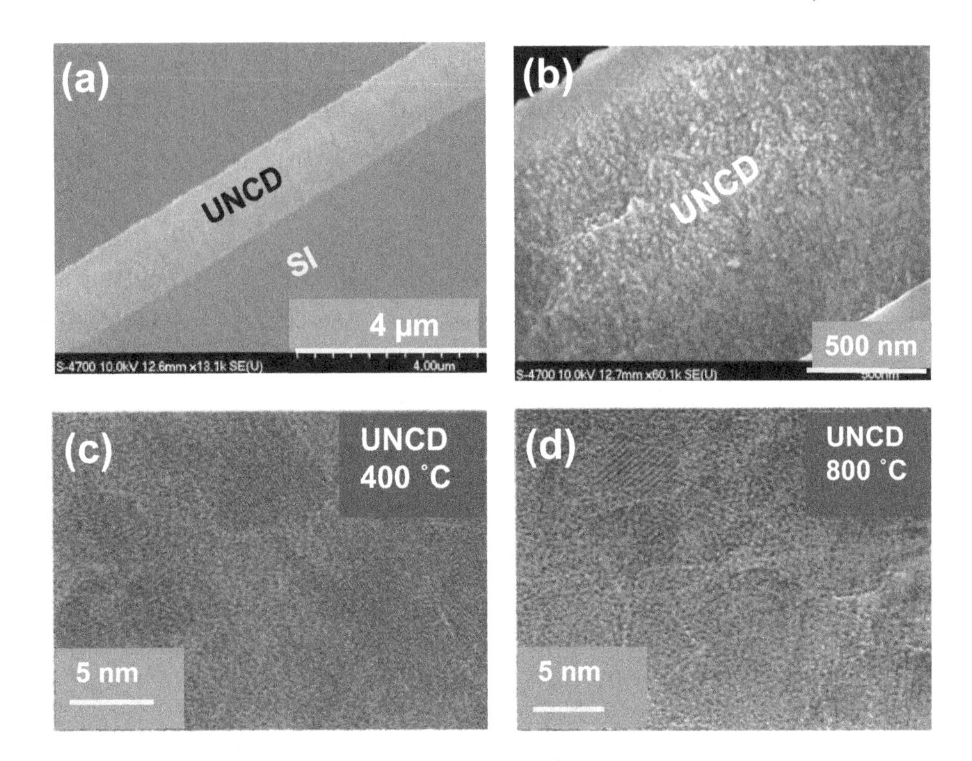

Figure 2.8 SEM images of polycrystalline diamond films grown with different hydrogen flow percentages in the Ar/CH_4 (1%) gas mixture, showing the microstructure changes induced by hydrogen incorporation into the films.

The comparisons of the nanostructures of UNCD coatings grown at 400 °C and 800 °C, as shown in Figure 2.8, and specifically the cross-section SEM (Figure 2.7e,f) indicate a growth rate of ~0.2 μm/h at 400 °C and ~0.25 μm/h at 800 °C, which indicates that the growth of UNCD films with the Ar-rich/CH_4 plasma chemistry is not very temperature-dependent.

The data shown in Figures 2.7 and 2.8 indicate that the UNCD films deposited at 400 °C have similar nanostructures to those of UNCD films deposited at higher temperatures. The UNCD films are very dense and pinhole-free. The SEM and HRTEM images of the very dense UNCD films indicate that they are extremely good candidate materials for hermetic encapsulating coatings for implantable Si microchips.

2.6 Characterization of Electrochemical Performance in Eye Humor–Like Saline Solution to Test Corrosion-Resistance of Hermetic UNCD Encapsulating Coating on Silicon

Resistance to electrochemical corrosion of the UNCD coating on Si was investigated via insertion of UNCD-coated strips (Figures 2.9a,b) in simulated eye humor saline.

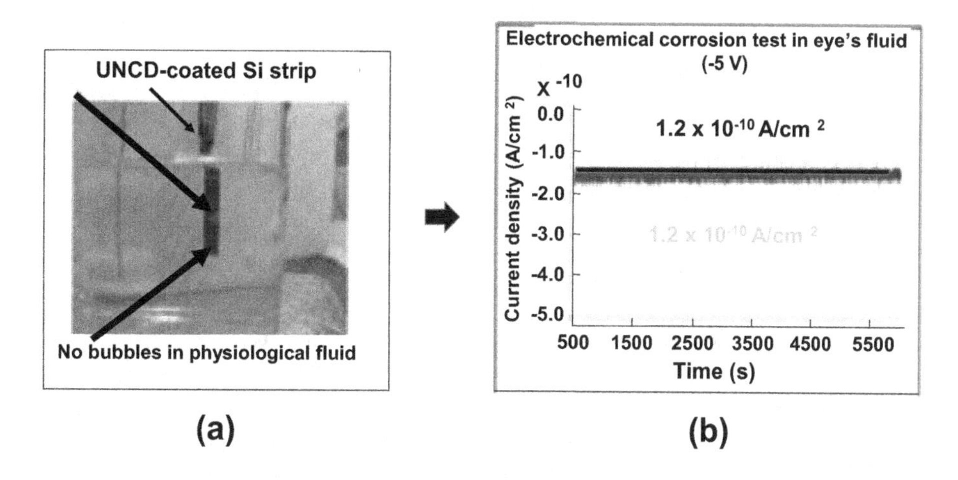

Figure 2.9 (a) Electrochemical test system with the eye humor-equivalent fluid and a UNCD-coated Si strip. (b) One of many electrochemical tests conducted on UNCD-coated Si strips to measure quantitatively the leakage current induced by the humor-equivalent solution. Notice the extremely low leakage current, indicating the outstanding performance of the UNCD coating as a corrosion-protective coating for an Si microchip.

The electrochemical tests involved cyclic voltammetry tests using three electrodes in a potentiostat (Solartron 1287A, Solartron Analytical). Phosphate buffered saline (PBS) was used as the electrolyte. A platinum rod was used as the counter electrode. An Ag/AgCl electrode was used as the reference electrode.

Electrochemical tests on UNCD-coated Si strips for UNCD coating grown at 400 °C, 600 °C, and 800 °C (Figure 2.10a) show that all UNCD films exhibit practically the same extremely low current density, indicating excellent corrosion protection to the Si underneath. Similar electrochemical tests in the same voltage range as for the test shown in Figure 2.10a done on UNCD-coated strips with hermetic UNCD and purposely scratched UNCD coatings show that the uncoated Si exhibits ≥4 orders of magnitude higher current density than even the scratched UNCD-coated Si (Figure 2.10b). Finally, Figure 2.10c shows that even the scratched UNCD coating can protect the underlying Si from extensive corrosion, while the hermetic UNCD coating practically eliminates any corrosion of the Si underneath.

Figure 2.10 shows that electrochemical tests of UNCD coatings on Si samples yielded relatively high leakage currents ($\sim 10^{-4}$ A/cm^2), which is not compatible with the needed functionality for several years inside human eyes. Introducing hydrogen into the Ar/CH_4 gas mixture used to grow the UNCD films resulted in the incorporation of hydrogen into the grain boundaries and saturated the dangling bonds, thus leading to a greatly decreased leakage current. UNCD films grown using $CH_4/Ar/H_2$ (1%) plasma exhibit the lowest leakage current ($\sim 7 \times 10^{-7}$ A/cm^2) in the fluids used to simulate the eye. This leakage is still on the borderline for the optimal performance of an Si microchip artificial retinal. However, it is expected that insertion of a biocompatible oxide layer (e.g., TiO_2 or Al_2O_3) on the Si surface, as an interface layer on which

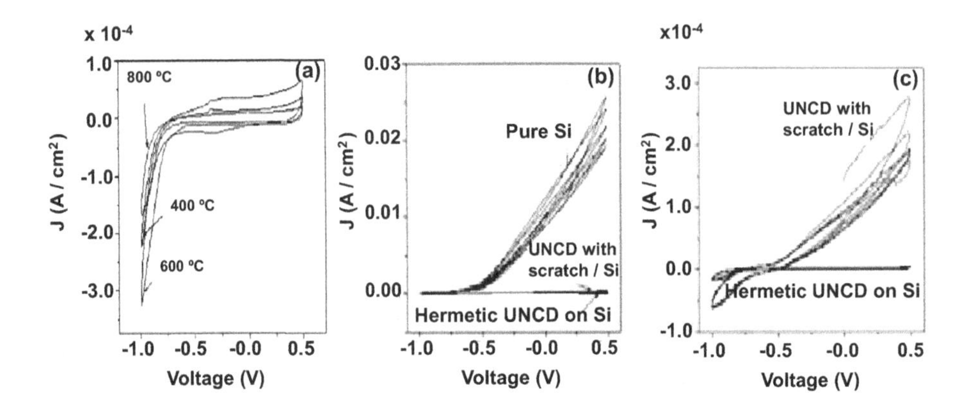

Figure 2.10 Cyclic voltammogram tests on UNCD-coated Si strips (a) for UNCD coating grown at different temperatures, with 1% H_2 flow in the MPCVD plasma, as described in the text (notice the extremely low current density, indicating excellent corrosion protection of the Si underneath); (b) for pure exposed Si, scratched UNCD-coated Si, and hermetic UNCD-coated Si; (c) for scratched UNCD-coated Si and hermetic UNCD-coated Si, showing the orders of magnitude lower current than for pure exposed Si (reprinted from *J. Biomed. Mater.*, vol. 77B, p. 273, 2006, (Figs. 2 and 3) in [36] with permission from Wiley Publisher).

the UNCD layer can be grown, may provide the means to reduce the leakage current to $\leq 10^{-10}$ A/cm^2, which would provide a reliable leakage current output for long-lasting artificial retina performance in the human eye.

In summary, the electrochemical tests on pure exposed Si and hermetic UNCD coatings on Si show that the hermetic UNCD coating provides excellent protection against chemical corrosion from eye fluids.

2.7 Characterization of UNCD/CMOS Integration

A critical issue related to the development of UNCD films as hermetic/biocompatible/corrosion-resistant coatings for encapsulation of Si-based complementary metal-oxide-semiconductor (CMOS) microchips implantable in the human eye or any other part of the human body is the demonstration that the CMOS devices perform to specification before and after being coated with UNCD films at 400 °C, which is the limit of the thermal budget of CMOS devices, beyond which they are destroyed. Therefore, systematic experiments were performed growing UNCD hermetic coatings using H_2 (1%) related flux chemistry during film growth, as explained above, on 200 mm commercial CMOS wafers (Figure 2.11a,b) kindly provide by Freescale. All CMOS devices performed with practically no degradation in output electrical parameters (Figure 2.11c) after being coated with UNCD films of ~1 μm, produced during four hours of deposition at 400 °C using the MPCVD growth process.

The data shown in Figure 2.11 from electrical tests performed on CMOS devices coated with UNCD films grown by MPCVD at 400 °C show that UNCD films grown

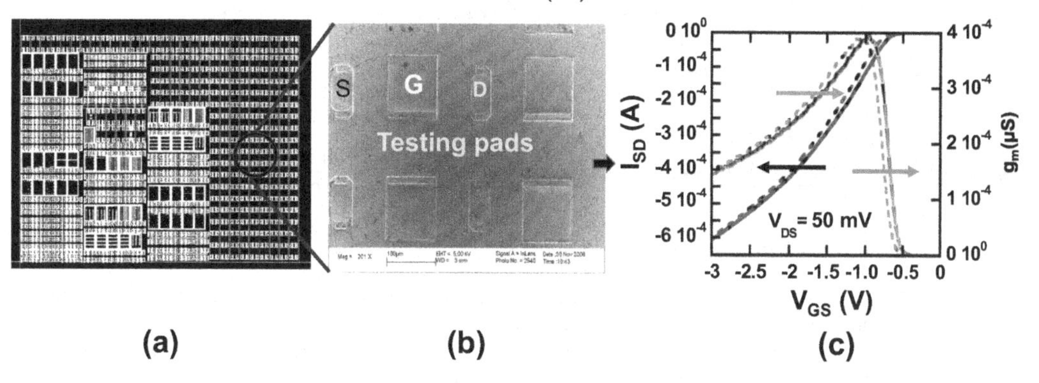

Figure 2.11 (a) Si-based CMOS devices on a commercial Si wafer; (b) SEM image of the CMOS region with test pads (source [S], gate [G], drain [D]) used to measure the performance of the Si CMOS devices before and after coating with UNCD films. (c) Measurement of CMOS device performance before and after coating with UNCD film for 4 h at 400 °C. There is negligible change in the electrical parameters of the CMOS device before (dotted curves) or after UNCD coating (full curves), which demonstrate that the UNCD growth process is extremely compatible with CMOS devices (200 mm CMOS wafers were supplied by S. Pacheco [Freescale] and the electrical characterization was performed by Prof. Z. Ma's group at University of Wisconsin-Madison) (reprinted from *Proc. SPIE*, vol. 7318, p. 17, 2009 (Figure 7) in [37] with permission from SPIE Publisher).

as encapsulating coating on Si-based microchips are fully compatible with the CMOS device technology, and enable encapsulation of Si CMOS devices implantable in the human body using UNCD.

2.8 *In-Vivo* Animal Tests of Hermetic/Biocompatible/Corrosion-Resistant UNCD Coatings Encapsulating Si Microchips for Artificial Retinas

Several Si microchips were implanted in rabbit eyes for durations between several months to about three years to test the biological performance in the presence of the eye's humor. Two type of samples were used for the *in-vivo* animal studies:

1. UNCD-coated Si samples encapsulated with hermetic/pinhole-free UNCD coatings, grown at 400 °C, where the Si was totally encapsulated by the UNCD layer;
2. Si samples coated with UNCD films with several pinholes due to the low nucleation density, to explore what defective UNCD coatings may do to the protection of the underlying Si.

The *in-vivo* animal tests were passive in the sense that the implants were just Si substrates without any CMOS circuits, so no power was applied to the Si substrate.

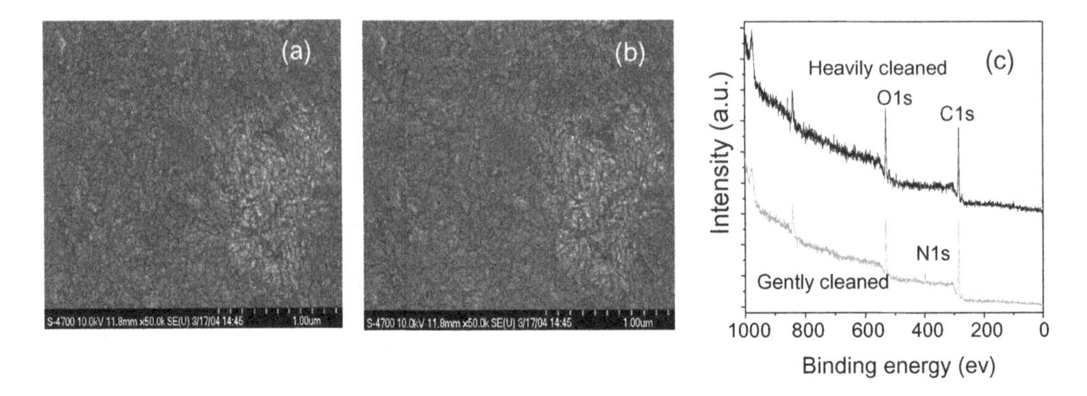

Figure 2.12 SEM images of the surface of a UNCD coating encapsulating an Si sample, before (a) and after (b) exposure to the saline fluid in a rabbit's eye for about three years. There is no degradation whatsoever of the surface topography of the UNCD coating after such a long implantation inside a rabbit's eye and exposure to the corrosive action of the eye's fluid (a small mark was made on the UNCD coating to be able to see the same region before and after implantation). (c) XPS analysis of the surface of the UNCD coating, corresponding to the SEM image shown in (b), showing no chemical reactions with the eye's fluid (reprinted from *MRS Bull.*, vol. 39, p. 621, 2014 (Fig. 3) in [24] with permission from Cambridge Publisher).

After six months to three years following implantation in several animals, no acute surface damage or evidence of biomaterial attachment was noted, as determined by SEM imaging of the surface of the UNCD coating before (Figure 2.12a) and after extraction (Figure 2.12b) of the samples from the rabbits' eyes, and X-ray photoelectron spectroscopy (XPS) analysis (Figure 2.12c), the latter to determine whether there had been any chemical modification due to biological molecules or fluids that may have reacted with the surface of the UNCD coating. XPS is an excellent technique to determine whether there has been any chemical reaction from the biological material in the rabbits' eyes on the surface of the UNCD coating (see the description of a comprehensive study of an extended variety of diamond films using XPS analysis in a recent publication by Veyan et al. [38]).

Some small reactions were observed in two samples with defective coatings, presumably due to reaction with the physiological fluid in the rabbits' eyes through the pinholes, demonstrating the importance of producing the densest possible UNCD coatings.

A very relevant outcome of the *in-vivo* animal studies of UNCD-coated Si samples, described above, is that no deterioration of the rabbit's eye retina was elicited by the UNCD coating. These results indicate that the UNCD coatings exhibit very promising bioinert performance in animal eyes, and tests in human eyes are thus warranted.

2.9 Challenges for Bioinert Microchip Encapsulation Hermetic Coatings

Three critical issues need to be taken into account when developing biocompatible/hermetic/corrosion-resistant coatings for encapsulation of implantable medical devices, and specifically in the case of the technology discussed in this chapter, for microchip retinal implants: (1) bioinertness and biocompatibility; (2) mechanical robustness; and (3) chemical inertness and electrical insulation.

Chemical inertness and electrical insulation are critical for encapsulation of retinal microchips in order to avoid electrical shorts between the independent connector lines to electrodes that are covered by the UNCD coating, and to avoid electrochemical reactions on the UNCD coating surface when exposed to the eye's fluid. Prior work by Auciello's group showed that the grain boundaries in the UNCD films consisted largely of sp^2-bonded carbon atoms and that electron transport occurred through those grain boundaries [35]. In the case of using UNCD coatings as an electron-emitting surface in field electron emitters for cold cathodes, this is a desirable property. However, in the case of application of UNCD films as hermetic coatings for encapsulation of microchips for retinal implants, electrical conductivity, even if small, can result in undesirable electrochemical reactions. The data presented in the review of the R&D done for the last 10 years by Auciello's group [24, 33–38] on this topic indicate that the insulation characteristics of UNCD layers can be greatly improved by engineering the grain boundaries of UNCD via elimination of dangling bonds in sp^2-coordinated carbon atoms, which results in elimination of sites that contribute to grain boundary–based electrical conduction in UNCD. The approach that can be used to eliminate dangling C atom bonds at the grain boundaries is to saturate them with hydrogen atoms. In this sense, theoretical calculations revealed that hydrogen exists in the grain boundaries in the form of C–H bonds. Moreover, the hydrogen addition increases the coordination of carbon and decreases the density of state in the diamond band gap. It can also decrease the whole concentration in p-type semiconductors, or neutralize dopants in n-type materials. The studies performed by Auciello's group confirmed that hydrogen incorporation into UNCD grain boundaries transforms UNCD coatings into better insulators with low enough leakage current to be compatible with the functionality required for a hermetic coating for retinal microchip implants. However, a critical issue under investigation is the possibility that sharp edges or corners of the coated microchip, which could concentrate electric fields and/or dust particles, resulting in pinholes can compromise the coating. To address these issues, Auciello's group is currently exploring hybrid oxide/UNCD coatings, where a highly insulating oxide such as biocompatible TiO_2 or Al_2O_3, or TiO_2/Al_2O_3 nanolaminates, are grown first on the Si chip, followed by a very thin metallic layer (this is used also to enhance the growth of the UNCD layer) on top of the oxide, but electrically isolated from the underlying Si substrate, followed by a UNCD layer to provide the biointerface to the retina.

2.10 Conclusions and Future Outlook

The extensive R&D work described in this chapter has shown that UNCD coatings grown on Si substrates at temperatures ≤400 °C using MPCVD with CH_4/Ar plasmas may provide a reliable hermetic/biocompatible/corrosion-resistant/bioinert coating technology for encapsulation of Si microchips implantable in the eye specifically and in the human body in general. SEM analysis showed that films grown at 400 °C exhibit similar surface morphology and nanostructure, characteristic of the best UNCD films demonstrated in the past by Auciello's group. Electrochemical tests of the plain UNCD coatings yielded relatively high leakage currents (~10^{-4} A/cm^2), which is not compatible with the needed functionality in the eye environment. Introducing hydrogen into the Ar/CH_4 mixture resulted in the incorporation of hydrogen into the grain boundaries and saturated the dangling bonds, thus leading to a greatly decreased leakage current. UNCD films grown using CH_4/Ar/H_2 (1%) plasma exhibit the lowest leakage current (~7×10^{-7} A/cm^2) in a saline solution simulating the eye environment. This leakage is incompatible with the functionality of the first generation of artificial retinal microchips. However, it is expected that when growing UNCD on top of the Si microchip passivated by a silicon nitride layer or oxide layers, also under investigation, the electrochemically induced leakage will be reduced by at least 1–3 orders of magnitude to the range of 10^{-10} A/cm^2, which is compatible with reliable, long-term implants in human eyes.

Acknowledgments

O. Auciello is grateful to M. Humayun (director of the DOE-funded Artificial Retina Project at Doheny Eye Institute – University of Southern California, 2000–2010) and to all the team members for several years of exciting scientific and technological development, which resulted in the development of a device (Argus II) that is currently on the market, restoring partial vision to people blinded by RP. O. Auciello also acknowledges the strong support from DOE Basic Energy Science and Bioengineering during the 2000–2010 period when R&D for the Artificial Retina Project was performed.

References

[1] D. T. Hartong, E. I. Berson, and T. P. Dryja, "Retinitis pigmentosa," *Lancet*, vol. 368 (9549), p. 1795, 2006.

[2] S. P. Daiger, L. S. Sullivan, and S. J. Bowne, "Genes and mutations causing retinitis pigmentosa," *Clin. Genet.*, vol. 84(2), p. 132, 2013.

[3] A. H. Milam, Z. Y. Li, and R. N. Fariss, "Histopathology of the human retina in retinitis pigmentosa," *Prog. Retin. Eye Res.*, vol. 17(2), p. 175, 1998.

[4] A. Santos, M. S. Humayun, E. de Juan, Jr., et al., "Preservation of the inner retina in retinitis pigmentosa: a morphometric analysis," *Arch. Ophthalmol.*, vol. 115, p. 511, 1997.

[5] W. A. Beltran, A. V. Cideciyan, A. S. Lewin, et al. "Gene therapy rescues photoreceptor blindness in dogs and paves the way for treating human X-linked retinitis pigmentosa," *Proc. Natl. Acad. Sci. USA*, vol. 109(6), p. 2132, 2012.

[6] N. Chadderton, S. Millington-Ward, A. Palfi, et al. "Improved retinal function in a mouse model of dominant retinitis pigmentosa following AAV-delivered gene therapy," *Mol. Ther.*, vol. 17(4), p. 593, 2009.
[7] J. W. B. Bainbridge, A. J. Smith, S. S. Barker, et al., "Effect of gene therapy on visual function in Leber's congenital amaurosis," *N. Engl. J. Med.*, vol. 358(21), p. 2231, 2008.
[8] A. M. Maguire, F. Simonelli, E. A. Pierce, et al., "Safety and efficacy of gene transfer for Leber's congenital amaurosis," *N. Engl. J. Med.*, vol. 358(21), p. 2240, 2008.
[9] V. Busskamp, S. Picaud, J. A. Sahel, and B. Roska, "Optogenetic therapy for retinitis pigmentosa," *Gene Ther.*, vol. 9(2), p. 169, 2012.
[10] C. A. Ku, S. M. Hariprasad, and M. E. Pennesi, "Gene therapy trial update: a primer for vitreo-retinal specialists," *Ophthalmic Surg. Lasers Imaging Retina*, vol. 47(1), p. 6, 2016.
[11] O. Foerster, "Beitrage zur pathophysiologie der sehbahn und der spehsphare," *J. Psychol. Neurol.*, vol. 39, p. 435, 1929.
[12] G. S. Brindley and W. S. Lewin, "The sensations produced by electrical stimulation of the visual cortex," *J. Physiol.*, vol. 196(2), p. 479, 1968.
[13] P. M. Lewis, H. M. Ackland, A. J. Lowery, and J. V. Rosenfeld, "Restoration of vision in blind individuals using bionic devices: a review with a focus on cortical visual prostheses," *Brain Res.*, vol. 1595, p. 51, 2015.
[14] A. M. Fekry, R. S. El-Kamel, and A. A. Ghoneim, "Electrochemical behavior of surgical 316L stainless steel eye glaucoma shunt (Ex-PRESS) in artificial aqueous humor," *J. Mater. Chem.*, vol. B4, p. 4542, 2016
[15] www.eurekanetwork.org/project/-/id/5558
[16] S. S. Lane and B. D. Kuppermann, "The implantable miniature telescope for macular degeneration," *Curr Opin. Ophthalmol.*, vol. 17 (1), p. 94, 2006.
[17] E. Zrenner, K. Ulrich Bartz-Schmidt, H. Benav, et al., "Subretinal electronic chips allow blind patients to read letters and combine them to words," *Proc. Roy. Soc. B*, vol. 287 (1711), p. 1489, 2010.
[18] D. Loudin, D. M. Simanovskii, K. Vijayraghavan, et al., "Optoelectronic retinal prosthesis: system design and performance," *J. Neural Eng.*, vol. 4 (1), p. S72, 2007.
[19] S. Ings, "Making eyes to see," in *The Eye: a Natural History*. London: Bloomsbury, p. 276, 2007.
[20] A. Rush and P. R. Troyk, "A power and data link for a wireless-implanted neural recording system," *Trans. Biomed. Eng.*, vol. 59 (11), p. 3255, 2012.
[21] Bionics Institute, "Bionic Vision Australia's progress of the bionic eye," July 23, 2012. www.bionicsinstitute.org/news/restoring-sight-australias-bionic-eye.
[22] J. L. Wyatt, Jr., S. Kelly, O. Ziv, et al., "The retinal implant project," MIT, 2011.
[23] M. Humayun, "Interim results from the international trial of Second Sight's visual prosthesis," *Ophthalmology*, vol. 119 (4), p. 779, 2012.
[24] O. Auciello, P. Gurman, M. B. Guglielmotti, et al., "Biocompatible ultrananocrystalline diamond coatings for implantable medical devices," *MRS Bull.*, vol. 39, p. 621, 2014.
[25] D. D. Zhou and R. J. Greenberg, "Microelectronic visual prostheses," in *Implantable Neural Prostheses 1*, D. D. Zhou and E. Greenbaum, Eds. New York: Springer.
[26] J. Meyer, "System on chip mass sensor based on polysilicon cantilevers," *Sens. Actuator*, vol. A97–98, p. 1, 2010.
[27] F. S. Cogan, D. Edell, A. Guzelian, Y. Liu, and R. Edell, "Plasma-enhanced chemical vapor deposited silicon carbide as an implantable dielectric coating," *J. Biomedical Mater. Res.*, vol. A67, p. 856, 2003.

[28] J. Seo, S. Kim, H. Chung, et al., "Biocompatibility of polyimide micro-electrode array for retinal stimulation," *Mater. Sci. Eng.*, vol. C24, p. 185, 2004.
[29] P. Heiduschka and S. Thanos, "Implantable bioelectronic interfaces for lost nerve functions," *Prog Neurobiol.*, vol. 55, p. 433, 1998.
[30] H. Hammerle, K. Kobuch, K. Kohler, et al., "Biostability of micro-photodiode arrays for subretinal implantation," *Biomaterials*, vol. 23, p. 797, 2002.
[31] M. Rojahn, "Encapsulation of a retina implant," PhD dissertation, University of Stuggart, 2003.
[32] J. D. Weiland, W. Liu, and M. S. Humayun, "Retinal prothesis," *Annu. Rev. Biomed. Eng.*, vol. 7, p. 361, 2005.
[33] O. Auciello," "Microchip embedded capacitors for implantable neural stimulators," in *Implantable Neural Prostheses: Techniques and Engineering Approaches*, D. D. Zhou and E. Greenbaum, Eds. Gland: Springer, p. 331, 2010.
[34] J. A. Carlisle, D. M. Gruen, O. Auciello, and X. Xiao, "A method to grow pure nanocrystalline diamond films at low temperatures and high deposition rates," US Patent #7,556,982, 2009.
[35] O. Auciello and A. V. Sumant, "Status review of the science and technology of ultrananocrystalline diamond (UNCD^{TM}) films and application to multifunctional devices," *Diam. Relat. Mater.*, vol. 19, p. 699, 2010.
[36] X. Xiao, J. Wang, J. A. Carlisle, et al., "In vitro and in vivo evaluation of ultrananocrystalline diamond for coating of implantable retinal microchips," *J. Biomed. Mater.*, vol. 77B, p. 273, 2006.
[37] A. V. Sumant, O. Auciello, H.-C. Yuan, et al., "Large area low temperature ultrananocrystalline diamond (UNCD) films and integration with CMOS devices for monolithically integrated diamond MEMS/NEMS-CMOS systems," *Proc. SPIE*, vol. 7318, p. 17, 2009.
[38] J-F. Veyan, E. de Obaldia, J. J. Alcantar-Peña, et al., "Argon atoms insertion in diamond: new insights in the identification of carbon C1s peak in X-ray photoelectron spectroscopy analysis," *Carbon*, vol. 134, p. 29, 2018.

3 Science and Technology of Ultrananocrystalline Diamond (UNCD™) Coatings for Glaucoma Treatment Devices

Alejandro Berra, Mario J. Saravia, Pablo Gurman, and Orlando Auciello

3.1 Introduction

Approximately 60 million people worldwide are affected by glaucoma, making it the leading cause of irreversible blindness worldwide [1]. Therefore, it is very important to understand the condition and the current treatments in order to determine what new science and technology developments are necessary to optimize devices to be used for glaucoma treatment. This chapter includes key sections describing the glaucoma condition, including the different types of glaucoma identified over the years, current medicine-based treatments for glaucoma, current device-based treatments, including key materials-related issues affecting device performance, and the new materials science and device design necessary to improve by at least an order of magnitude the performance of current glaucoma treatment devices.

3.1.1 Glaucoma Condition

Glaucoma Process. Glaucoma is a condition related to the clogging of the tubes in the trabecular mesh of the eye (Figure 3.1a), which continuously evacuate the aqueous humor from the inner region of the eye toward the outside (Figure 3.1a) to keep the inner eye pressure constant (Figure 3.1b) to avoid increased intraocular pressure (IOP), which can destroy the optic nerve, leading to blindness (Figure 3.1b).

Other causes of glaucoma, although less common, include blunt/chemical injury in the eye, eye infection, clogging of blood vessels or eye inflammation, and eye surgery done to correct some other eye condition. Glaucoma generally occurs in both eyes, although each is affected differently.

Glaucoma is the second leading cause of blindness after cataracts. Glaucoma affects 1 in 200 people aged 50 and younger, and 1 in 10 over the age of 80, worldwide. Early detection of glaucoma can arrest or slow progression via pharmacological and surgical means. Further details about glaucoma can be found in [1].

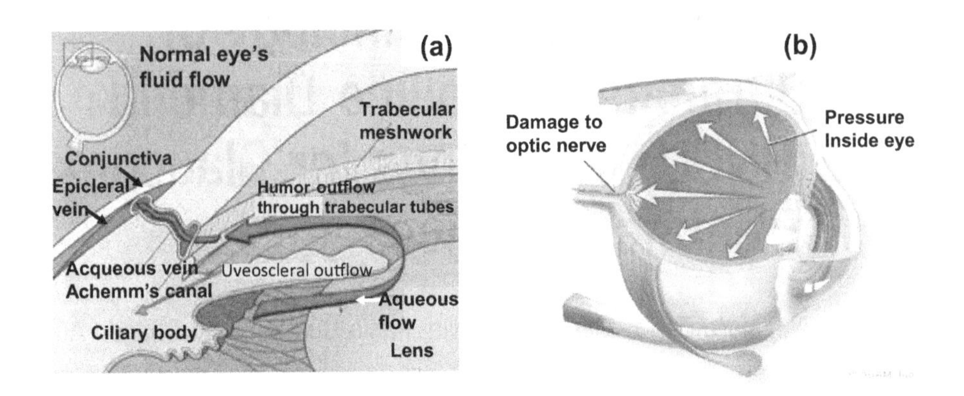

Figure 3.1 (a) Schematic showing key components of the human eye's trabecular region that drains humor from the inner part of the eye to the exterior to keep the inner eye pressure stable. (b) Increased pressure in the human eye induced by clogging of the trabecular tubes, which inhibits fluid drain, resulting in inner eye pressure, which can finally destroy the optic nerve, potentially resulting in blindness.

Glaucoma Conditions and Symptoms. Various symptoms have been identified in relation to glaucoma [2]. *Open-angle glaucoma* is the most common form of the disease. The drainage angle formed by the cornea and iris remains open, but the trabecular meshwork is partially blocked. This causes pressure in the inner part of the eye to gradually increase. This pressure damages the optic nerve. This condition occurs very slowly, and the person affected by it may not be aware of the problem and may start losing their vision. This condition has two main symptoms:

- patchy blind spots in the peripheral or central vision, frequently in both eyes; and
- tunnel vision in the advanced stages.

Angle-closure glaucoma, also called closed-angle glaucoma, is induced when the iris of the eye bulges forward, narrowing or blocking the eye's fluid drainage angle formed by the cornea and iris. As a result, fluid cannot circulate through the eye, resulting in deleterious pressure increase. Some people may have narrow drainage angles, putting them at increased risk related to this condition. Angle-closure glaucoma may occur suddenly (acute angle-closure glaucoma) or gradually (chronic angle-closure glaucoma). In general, acute angle-closure glaucoma represents a medical emergency. Acute angle-closure glaucoma features various symptoms:

- severe headache
- eye pain
- nausea and vomiting
- blurred vision
- halos around lights
- eye redness.

In *normal-tension glaucoma* the optic nerve of the eye is damaged even though the eye pressure is within the normal range. The underlying reason or mechanism for this condition is not yet known. The individual patient may have a sensitive optic nerve or may have less blood being supplied to the optic nerve. This limited blood flow could be caused by atherosclerosis – the build up of fatty deposits (plaque) in the arteries – or other conditions that impair blood circulation [2].

Pigmentary glaucoma involves the appearance of pigment granules in the iris, building up in the drainage channels and slowing or blocking fluid exiting the eye. Activities such as jogging sometimes stir up the nucleation and growth of pigment granules in the trabecular meshwork, causing intermittent pressure elevations [2].

Risk Issues Related to Glaucoma. Considering that chronic forms of glaucoma can destroy vision before any signs or symptoms are detected, people should be aware of risk factors, as described below:

- having high inner eye Intraocular Pressure (IOP);
- being over 60 years of age;
- being black, Asian, or Hispanic;
- having a family history of glaucoma;
- having certain medical conditions, such as diabetes, heart disease, high blood pressure, or sickle cell anemia (this condition affects mainly people of color);
- having corneas that are thin in the center;
- being extremely nearsighted or farsighted;
- having had an eye injury or certain types of eye surgery; and
- taking corticosteroid medications, especially eye drops, for a long time.

3.1.2 Prevention to Stop or Delay Glaucoma Development Until Positive Treatment Implementation

Some key self-care steps can help people detect glaucoma in the early stages:

- **Regular dilated eye examinations.** Regular dilated eye exams can help detect glaucoma development in the early stages, before significant damage occurs. As a general rule, the American Academy of Ophthalmology recommends having a comprehensive eye exam every 5–10 years for people younger than 40 years old; every 2–4 years for people 40–54 years old; every 1–3 years for people 55–64 years old; and every 1–2 years for people older than 65. In addition, people at risk of glaucoma need more frequent screening. People's personal doctors can recommend the right screening schedule.
- **Know the family's eye health history.** Glaucoma tends to run in families, and if you're at increased risk you may need more frequent screening.
- **Exercise safely.** Regular, moderate exercise may help prevent glaucoma by reducing eye pressure. Discussion with one's doctor can provide information to establish appropriate exercise programs.
- **Take prescribed eye drops regularly.** Glaucoma eye drops can significantly reduce the risk of high eye pressure progressing to glaucoma. To be effective, eye

drops prescribed by doctors need to be used regularly even if no further symptoms are detected.

- **Wear eye protection.** Serious eye injuries can lead to glaucoma. Thus, eye protection should be used when handling power tools or playing high-speed racket sports.

3.1.3 Current Glaucoma Treatments

Current treatments for glaucoma involve lowering the IOP, achieved via eye drops, laser beam exposure of the eye, or incisional surgery. Eye drops are the most common treatment modality, and, over time, patients may need to take multiple types of eye drops to stop progression of a disease that typically has no symptoms. The drops may lower IOP by reducing fluid formation or encouraging better drainage of fluid from the eye via one or both of two pathways: the trabecular meshwork/Schlemm's canal (TM/SC) and the uveoscleral pathway (Figure 3.1a).

Currently, the US market includes five families of eye drops used to lower IOP [3]. In the near future, new compounds that reduce IOP may become available, as described in the following subsection.

3.1.4 New Glaucoma Eye Drops on the Horizon

One new compound nearing commercial availability is Latanoprostene Bunod 0.024% (LBN) [3]. This compound reduces IOP by enhancing fluid outflow through both the TM/SC and uveoscleral pathways. It has two components: a prostaglandin – latanoprost – which reduces IOP by increasing uveoscleral outflow; and a nitric oxide–donating component that enhances drainage via the TM/SC pathway. Once-daily LBN performed well in clinical trials, showing superiority to both twice-daily timolol and once-daily latanoprost. The most common side-effect was mild redness of the eyes.

Rhopressa (Netarsudil 0.02%) is another molecule taken daily, involving two action mechanisms and two targets. Rhopressa targets rho-kinase (ROCK) and a norepinephrine transporter (NET), inhibiting ROCK and enhancing fluid outflow through the conventional pathway. The second target, NET, reduces the production of fluid in the eye. This compound was as effective as twice-daily timolol in a clinical trial. The most common side-effect was mild redness of the eyes.

A combination product, Roclatan, is a once-daily combination of netarsudil 0.02% and latanoprost 0.005%. This product potentially acts on both outflow pathways and also reduces production of fluid in the eye. Results from clinical trials for these products were available in 2017, showing initial positive results.

Another compound in clinical trials is trabodenoson, based on opening the conventional outflow pathway, through the trabecular meshwork via a unique mechanism. In addition, trabodenoson has been combined with latanoprost to further improve the drainage of eye fluid. Trials of this combination are moving forward.

3.1.5 Selective Laser Trabeculoplasty Treatment

Selective laser trabeculoplasty involves using short pulses of a low-energy laser beam to excite specific cells in the drainage system of the eye to stimulate improved function [4]. It is typically used when other glaucoma treatment options, including eye drops, are not lowering the eye pressure enough or are causing significant side-effects. It can also be used as an initial glaucoma treatment option.

3.1.6 Flexible Glaucoma Drainage Device

One type of incisional surgery to treat glaucoma involves using a tube shunt connected to a flexible pump as a flexible glaucoma drainage device that is implanted in the eye to pump fluid from the inner region of the eye (Figure 3.2a) to an external reservoir. The commercial Ahmed glaucoma valve (AGV) (Figure 3.2a) is currently extensively used to enable eye fluid drainage. The Ahmed device is made of silicone or polypropylene, a material that will not break down when inserted on the outside wall of the eye (Figure 3.2b). The entire implant is covered with the eye's own external covering. The Ahmed shunt itself is shaped like a miniature computer mouse with a tube at the end. The tube portion enters the front of the eye, or anterior chamber, while the rest of the implant sits on the outside surface of the eyeball (Figure 3.2b), underneath the conjunctiva, and is covered by the eyelid. The fluid that collects is then absorbed by the eye's own veins and transported out of the eye cavity. The Ahmed valve implant procedure is done as outpatient surgery.

Tube shunts have been and are used to control eye pressure in patients for whom traditional eye surgery to relieve fluid pressure has failed, or in patients who have had previous surgeries or trauma that caused substantial scarring of the conjunctiva. Tube shunts have also been successfully used to control eye pressure in patients suffering

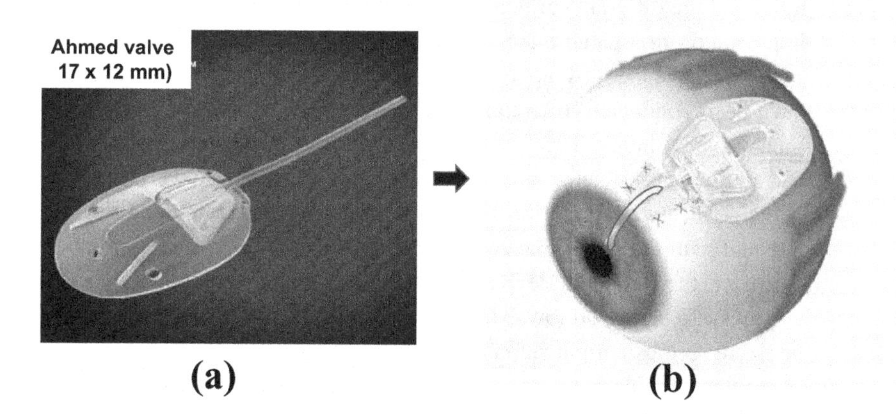

Figure 3.2 (a) A commercial Ahmed valve for drainage of inner eye fluid to control glaucoma. (b) An Ahmed valve positioned on the outside surface of the eye, with the flexible tube inserted in the inner part of the eye for fluid drainage.

with other types of glaucoma, such as glaucoma associated with uveitis or inflammation, neovascular glaucoma (associated with diabetes or other vascular eye diseases), pediatric glaucoma, traumatic glaucoma, and others.

Clinical trials performed in a multicenter joint program (the Tube Versus Trabeculectomy Study) involving patients previously subjected to other surgeries, including trabeculectomy and/or cataract surgery, showed that tube shunt surgery had a higher success rate after about five years compared to trabeculectomy, with similar reductions in eye pressure and need for supplemental glaucoma medication.

The information presented above supports the fact that in recent years some surgeons have been using tube shunts or glaucoma drainage devices as first-line surgery, bypassing trabeculectomy as the first-choice surgery. Generally, trabeculectomy is considered to be a better approach to reduce eye pressure than tube shunt surgery, but involves greater risks. There are several types of tube shunt available for surgeons to choose from. The AGV is the most commonly used type of shunt.

The valve operation of a glaucoma drainage device limits the flow of eye fluid to one direction, which puts a theoretical limit to how low the eye pressure can fall. However, in many patients supplemental glaucoma medication such as eye drops are still required after an Ahmed valve is implanted to keep the eye's pressure within normal pressure conditions.

Sometimes the tube shunt is not attached to a valve (e.g., the Baerveldt or Molteno shunts). In this case, some scarring is initially required before the tube opens, otherwise the eye pressure would be too low. The tube opens when a special suture, used by the surgeon to tie the tube, dissolves over time (or another type of suture that the surgeon removes approximately 4–6 weeks after surgery). The risk of the eye pressure falling too low, while rare, is somewhat higher with these devices than with the Ahmed valve, but non-valved shunts sometimes work better for attaining lower eye pressure.

Ahmed Device vs. Baerveldt Shunt. Because of the different sizes of drainage tubes and the different surgical techniques, the selection of a particular tube shunt for implantation in a patient depends on the surgeon's preference. In this sense, to help surgeons make the most appropriate choice, two multicenter randomized clinical trials have been conducted to examine the performance of Ahmed (valved) vs. Baerveldt (non-valved) shunts. These studies demonstrated that these two types of tube shunt had similar surgical success at five years [5]. The results showed that both devices are effective at lowering eye pressure, but the Baerveldt device exhibited lower failure rate and required fewer glaucoma medications to lower eye pressure after five years. However, the Baerveldt device exhibited more serious complications associated with eye pressure being too low. The vast majority of complications are short-lived or can be fixed; serious complications are rarer.

Another problem to be considered, although rare, is infection inside the eye induced by tube shunt surgery, which can occur months to years after the actual operation, sometimes requiring shunt removal.

Any tube shunt procedure may fail over time due to the natural healing tendencies of the eye. The body will always react to a foreign object – in this case the tube shunt – and scarring around the plate is not uncommon. Thus, the patient may need to resume taking glaucoma medication, or even further surgery. On the flip side, sometimes tube shunts, especially non-valved implants, can result in too low eye pressure, or hypotony. Some patients with hypotony will have vision impairment. The tube may have to be revised or removed if hypotony persists.

A problem that is not yet solved with the current polymer-based glaucoma valves shown in Figure 3.2 is the eye's biological reaction to the surface of the implant due to protein layer adhesion to the hydrophilic (eye fluid adsorption) surface of the polymer, resulting in eye inflammation. Macrophages from the eye are attracted by chemical signals from the eye to the implant surface, where they release growth factors that induce growth of fibroblast cells that yield collagen production, which results in development of a fibrous capsule around the implant (see Figure 3.3a). Subsequently, complete fibrosis occurs on the surface of the valve, isolating the implant from the rest of the eye. This fibrous capsule induces material degradation effects, not only on glaucoma valves discussed here, but also on other polymer-based medical implants, resulting in degraded sensor performance or transport of drugs in drug-delivery devices. Several approaches have been investigated to overcome these problems, but with no proven success yet. The fibrosis issue is critical as it is the reason for the limited five-year lifetime of current polymer-based devices, which have been shown to have a failure rate of about 31.8% worldwide [6].

Failure of polymer-based glaucoma devices is mainly due to fibrotic encapsulation of the shunt at the implant site. Studies have shown that fibrotic encapsulation results in increased outflow resistance, leading to inadequate control of IOP [7–9]. The use of antimetabolites (e.g., mitomycin C [MMC] and 5-fluorouracil [5-FU]) showed

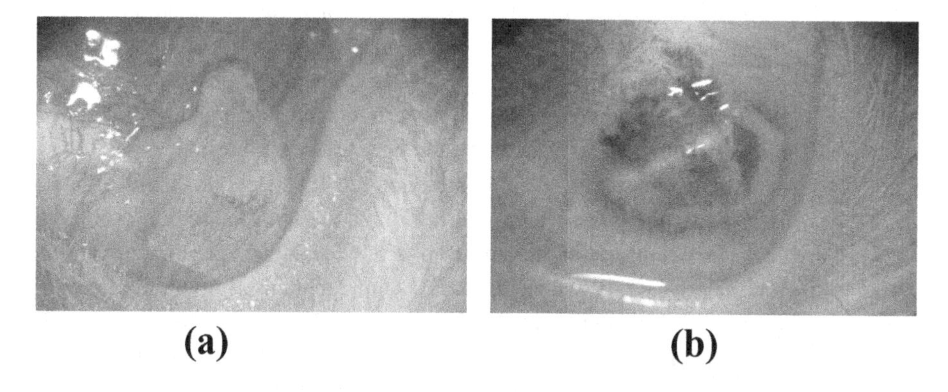

Figure 3.3 (a) Optical picture of a commercial AGV valve implanted in a rabbit's left eye, showing the fibrosis generated in ~24 h due to the hydrophilic nature of the polymer surface of the AGV valve; (b) Another commercial AGV valve coated with ~0.2 μm of a UNCD coating after about 12 months implantation in the other eye of the same rabbit in which the uncoated AGV (a) was implanted, showing ability of the powerful surface property of the UNCD coating to eliminate biomolecule adhesion (reprinted from *MRS Bull.*, vol. 39, p. 621, 2014 (Fig. 5) in [30] with permission from Cambridge Publisher).

improvements in success rates of trabeculectomy in clinical trials and decades of clinical practice [10–12]. Unfortunately, many studies revealed that use of MMC in glaucoma drainage devices implantation does not increase success rates [13, 14]. In addition, other adverse effects were identified, such as infection, wound dehiscence, tube extrusion, and persistent fibrotic encapsulation, further limiting adjunctive use of antimetabolites. The information presented above sparked investigations into safer, more effective agents to improve survivability of glaucoma drainage devices. Dynamic homeostasis of the extracellular matrix is maintained by biochemical and biophysical signaling. Disruption of this homeostasis, as in injury, leads to wound healing and promotion of the fibrotic pathway. Molecular signaling, mainly due to transformation growth factor b (TGF-b), drives trans-differentiation of fibroblasts into myofibroblasts [15]. Myofibroblasts subsequently upregulate smooth muscle actin (SMA) fibers, as well as several integrins and growth factors, which induce deposition of a dense fibrotic collagen capsule [16]. Several studies have focused on investigating agents targeting specific molecules in the fibrosis cascade. Inhibition of TGF-b showed promise in animal studies, but a clinical study in humans revealed that a monoclonal antibody (CAT-152) was not efficient in preventing trabeculectomy failure [17]. Preliminary studies investigating antitransforming growth factor b 2 (anti-TGFb 2), antivascular endothelial growth factor (anti-VEGF), antiplacental growth factor (anti-PIGF), inhibitors of platelet collagen interaction (saratin), and application of amniotic membranes have all shown relative success, but did not eliminate the fibrosis problem.

In relation to the failure of polymer-based glaucoma devices, medical device coatings have emerged as feasible materials to provide a biochemical solution to the adverse reactions at and around foreign implant sites. R&D is being focused on coatings that may exhibit antithrombogenic properties. Specifically, hydrophilic coatings are being used in endovascular catheters to improve mobility, prevent friction, and abate particulate formation in the hope of reducing inflammatory response [18]. Several studies revealed that heparin may inhibit early inflammatory signaling by blocking the aforementioned chemokines, integrins, growth factors, and receptors [19]. These techniques, including heparin-coated intraocular lenses and slow-release antimetabolite-coated glaucoma drainage devices, were investigated for ophthalmic use to reduce postoperative inflammation [20, 21]. Beyond chemical factors, mechanical force on the extracellular matrix is a major factor in the fibroproliferative process. Mechanical tension in the material acts through transmembrane integrins to free sequestered TGF-b, which promotes and maintains a myofibroblast response [22]. Myofibroblast wound contraction induces higher tension, resulting in positive feedback and dense scar formation [23]. Medical devices with 3D micropatterned surfaces mimic the extracellular matrix; thus, researchers are exploring laser beam bombardment of medical device surfaces to induce 3D patterning, which may help create an antiproliferative effect. Studies of healing ligaments and tendons show the collagen matrix is not aligned as it is in normal tissues [24]. Micropatterned surfaces provide mechanical-based guides, inducing cellular growth orientation, morphology, and migration during scar formation [25, 26]. Studies of post-infarct myocardial scar formation have shown mechanical guide effects.

Based on the information presented above, modified AGVs were recently produced with different coatings in an attempt to reduce fibrotic encapsulation and lower outflow resistance [27]. The research involved using four samples:

1. a control AGV (AGV Model FP7) consisting of a silicone plate and tube;
2. an AGV FP7 with a heparin plate coating;
3. an AGV FP7 implant with a hydrophilic plate coating; and
4. an AGV FP7 implant with a micropatterned plate surface.

The implants were inserted in rabbits' eyes for six weeks, followed by detailed microscopic and biological analysis. The edge of fibrosis was measured from the inner edge of the bleb to the transition point at the conjunctival interface. For hematoxylin and eosin (H&E) and trichrome staining, the transition point was determined to be at the boundary of tightly bound collagen fibers and loosely organized collagen fibers. For SMA stain, the transition point was determined to be when the brown immunohistochemistry (ICH) stain faded to the background.

The results of these studies [27] suggest that a device with a micropatterned plate surface and heparin plate coating may reduce postoperative fibrosis following glaucoma drainage device surgery. The micropatterned plate surface and heparin plate coating showed significantly less fibrosis and less outflow resistance than the control. However, although the AGV with a hydrophilic plate coating outperformed the control in reducing outflow resistance, it did not have a significantly different degree of fibrosis compared to the control.

An important issue to be considered in relation to the studies published in [27] is that the coating used was hydrophilic, which induces eye fluid adhesion on the surface of the coating, and thus adsorption of proteins and other biomolecules that may induce the formation of fibrosis, as observed.

Based on all the information presented so far, R&D was performed by the authors of this chapter to investigate the properties of a unique biocompatible/hydrophobic/antifouling coating named ultrananocrystalline diamond (UNCD) [28], which has been shown to provide probably the best biocompatibility, since it is made of C atoms (the element of life in human DNA, cells, and molecules), as the coating for glaucoma valves, providing a solution to the materials properties required for the coating of glaucoma valves.

3.2 Low-Cost, Best Biocompatible, Extremely Hydrophobic, and Antifouling UNCD Coating for Encapsulation of Commercial Polymer-Based Glaucoma Valves for Orders of Magnitude Performance Improvement

Current commercial glaucoma treatment devices inserted in the eye to drain fluid are based mainly on polymers and are rather large (~1.7 × 1 cm, as shown in Figure 3.2a). The implant surgery requires two incisions, one to insert the body of the valve on the sclera and the other to insert the polymer tube inside the eye to drain the fluid (see Figure 3.2b).

A problem that is not yet solved with current glaucoma treatment devices is the eye's reaction to the implant due to protein layer adhesion to the hydrophilic surface of the polymer, resulting in eye inflammation. Macrophages are attracted by chemical signals from the eye to the implant surface, where they release growth factors and induce growth of fibroblast cells that yield collagen production that results in development of a fibrous capsule around the implant. Subsequently, complete fibrosis occurs on the surface of the valve, isolating the implant from the rest of the body. This fibrous capsule induces material degradation effects, not only on glaucoma valves discussed here, but also on other medical implants, resulting in degraded sensor performance or transport of drugs in drug-delivery devices. Several approaches have been investigated to overcome these problems, but without proven success.

Thus, this section focuses on presenting a brief description of animal studies done in a joint collaboration between researchers at the University of Texas-Dallas and the University of Buenos Aires and Hospital Austral in Buenos Aires, Argentina, directed at investigating the effectiveness of a unique biocompatible/super-hydrophobic UNCD coating in eliminating the development of fibrotic capsules on the surface of commercial glaucoma silicone-based valves, which over time degrade the performance of the valve. The UNCD coating was grown using the process described in detail in Chapter 1 and elsewhere in a review article [28]. The animal studies were performed following NIH and ANMAT (Argentina's FDA equivalent) guidelines for animal care. The studies involved two rabbits, each receiving one uncoated AGV drainage device (Figure 3.3a) on one eye, and one coated with a low-temperature (~400 °C) UNCD film (Figure 3.3b) on the other eye in order to avoid ambiguities from using different animals in one experiment. The uncoated valve developed fibrosis (Figure 3.3a) due to eye protein adhesion within about 24 h after implantation, while the UNCD-coated device remained clear even after several months of implantation (Figure 3.3b), thus demonstrating the power of UNCD films as bioinert/biocompatible encapsulating coatings to improve by orders of magnitude the lifetime of the implanted polymer-based glaucoma valve.

The elimination of protein adhesion on the surface of the UNCD-coated AGV drainage device is attributed to the fact that the surface of UNCD films can be made super-hydrophobic (no water adhesion) [29] due to F atom termination of the surface via plasma treatment of as-grown UNCD films [29], and in addition providing the best biocompatible material due to being formed by C atoms [30–32].

The results presented in this section indicate that UNCD coatings may produce a new generation of transformational flexible polymer-based glaucoma treatment valves with orders of magnitude superior performance compared to current commercial valves.

3.3 Alternative Low-Cost, Best Biocompatible, Extremely Hydrophobic, and Antifouling UNCD-Coated Mesh Implantable in the Trabecular Region of the Eye to Drain Fluid

An alternative approach was explored to determine the feasibility of inserting a glaucoma drainage device directly in the trabecular tube area of the eye, thus eliminating connection of the device with the outside world, as in the case of AGV devices.

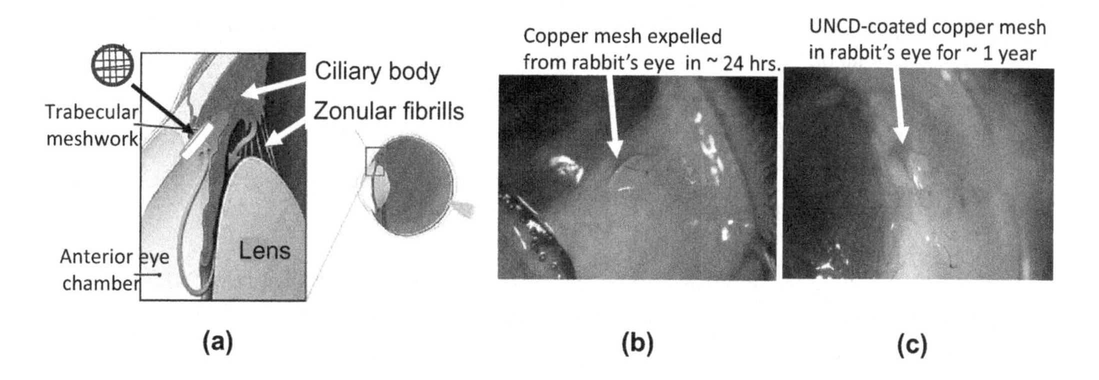

Figure 3.4 (a) A schematic showing the concept of inserting a mesh with a large arrays of holes in the eye's trabecular region to provide a pathway for eye fluid drainage. (b) A copper mesh inserted in the trabecular region of the right eye of a rabbit; the rabbit eye detected the poison represented by copper for the biological animal body, inducing rejection in a short time (see the copper grid driven through the eye tissue to the outside world). (c) A UNCD-coated copper grid implanted in the left eye of the same rabbit, showing total biological inertness for one year following implantation.

The approach described in this section involved inserting a circular mesh (ø = 3 mm) (Figure 3.4a) used in transmission electron microscopy (TEM) for sustaining films for TEM studies, with a large arrays of holes to provide a pathway for eye fluid drainage. The mesh is formed by an array of micron-scale diameter wires attached to a circular narrow holder to provide the mesh structure to place films for high-resolution TEM studies. In order to prove the unique biocompatibility of the UNCD coating for encapsulation of glaucoma drainage devices for insertion in the human eye, a copper grid was implanted in one rabbit's eye as a control, knowing that copper is poison for the biological environment of the human eye, thus eliciting rejection (Figure 3.4b). A UNCD-coated copper grid was implanted in the second eye of the same rabbit to avoid reactions to different rabbits' biological environments, which may compromise the validity of the test. The UNCD-coated copper mesh was implanted in the rabbit's eye without eliciting any rejection (Figure 3.4c) for one year following implantation, demonstrating the unique biocompatibility of the UNCD coating.

The results shown above on the performance of UNCD-coated AGV devices and copper grids implanted in rabbits' eyes show clearly the unique superior performance expected for UNCD-coated glaucoma drainage valves in human eyes. Clinical trials in humans are projected for the near future.

3.4 Conclusions

This chapter's description of glaucoma and the problems related to current treatments either using biological-related medicines or commercial glaucoma drainage valves, and the R&D described in relation to the development of a transformational low-cost and uniquely biocompatible/bioinert UNCD coating to encapsulate commercial polymer-based valves demonstrated that UNCD coating for encapsulation of commercial glaucoma drainage valves provides the pathway for a new revolutionary drainage valve technology for the treatment of glaucoma.

Acknowledgments

O. Auciello acknowledges the support from the University of Texas-Dallas through the Distinguished Endowed Chair Professor grant. A. Berra acknowledges support from CONICET-Argentina. M. Saravia acknowledges support from the Hospital Austral, where the animal studies were performed.

All authors acknowledge the great contribution to the R&D reported in this chapter of Dr. Pablo Gurman (MD-Argentina).

References

[1] C. Cook and P. Foster, "Epidemiology of glaucoma: what's new?," *Can. J. Ophthalmol.*, vol. 47, p. 223, 2012.
[2] www.mayoclinic.org/diseases-conditions/glaucoma/symptoms-causes/syc-20372839.
[3] www.glaucoma.org/treatment/new-medical-therapies-for-glaucoma.php.
[4] www.ivantisinc.com/glaucoma/treatmentoptions/?utm.sourceGoogle&utm.medium=PaidSearch at utm.vcampaign=Patient$utm.term=Glaucoma-Treatment&utm.content=Options FAD& gclid= EAIaIQobCh MI7uXn2f-Y6wIVCNbACh2XlAeFEAAYAiAAEgJDvPD_Bw
[5] www.brightfocus.org/glaucoma/article/glaucoma-surgery-series-tube-shunt-drainage
[6] S. J. Gedde, J. C. Schiffman, W. J. Feuer, et al. "Treatment outcomes in the tube versus trabeculectomy (TVT) study after 5 years of follow-up," *Am. J. Ophthalmol.*, vol. 153, p. 789, 2012.
[7] A. A. Sharkawy, B. Klitzman, G. A. Truskey et al. "Engineering the tissue which encapsulates subcutaneous implants I: diffusion properties," *J. Biomed. Mater. Res.*, vol. 37, p. 401, 1997.
[8] A. Mahale, M. W. Othman, S. Al Shahwan et al., "Altered expression of fibrosis genes in capsules of failed Ahmed glaucoma valve implants," *PLoS One*, vol. 10, p. E0122409, 2015.
[9] H. B. Thieme, L. Choritz, C. Hofmann-Rummelt, et al., "Histopathologic findings in early encapsulated blebs of young patients treated with the Ahmed glaucoma valve," *J. Glaucoma.*, vol. 20, p. 246, 2011.
[10] The Fluorouracil Filtering Surgery Study Group, "Five-year follow-up of the fluorouracil filtering surgery study," *Am. J. Ophthalmol.*, vol. 121, p. 349, 1996.
[11] J. S. Cohen, L. J. Greff , G. D. Novack, et al., "A placebo controlled, double-masked evaluation of mitomycin C in combined glaucoma and cataract procedures," *Ophthalmology.*, vol. 103, p. 1934, 1996.
[12] P. J. Lama and R. D. Fechtner, "Antifibrotics and wound healing in glaucoma surgery," *Surv. Ophthalmol.*, vol. 48, p. 314, 2003.

[13] V. P. Costa, A. Azuara-Blanco, P. A. Netland, et al., "Efficacy and safety of adjunctive mitomycin C during Ahmed glaucoma valve implantation: a prospective randomized clinical trial," *Ophthalmology*, vol. 111, p. 1071, 2004.
[14] J. A. Prata, Jr, D. S. Minckler, A. Mermoud, et al. "Effects of intraoperative mitomycin-C on the function of Baerveldt glaucoma drainage implants in rabbits," *J. Glaucoma*, vol. 5, p. 29, 1996.
[15] B. Hinz, D. Mastrangelo, C. E. Iselin, et al., "Mechanical tension controls granulation tissue contractile activity and myofibroblast differentiation," *Am. J. Pathol.*, vol. 159, p.1009, 2001.
[16] J. J. Tomasek, G. Gabbiani, B. Hinz, et al., "Myofibroblasts and mechano-regulation of connective tissue remodeling," *Nat. Rev. Mol. Cell. Biol.*, vol. 3, p. 349, 2002.
[17] P. Khaw, F. Grehn, G. Hollo, et al., "CAT-152 0102 Trabeculectomy Study Group, 161, A phase III study of sub-conjunctival human anti-transforming growth factor beta (2) monoclonal antibody (CAT-152) to prevent scarring after first-time trabeculectomy," *Ophthalmology*, vol. 114, p. 1822, 2007.
[18] A. Niemczyk, M. El Fray, and S. E. Franklin, "Friction behavior of hydrophilic lubricious coatings for medical device applications," *Tribol. Int.*, vol. 89, p. 54, 2015.
[19] G. Cassinelli and A. Naggi, "Old and new applications of non-anticoagulant heparin," *Int, J. Cardiology*, vol. 212, p. S14, 2016.
[20] C. L. Lin, A. G. Wang, J. C. K. Chou, et al., "Heparin-surface modified intraocular lens implantation in patients with glaucoma, diabetes, or uveitis," *J. Cataract Refract Surg.*, vol. 20, p. 550, 1994.
[21] S. Maedel, N. Hirnschall, Y. A. Chen et al., "Effect of heparin coating of a foldable intraocular lens on inflammation and capsular bag performance after cataract surgery," *J. Cataract Refract. Surg.*, vol. 39, p.1810, 2013.
[22] M. Xue and C. Jackson, "Extracellular matrix reorganization during wound healing and its impact on abnormal scarring," *Adv. Wound Care*, vol. 4, p. 119, 2015.
[23] D. Duscher, Z. N. Maan, V. W. Wong et al., "Mechanotransduction and fibrosis," *J. Biomech.*, vol. 47, p. 1997, 2014.
[24] J. H. Wang, F. Jia, T. W. Gilbert, et al., "Cell orientation determines the alignment of cell-produced collagenous matrix," *J. Biomech.*, vol. 36(1), p. 97, 2003.
[25] J. R. Gamboa, S. Mohandes, P. L. Tran, et al., "Linear fibroblast alignment on sinusoidal wave micropatterns," *Colloids Surf B. Biointerfaces*, vol. 104, p. 318, 2013.
[26] R. B. Dickinson, S. Guido, and R. T. Tranquillo, "Biased cell migration of fibroblasts exhibiting contact guidance in oriented collagen gels," *Ann. Biomed Eng.*, vol. 22(4), p. 342, 1994.
[27] N. A. Fischer, M. Y. Kahook, S, Abdullah, et al., "Effect of novel design modifications on fibrotic encapsulation: an in vivo glaucoma drainage device study in a rabbit model," *Ophthalmol. Ther.*, 2020. https://doi.org/10.6084/m9.figshare.11882403.
[28] O. Auciello and A. V. Sumant, "Status review of the science and technology of ultra-nanocrystalline diamond (UNCD^{TM}) films and application to multifunctional devices," *Diam. Relat. Mater.*, vol. 19, p. 699, 2010.
[29] A. Gabriela Montano-Figueroa, J. J. Alcantar-Peña, P. Tirado et al., "Tailoring of poly-crystalline diamond surfaces from hydrophilic to super-hydrophobic via synergistic chemical plus micro structuring processes," *Carbon*, vol. 139, p. 361, 2018.
[30] O. Auciello , P. Gurman , M. B. Guglielmotti, et al., "Biocompatible ultrananocrystalline diamond coatings for implantable medical devices," *MRS Bull.*, vol. 39, p. 621, 2014.

[31] O. Auciello, P. Gurman, A. Berra, M. Zaravia, and R. Zysler, "Ultrananocrystalline diamond (UNCD) films for ophthalmological applications," in *Diamond Based Materials for Biomedical Applications*, R. Narayan, Ed. Cambridge: Woodhead Publishing, p. 151, 2013.

[32] O. Auciello, "Novel biocompatible ultrananocrystalline diamond coating technology for a new generation of medical implants, devices, and scaffolds for developmental biology," *Biomater. Med. Appl. J.*, vol. 1 (1), p. 1000103, 2017.

4 Science and Technology of Novel Integrated Biocompatible Superparamagnetic Oxide Nanoparticles Injectable in the Human Eye and External Ultrananocrystalline Diamond (UNCD™)-Coated Magnet for a New Retina Reattachment Procedure

Mario J. Saravia, Roberto D. Zysler, Enio Lima, Jr., Pablo Gurman, and Orlando Auciello

4.1 Introduction

4.1.1 Biology and Function of the Human Eye and the Retina

The human eye and the key component, the retina (Figure 4.1), comprise one of the key parts of the human body. The eye, shown in Figure 4.1(a) and the schematic in Figure 4.1(b), captures images and light from the outside world, focusing the images through the lens onto the retina, where light induces optical excitation of the rods (responsible for dark vision) and the cones (responsible for color vision) (see the schematic in Figure 4.1c), which transform the optical excitation into electronic charges that excite the bipolar cells, which in turn amplify the electric charges and inject them into the ganglion cells. The ganglion cells transfer the electronic charges to the brain through their axons, which form the optic nerve. The retina is the neurosensory tissue located on the back wall of the human eye, involving several layers and functions:

1. an inner limiting membrane involving the so-called Muller's cells;
2. a nerve fiber layer formed by the axons linked to the ganglion cells;
3. a ganglion cell layer containing the nuclei of ganglion cells, the axons of which become the optic nerve fibers which are connected to the vision brain cells and some amacrine cells [1];
4. an inner plexiform layer, which contains the synapse between the bipolar cell axons and the dendrites of the ganglion and amacrine cells [1];

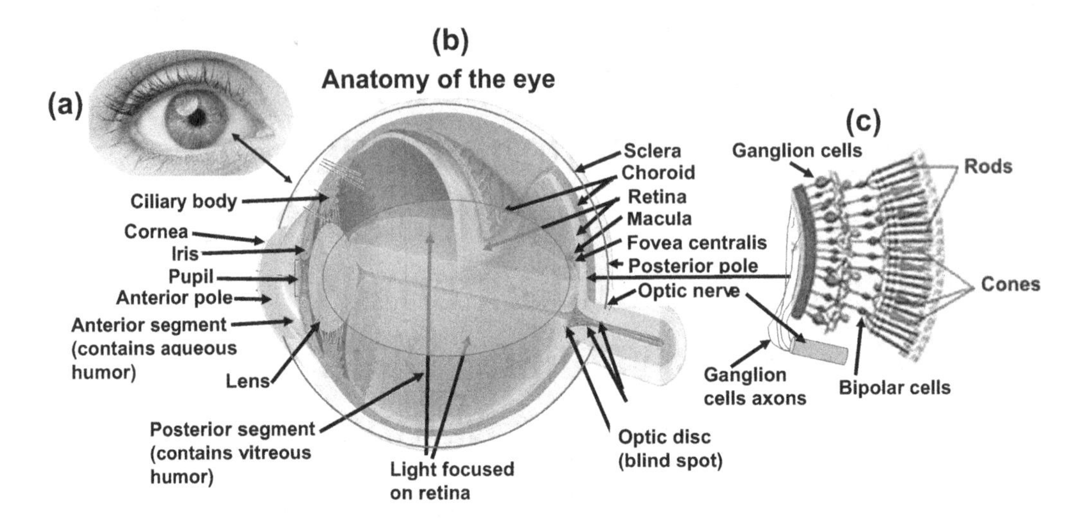

Figure 4.1 (a) A picture of the human eye. (b) A schematic of the human eye showing key components that produce vision. (c) A schematic of the retina showing the key components (rods and cones) responsible for capturing light and translating it into electronic charges, injected into bipolar cells, which amplify the charges and inject them into ganglion cells, the axon of which form the optical nerve, to transfer electronic charges to brain cells to produce vision.

5. an inner nuclear layer which contains the nuclei and surrounding cell bodies (perikarya) of the amacrine cells, bipolar cells, and horizontal cells [1];
6. an outer plexiform layer involving projections of rods and cones ending in the rod spherule and cone pedicle, respectively, making synapses with dendrites of bipolar cells and horizontal cells [1] in the macular region, known as the *fiber layer of Henle*;
7. an outer nuclear layer containing cells bodies of rods (responsible for dark vision) and cones (responsible for color vision);
8. an external limiting membrane layer separating the inner segment portions of the photoreceptors from their cell nuclei;
9. an inner segment/outer segment layer featuring inner segments and outer segments of rods and cones, with the outer segments containing a highly specialized light-sensing apparatus [2, 3]; and
10. the retinal pigment epithelium, which is a single layer of cuboidal epithelial cells closest to the choroid, providing nourishment and supportive functions to the neural retina, with the black pigment melanin in the pigment layer preventing light reflection throughout the globe of the eyeball; this behavior is extremely important for clear vision [4, 5].

These layers provide four main vision-related actions:

1. photoreception by photoreceptors;
2. photoreceptors transform the received photons into electronic charges;

3. electronic charges are transmitted to the bipolar cells, which amplify the charges and inject them into the ganglion cells;
4. the ganglion cells transmit the electronic charges through the bundle of ganglion cell axons (optic nerve) to the lateral geniculate body, a visual relay station in the diencephalon (the rear of the forebrain), and also they transmit the charges to the superior colliculus, the suprachiasmatic nucleus, and the nucleus of the optic tract to produce images in the brain [6].

In human adults, the retina comprises about 72% of a sphere about 22 mm in diameter, containing about seven million cones and 75–150 million rods. The optic disc, the area of the retina called "the blind spot" (Figure 4.1b) because it does not have photoreceptors, is located at the optic papilla, where the optic nerve fibers exit the eye. The optic disc looks like an oval white area of 3 mm^2. Close to this disc is the macula, which in the center has the fovea, a pit that is responsible for humans' sharp central vision, but is actually less sensitive to light because it lacks rods. Human and some mammals have one fovea, as opposed to certain bird species, such as hawks, which have two foveae, and dogs and cats, which do not have a fovea but rather a central band known as the visual streak. Around the fovea the central retina extends for about 6 mm, followed by the peripheral retina. The farthest edge of the retina is defined by the retina component named the ora serrata. The distance from the ora to the macula, which is the most sensitive area along the horizontal meridian, is about 32 mm.

The retina is a thin layer of light-sensitive tissue on the back wall of the eye (Figure 4.1b). The optical system of the eye focuses light on the retina much like light is focused on the film or sensor in a camera. The retina translates that optical image into electrical charges (impulses) and sends them to the brain via the optic nerve. The information presented here shows that the retina is a key component of the human eye, enabling vision.

4.1.2 Retina Trauma (Detachment): Symptoms and Current Treatments

4.1.2.1 Retina's Detachment Processes

Occasionally, injury or trauma to the eye or head may cause a small tear in the retina. The tear allows vitreous fluid to seep through it, getting under the retina and peeling it away like a bubble in wallpaper. Thus, when the retina suffers deleterious biological or structural changes, it may lead all the way to blindness. In this sense, when the retina detaches from the back wall of the eye due to concussion or a biological process, the sensory retinal tissue layer is separated from the underlying pigmented epithelium, with accumulation of subretinal fluid. The most common cause of retinal detachment is hematogenous, a process resulting from a break or tear of the sensory part of the retina from the rest of the eye tissue, resulting in partial or total loss of vision, and thus a medical emergency. The detached retina separates from the back wall of the eye and is removed from its blood supply and source of nutrition. The retina will degenerate and lose its ability to function if it remains detached. Central vision will be lost if the macula remains detached. The types of retinal detachment are as follows:

- Rhegmatogenous retinal detachment. A rhegmatogenous retinal detachment occurs due to a break in the retina (called a *retinal tear*) that allows fluid to pass from the vitreous space into the subretinal space between the sensory retina and the retinal pigment epithelium. Retinal breaks are divided into three types: holes, tears, and dialyses. Holes form due to retinal atrophy, especially within an area of lattice degeneration. Tears are due to vitreoretinal traction. Dialyses, which are very peripheral and circumferential, may be either tractional or atrophic, such that the atrophic form most often occurs as idiopathic dialysis of young people.
- Exudative, serous, or secondary retinal detachment. An exudative retinal detachment occurs due to inflammation, injury, or vascular abnormalities that results in fluid accumulating underneath the retina without the presence of a hole, tear, or break. In evaluation of retinal detachment, it is critical to exclude exudative detachment as surgery will make the situation worse. Although rare, exudative retinal detachment can be caused by the growth of a tumor on the layers of tissue beneath the retina, namely the choroid. This cancer is called a choroidal melanoma.
- Tractional retinal detachment. A tractional retinal detachment occurs when fibrous or fibrovascular tissue, caused by an injury, inflammation, or neovascularization, pulls the sensory retina from the retinal pigment epithelium.

4.1.2.2 Signs and Symptoms Associated with Retinal Detachment

Retinal detachment is commonly preceded by a posterior vitreous detachment, which gives rise to various symptoms:

- flashes of light (photopsia), which are very brief in the extreme peripheral vision;
- sudden dramatic increases in the number of “floaters,” which are visible deposits within the eye’s vitreous humor, with different sizes, shapes, consistency, refractive index, and motility [7]. The vitreous humor usually starts out transparent, but imperfections may gradually develop as people age. The common type of floater, present in most people’s eyes, is associated with degenerative changes in the vitreous humor. The perception of floaters, annoying to some people, is known as mydesopsia, or less common as *myodaeopsia*, *myiodeopsia*, or *myiodesopsia*. It is not often treated, except in severe cases, where vitrectomy (surgery), laser vitreolysis, and medication may be effective;
- a ring of floaters or hairs just to the temporal side of the central vision; and
- a slight feeling of heaviness in the eye.

Although most posterior vitreous detachments do not progress to retinal detachments, those that do produce the following symptoms:

- a dense shadow that starts in the peripheral vision and slowly progresses toward the central vision;

- the impression that a veil or curtain has been drawn over the field of vision;
- straight lines (e.g., on a scale, the edge of a wall, the road) that suddenly appear curved; and
- central vision loss.

A minority of retinal detachments result from trauma, including blunt blows to the orbit, penetrating trauma, and concussions. A retrospective Indian study of more than 500 cases of rhegmatogenous detachments found that 11% were due to trauma, and that gradual onset was the norm, with over 50% presenting more than one month after the inciting injury [8].

4.1.2.3 Risk Factors and Prevention of Retinal Detachment

Risk factors for retinal detachment include myopia, retinal tears, trauma, family history, and complications from cataract surgery [9]. Retinal detachment can be controlled in cases when the warning signs [10] are noticed early. The most effective means of prevention and risk reduction is by appropriate education of people for identification of the initial signs, and inducing them to seek medical attention if they suffer from symptoms suggesting the occurrence of posterior vitreous detachment. Early examination allows detection of retinal tears that can be treated with laser or cryotherapy. This reduces the risk of retinal detachment in those who have tears from around 1:3 to 1:20. For this reason, the governing bodies in some sports require regular eye examinations.

Trauma-related cases of retinal detachment may be related to high-impact sports (e.g., boxing, karate, kickboxing, the so called football (strong contact sport) played in the USA), or high-speed sports (e.g., automobile racing, sledding). Although some doctors recommend avoiding activities that increase pressure in the eye (e.g., diving and skydiving), there is little evidence to support that retinal detachment may be substantially induced by these activities. Nevertheless, ophthalmologists generally advise patients with a high degree of myopia to avoid activities with the potential for trauma, increasing pressure on or within the eye itself, or including rapid acceleration and deceleration.

Intraocular pressure spikes may occur during activities accompanied by the Valsalva maneuver, including weightlifting. An epidemiological study suggested that manual heavy lifting at work may be associated with increased risk of rhegmatogenous retinal detachment [11]. This study also indicates that obesity may increase the risk of retinal detachment. In addition, genetic-based factors may induce local inflammation and photoreceptor degeneration [12].

4.1.2.4 Diagnosis of Retinal Detachment

Retinal detachment can be examined by fundus photography or ophthalmography. Fundus photography generally requires a substantially larger instrument than ophthalmography, but it has the advantage of enabling image examination by a specialist at another location and/or time, as well as providing photo documentation for future reference. Modern fundus photographs generally recreate considerably larger areas of the fundus than can be seen at any one time with handheld ophthalmoscopes.

4.1.2.5 Current Treatment of Retinal Detachment

There are several methods for treating retinal detachment, each of which depends on finding and closing the breaks that have developed in the retina. All three of the current procedures, described below, follow the same three general principles:

1. find all retinal breaks;
2. seal all retinal breaks; and
3. relieve present (and future) vitreoretinal traction.

- **Cryopexy and laser photocoagulation:** Cryotherapy (freezing of the retina) [13] or laser photocoagulation [14] are used to treat a small area of retinal detachment so that the detachment does not spread.
- **Scleral buckle surgery:** This surgery is an established procedure whereby the eye surgeon sews one or more flexible silicone bands to the sclera (the white outer coat of the eyeball). The bands push the wall of the eye inward against the hole produced in the retina, closing the break or reducing fluid flow through it and reducing the effect of vitreous traction, thus allowing reattachment of the retina. In this procedure, cryotherapy (eye wall freezing) is applied around the retinal breaks prior to placing the buckle. Often, subretinal fluid is drained as part of the buckling procedure. The buckle remains in place. The most common side-effect of a scleral operation is a myopic shift – that is, the affected eye will be more short-sighted after the operation. Radial scleral buckle is the most appropriate procedure to treat U-shaped or fish-mouth tears and posterior breaks.
  A circumferential scleral buckle is used to treat multiple breaks, anterior breaks, and wide breaks. Encircling buckles are used to treat breaks extended to more than two quadrants of the retinal area, lattice degeneration located on more than two quadrants of the retinal area, undetectable breaks, and proliferative vitreous retinopathy.
- **Pneumatic retinopexy:** This treatment, extensively applied today, is generally performed in the doctor's office under local anesthesia. In the pneumatic retinopexy procedure to reattach the retina, a gas bubble (SF_6 or C_3F_8 gas) is injected into the eye (Figure 4.2), followed by a laser or freezing treatment applied to the retinal hole. The patient's head is then positioned so that the bubble rests against the retinal hole. Patients may have to keep their heads tilted for several days to keep the gas bubble in contact with the retinal hole. The surface tension of the air–water interface seals the hole in the retina and allows the retinal pigment epithelium to pump the subretinal space dry and suck the retina back into place.

The strict positioning requirement of the pneumatic retinopexy process, including the laser or cryopexy treatments, makes the treatment of the retinal holes and detachment that occurs in the lower part of the eyeball impractical.

Vitrectomy. This process is increasingly being used for treatment of retinal detachment. It involves the removal of the vitreous gel and is usually combined with filling

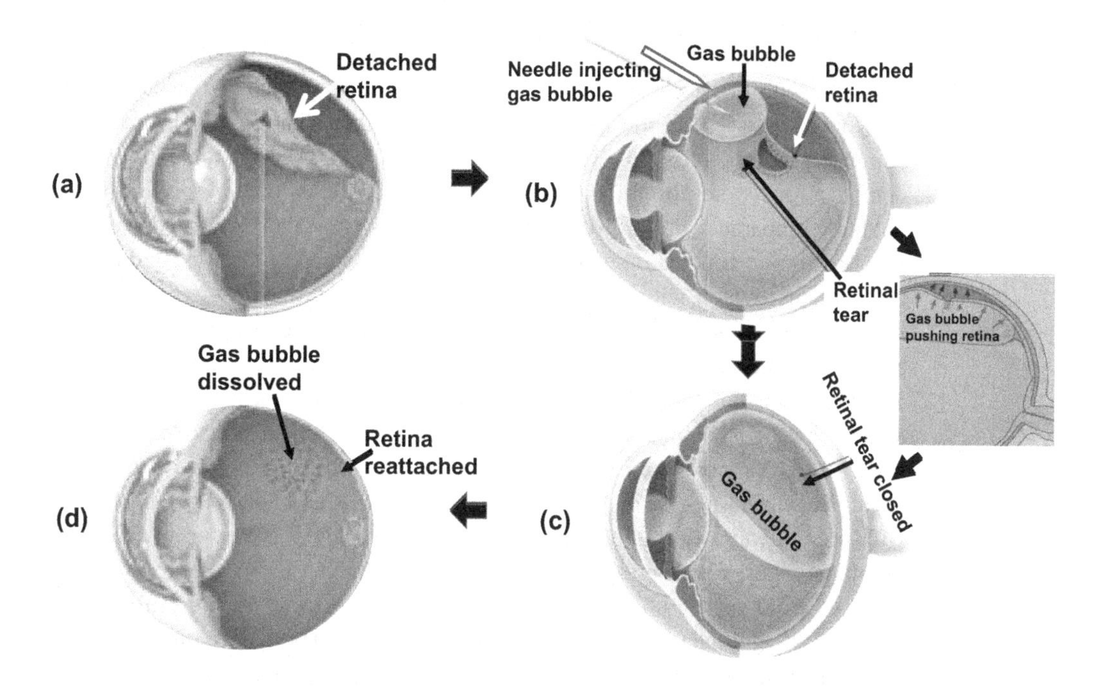

Figure 4.2 (a) Schematic of the human eye with a detached retina. (b–d) Using a gas (SF_6 or C_3F_8 gas) bubble inserted inside the eye to push the retina back onto the eye wall. The gas bubble pushes the retina back onto the eye wall and laser or cryopexy treatments induce chemical reactions of retina atoms/molecular bonds with the eye wall. The gas bubble is removed after the retina is reattached. Serious problems related to the pneumatic retinopexy shown in this figure are: (1) it does not work well for inferior retinal detachment; (2) it induces temporarily impaired vision; and (3) it has a low rate of success.

the eye with either a gas bubble (SF_6 or C_3F_8 gas) or silicone oil. An advantage of using gas in this operation is that there is no myopic shift after the operation and the gas is absorbed within a few weeks. If silicone oil (PDMS) is used instead of a gas bubble, it needs to be removed after a period of 2–8 months, depending on the surgeon's preference. Silicone oil is more commonly used in cases associated with proliferative vitreoretinopathy (PVR). The literature indicates that silicone oil is the most common choice of tamponade process used when managing a giant retinal tear (GRT). However, there is concern related to potential detrimental effects on vision. Detailed studies of gas vs. oil bubble treatment for retina reattachment have been performed by independent groups:

1. A Moorfield's Eye Hospital group performed systematic studies for GRT over a five-year period (2005–2010) [15]. In addition, the group analyzed a subgroup of fovea-sparing retinal detachments treated by pars plana vitrectomy (PPV) with either silicone oil or gas tamponade methods. Subgroup analysis of

fovea-sparing retinal detachments (silicone oil $n = 49$; gas $n = 15$) revealed visual loss ($\geq$2 Snellen lines of vision) in 49.0% ($n = 24$) of patients managed with oil, compared to 13.3% ($n = 2$) of gas patients ($p = 0.019$). In all, 73.3% ($n = 11$) in the gas group achieved a final vision of 6/12 or better, compared to 36.7% ($n = 18$) in the oil group ($p = 0.031$). Postoperative complications were absent in 66.7% ($n = 10$) of patients treated with a gas bubble, compared with 14.3% ($n = 7$) of patients ($p = 0.002$) treated with an oil bubble. The conclusion was that eyes with fovea-sparing GRT-related retinal detachments treated with a gas bubble yielded a better visual outcome with fewer postoperative complications and no significant difference in anatomical success.

2. Systematic studies by Moharran et al. [16], involved treatment of retinal detachment of 88 human eyes, with 48 retinas reattachment processes involving C_3F_8 gas bubbles and 40 using silicone oil as a tamponading agent. The mean age of patients was 39 years. All eyes underwent 23G PPV with no adjuvant scleral buckling and phacovitrectomy was performed for all phakic eyes. Final retinal reattachment was achieved in 86 eyes (97.7%). One eye from each group had recurrent retinal detachment. Postoperative vision was significantly better in the gas group ($p = 0.008$). Prolonged increase of intraocular pressure (IOP) developed in six eyes in the silicone oil group and five eyes in the gas group. Prolonged uveitis developed in four eyes in the gas group and six eyes in the oil group ($p = 0.04$). Epiretinal membranes (ERM) developed in ten eyes in the gas group and nine eyes in the oil group. This group found no significant difference between gas and oil bubble treatments of retinal detachment regarding postoperative glaucoma or ERM formation.
3. Ratanasukon et al. [17] investigated the treatment of patients with HIV-related retinal detachment, treated with PPV with silicone oil and gas endotamponade between January 2003 and June 2005. They determined that PPV with silicone oil and gas tamponade provided the same reattachment rate and visual results between the two tamponade groups. However, they determined that use of gas tamponade may be more effective in treatment of patients with highly active antiretroviral therapy (HAART).

Information from the literature and hospitals worldwide indicates that vitrectomy is the most commonly performed operation for the treatment of retinal detachment. However, a general disadvantage observed in most cases is that a vitrectomy always leads to more rapid progression of a cataract in the operated eye.

Results of Surgery. The results of surgery to reattach the retina indicate that 85% of cases are successful with one operation, with the remaining 15% requiring two or more operations. After treatment, patients gradually regain their vision over a period of a few weeks, although the visual acuity may not be as good as it was prior to the detachment, particularly if the macula was involved in the area of the detachment. However, if left untreated, total blindness could occur in a few days.

4.2 New Revolutionary Nanotechnology-Based Retina's Reattachment Procedure

The advent of nanotechnology in recent years is considered the technological revolution of the twenty-first century. Nanotechnology deals with manipulation and control of matter at a scale of 1–100 nm, pursuing the development of new products by taking advantage of new physical, mechanical, and chemical properties that arise as the scale goes down from molecular to atomic.

Nanoparticles are among the first true nanotechnology products already on the market. Among several nanoparticle systems available today, magnetic nanoparticles have drawn a lot of attention [18–22] from industry, academia, and national laboratories because of the great variety of applications. Among these applications, biotechnology and medicine [18–22] are two fields where magnetic nanoparticles have found a wide variety of uses, including bio-separation, contrast agents for MRI, therapeutic agents in hyperthermia and drug delivery, and recently as nanoactuators and nanosensors for lab-on-a-chip devices.

In the work described in this chapter, the focus of recent R&D has been the use of superparamagnetic nanoparticles for the treatment of retinal detachment to overcome the drawbacks of the current reattachment procedures described above.

The work described in this chapter demonstrated how, by taking advantage of superparamagnetic properties of magnetic nanoparticles, a sealing mechanism for reattaching the retina is achieved by using an external magnetic field induced by a magnet temporarily implanted on the outer wall of the human eye, such that the magnetic field generated by the magnet produces a torque in the inner part of the nanoparticles [22], increasing their magnetization and creating a nanoscale force capable of attracting the nanoparticles to agglomerate, impacting on the detached retina, and pushing it back onto the inner wall of the eye for the period of time necessary to be fully reattached via two different procedures:

1. **Laser photocoagulation process:** In this process an excimer laser beam (UV 192 nm wavelength) is directed through the eye's pupil and lens, focused on the retina, breaking chemical bonds of biological material in the eye and inducing formation of molecules which diffuse into the retina tissue, resulting in denaturing and coagulation of the retina cells to produce reattachment of the retina on the inner wall of the eye. In addition, a key component of the mechanism of retinal photocoagulation is the oxygenation of the tissue [23], such that oxygen improves the hypoxia induced by capillary nonperfusion or ischemia, reversing the consequences of hypoxia – that is, VEGF formation and vasodilation, new vessel formation, and edema to complete the retina reattachment.
2. **Cryotherapy (freezing of the retina):** This process uses cryopexy (intense cold [–40 °C] applied to the retina from the eye's outer wall) using freon, nitrous oxide, or carbon dioxide gas as the cooling agent to decrease cellular metabolism in the retina and promote cellular survival as the means of the process of retina reattachment [24].

4.2.1 Materials and Methods

4.2.1.1 Superparamagnetic Iron Oxide (Fe_3O_4) Nanoparticles (SPION-Fe_3O_4)

Summary of R&D and Biomedical Use of SPION-Fe_3O_4

In recent years, SPION-Fe_3O_4 has been extensively investigated for biological applications due to studies that indicated that these nanoparticles have very low toxicity [19–22] and high magnetic moment saturation at room temperature. SPION-Fe_3O_4 has been investigated for the following applications: (1) as a contrast medium to obtain images using magnetic resonance [25, 26]; (2) for solutions in magnetic hyperthermia fluids [27, 28]; (3) as carriers for drug delivery [29–31]; and (4) for ophthalmological diagnosis and treatment [21, 32–38].

Key properties of SPION-Fe_3O_4 are: (1) their magnetic moment induces strong interactions with external magnetic fields, enabling controlled magnetic field-induced motion of the particles through control of particle size, with both parameters contributing to tailoring the force exerted on the particles; (2) the superparamagnetism of SPION-Fe_3O_4 is a critical property related to the fact that they become magnetic only when exposed to a magnetic field, which attracts them, inducing their agglomeration to push the retina (see the description of the process in the sections below); once the field is removed, the particles are dispersed and can be adsorbed into the biological tissue without secondary effects, thus eliminating the need for particle extraction from the eye, and thus a better post-reattachment procedure compared with the gas or oil bubble treatment.

A key issue related to using SPION-Fe_3O_4 for retina reattachment is that these particles have been approved by the FDA in the USA and the European Medicines Agency for use in humans, including retina reattachment [39, 40], nuclear magnetic resonance (NMR) imaging [39–44], and anemia treatment [20]. For the specific case of using SPION-Fe_3O_4 for imaging contrast in NMR, the FDA approved the use of ferumoxides, which are SPION-Fe_3O_4 encapsulated with dextran (also known as Feridex®) in liquid solution, commercialized in the USA by Bayer Health Care Pharmaceuticals Inc., and in Europe by Bayer (Belgium, Switzerland, Germany, Spain; Bayer Animal Health, Luxembourg; Bayer Schering, Greece, Denmark, Norway; Bayer Yakuhin, Japan; Schering, Czech Republic, Italy, Romania, Serbia, Slovenia; and Schering-Plough, Israel).

Fabrication and Characterization of SPION-Fe_3O_4

SPION-Fe_3O_4 was synthesized by a chemical route according to the method described in [33, 45]. Briefly, SPION-Fe_3O_4 was prepared by the high-temperature decomposition of Fe(III)acetylacetonate ($Fe(acac)_3$) in the presence of surfactants and of a long-chain alcohol. In order to obtain the desired particle size, which is critical to get into the superparamagnetic regime, the surfactant:precursor molar ratio was precisely controlled, obtaining 18 nm nanoparticles with a very narrow size distribution. Increasingly relevant applications of SPION-Fe_3O_4 in biology and medicine require strict control of their morphological, physicochemical, and magnetic properties. Usually, the tuning of the SPION-Fe_3O_4's magnetic properties is achieved by controlling the size and crystallinity of the nanoparticles. It is known that the peroxidase-like activity of the ferrite nanoparticles can catalyze the decomposition of H_2O_2, producing reactive oxygen species (ROS). The presence or not of this activity is crucial for different applications of SPION-Fe_3O_4 in medicine: ROS generation helps in tumor treatment but, on the other hand, the free radical presence is undesirable in nanoparticle-assisted drug delivery. Several works have been dedicated to understanding the catalytic activity of SPION-Fe_3O_4, in particular the Fenton-like activity of ferrite nanoparticles (MFe_2O_4, M = Fe_2^+/Fe_3^+, Ni, and Mn) with mean diameter in the 10–12 nm range in different media has been characterized [32, 46]. This peroxidase-like activity was investigated via electron paramagnetic resonance (EPR) using the spin-trap 5,5-dimethyl-1-pyrroline N –oxide (DMPO) at different pHs (4.8 and 7.4), temperatures (25 and 40 °C), and buffer media, enabled identification of an enhanced amount of hydroxyl (•OH) and perhydroxyl (•OOH) radicals in the SPION- Fe_3O_4 solution compared to a blank solution. The studies showed that [•OH] is the dominant radical formed for Fe_3O_4, which is strongly reduced with the concomitant oxidation of Fe^{2+} or its substitution by Ni or Mn. A comparative analysis of the EPR data against *in-vitro* production of ROS in microglial BV2 cell culture provided additional insights regarding the catalytic activity of SPION-Fe_3O_4, which should be considered for biomedical uses. Therefore, prior to the use of nanoparticles in ophthalmological treatment, it is necessary to eliminate the catalytic effect of Fe^{2+} with controlled surface oxidation (from Fe^{2+} to Fe^{3+}, which is more harmless). EPR experiments were carried out on these "passivated" samples, showing that they do not show appreciable radicals (•OH and •OOH) production activity and are safe for use in the treatment of retinal detachment. This "passivation" does not produce a significant change in the magnetic properties of the SPION-Fe_3O_4. Other groups also demonstrated processes for fabrication of SPION-Fe_3O_4 [47, 48].

A suspension of the SPION-Fe_3O_4 (with appropriate biofunctionalization in the surface) in saline solution was prepared [33, 45]. The color of the solution was black, and the SPION-Fe_3O_4 remained in the liquid phase without further precipitation.

Figure 4.3a shows a picture of a SPION-Fe_3O_4 solution in a container; Figure 4.3b and c show transmission electron microscopy (TEM) and high-resolution TEM (HRTEM) images, respectively, of SPION-Fe_3O_4, revealing the nanostructure and dimensions.

The superparamagnetic characteristics and magnetization value per particle were obtained by measuring magnetization vs. magnetic field and temperature, using tailored magnetometers (see [33]). These measurements are critical, since the SPION-Fe_3O_4 nanoparticles' magnetization value determines the force (pressure) that the nanoparticles exert on the detached retina to push it back onto the eye's inner wall for reattachment. The value of the magnetic moment is proportional to the value of magnetization saturation (Ms) of SPION-Fe_3O_4; that is: $M = Ms{\cdot}V$, where V is the mean volume of the SPION-Fe_3O_4s and Ms is the normalized magnetization saturation value of this material. Measurements of magnetization vs. magnetic field are shown in Figure 4.4, which reveals a saturation magnetization $Ms = 42$ emu/g (see extrapolation of the straight black line up to the magnetization axis).

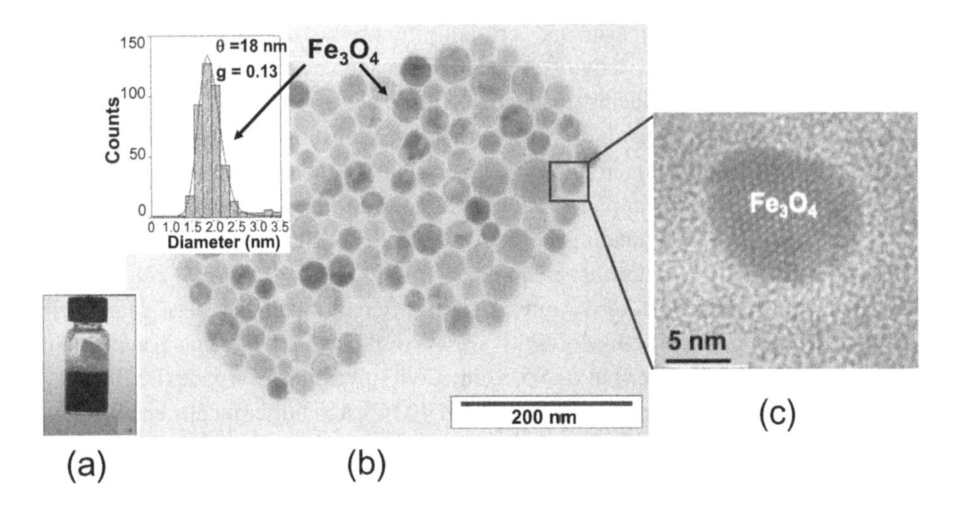

Figure 4.3 (a) SPION-Fe_3O_4 nanoparticles in solution. (b) Medium- and (c) high-resolution TEM images of SPION-Fe_3O_4 nanoparticles.

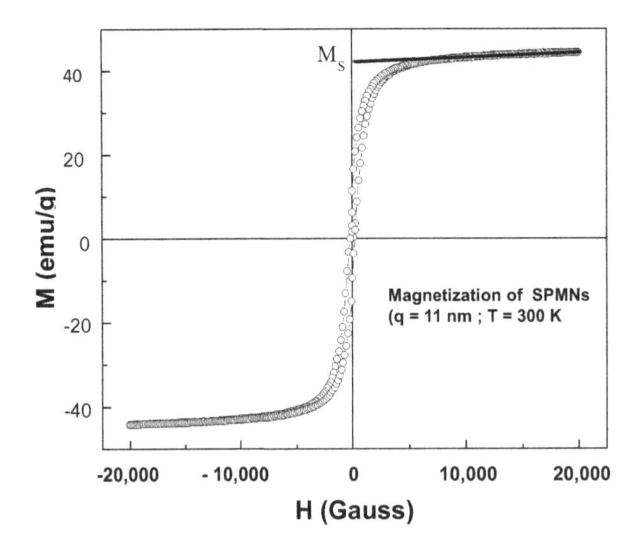

Figure 4.4 Magnetization vs. magnetic field at 300 K, measured for SPION-Fe_3O_4 nanoparticles.

Characterization of Magnetic Properties of Magnets

Neodymium-based magnets (1 mm thick and semicircular with 5 mm diameter – see Figure 4.5a) were purchased from a supplier. In order to get some statistics, the magnetization of five magnets was measured (Figure 4.5b,c). It was determined that the magnetization is established along the width of the magnet (see Figure 4.6a). The vertical edges of the magnets revealed magnetic field near $H = 0$ Gauss. On the other hand, the circular edges yielded the magnetic fields shown in Figure 4.5c.

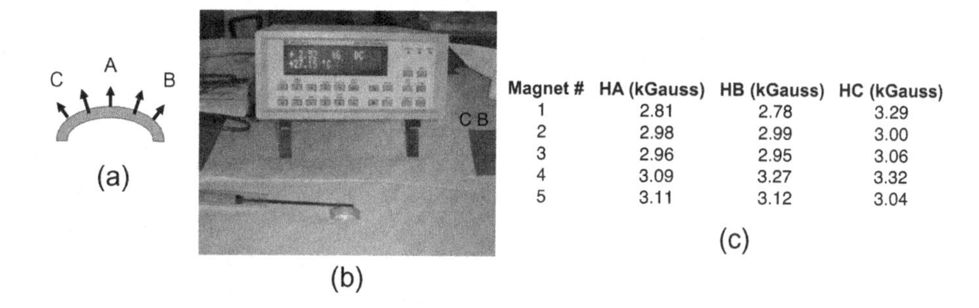

Magnet #	HA (kGauss)	HB (kGauss)	HC (kGauss)
1	2.81	2.78	3.29
2	2.98	2.99	3.00
3	2.96	2.95	3.06
4	3.09	3.27	3.32
5	3.11	3.12	3.04

Figure 4.5 (a) Schematic of the magnet. (b) Instrument for measuring magnetization. (c) Values of magnetization for five magnets with equal dimensions, showing relatively good reproducibility of magnetization values (see also [34]).

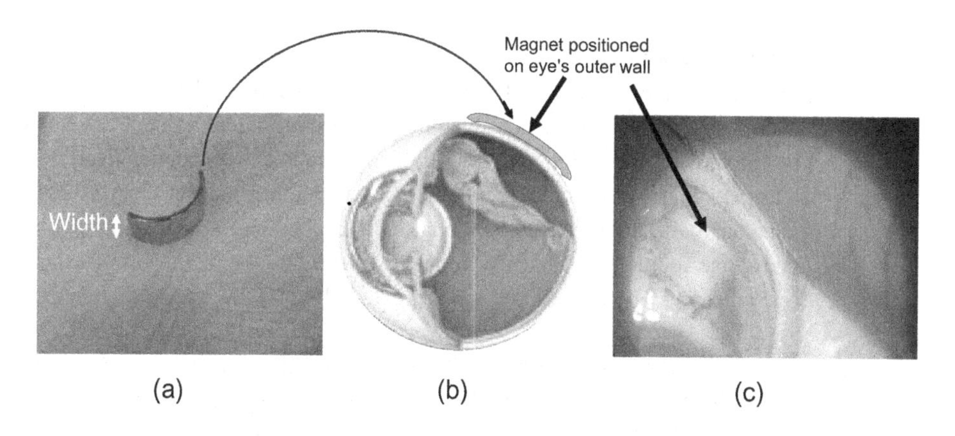

Figure 4.6 (a) A semicircular neodymium-based magnet (1 mm thick and 5 mm diameter). (b) A schematic showing the positioning of the magnet on the surface of the eye's outer wall, near the detached retina. (c) A magnet positioned on the surface of the eye's outer wall in a rabbit used for animal studies.

The mean value of the magnetic field, as shown in Figure 4.5c, is 3.1 kGauss. This magnetic field value is compatible with the values provided by simulations of magnetic fields for magnets, which can produce the required force for **SPION**-Fe_3O_4 to push the retina into place.

Neodymium-based magnets, coated with a layer of nickel (Ni), were purchased from an industrial manufacturer (Figure 4.6a). The' magnet is positioned, with stitches, temporarily on the surface of the eye's outer wall on the area where the retina is detached (Figure 4.6b,c). The magnet is coated either with biocompatible Ti or with a UNCD layer, the latter being preferably because of the demonstrated superior biocompatibility of UNCD coating in connection with the eye, as was demonstrated in prior work by Auciello et al. [47–49] and Weiland et al. [50], involving insertion of UNCD-coated silicon microchips into rabbits' eyes, implanted on the inner side of the retina wall surface as part of R&D to develop an artificial retina [47, 48] to restore partial vision to people blinded by genetically induced degeneration of the retina photoreceptors (see Chapter 2).

Work in progress by Dr. M. Saravia in Argentina is focused on testing whether a Ti-coated magnet would work without undesirable biological effects when positioned on the surface of the outer wall of a human eye during the 10–15 days needed for retina reattachment. A Ti coating, which can be grown at room temperature, would be easier and lower cost than a UNCD coating because the UNCD growth temperature (~400 °C) would require remagnetizing the magnet after the UNCD coating process.

4.2.2 *In-Vivo* Animal Studies for SPION-Fe_3O_4-based Retina Reattachment

Animal studies were performed using 10 rabbits, 5 of them being albinos NZ and 5 pigmented.

4.2.2.1 Surgically Induced Retina's Detachment

In order to induce retinal detachment in the animals, all rabbits were subjected to anesthesia using a mixture of ketamine hydrochloride and xylazine. A vitrectomy was performed in the right eyes of eight NZ rabbits. A conjunctival 360° aperture was performed, resulting in a conjunctiva separation from the sclera. A retinal tear was generated with a 23G cannula in a place correspondent to the external implant location, and saline solution was injected through the tear until the retina was detached for at least one quadrant. The magnet was sutured externally on the superior sclera of the right eye.

4.2.2.2 Magnet Attachment on External Wall of the Eye and SPION-Fe_3O_4 Solution Injection into the Eye for Retina Reattachment

SPION-Fe_3O_4 (20 nm in diameter) particles, in solution, were injected around the lesion in seven eyes. Three eyes remained without injection for control. Three additional rabbits were injected without surgery to evaluate inflammatory response to SPION-Fe_3O_4. Images of the fundus and optical coherence tomography (OCT) were obtained at days 3, 7, 14, and 21 after surgery. The 13 rabbits were sacrificed on day 21 and their eyes were fixed for retinal histology with hematoxylin and eosin (H&E) to assess for signs of inflammation, toxicity, or damage. All the experiments were carried out following the guidelines for animal care of the Association for Research in Vision and Ophthalmology (ARVO) in Argentina.

Results from Animal Retina Reattachment Studies via Combined Magnetic Field Created by a Magnet on the Eye's Outer Wall and Injection of SPION-Fe_3O_4 Solution into the Eye

Retinas were attached in the eyes of seven rabbits via injection of SPION-Fe_3O_4 solution, with magnets located on the surface of the outer wall of the eyes. Retinas remained detached in the eyes of rabbits where no SPION-Fe_3O_4 solution was injected. Neither signs of cataract, inflammation, or toxicity were detected in the histology of the three groups of rabbits. No SPION-Fe_3O_4 particles were detected beyond the internal limiting membrane of the retina. No signs of structural damage were observed.

Retinography and OCT Analysis of Retinas with Adhered SPION-Fe_3O_4

After three weeks of the retina reattachment procedure with the SPION-Fe_3O_4 particles, retinography-based studies of the rabbits' eyes showed SPION-Fe_3O_4 particles distributed in a semicircle (Figure 4.7a–c), which is correlated with the magnetic field distribution produced by the magnets outside the eye with the same semicircular geometry (see Figure 4.6). The OCT analysis (Figure 4.7a,b) revealed that the SPION-Fe_3O_4 particles were over the surface of the retina without penetrating it. The curve shown in Figure 4.7d correlates with the SPION-Fe_3O_4 particle distribution observed in Figure 4.7b,c.

In spite of the fact that SPION-Fe_3O_4 nanoparticles remain in the vitreous cavity for up to four months, the continuous observation of the treated eyes of the rabbits showed neither signs of inflammation nor toxicity, which strongly indicates that SPION-Fe_3O_4 nanoparticles are not biodegraded and can remain in the eye without eliciting toxic effects. Studies to obtain further data to confirm the initial results are in progress.

4.2.2.3 Histological Evaluation of the Animals' Retinas Exposed to SPION-Fe_3O_4 Nanoparticles and Magnets Coated with Gold

Thirty days after the SPION-Fe_3O_4 nanoparticles procedures for animal retina reattachment, euthanasia was completed to perform histological evaluation in the ocular lobe, specifically in the area where the magnet was positioned on the outer wall of the eye. The samples were chemically treated to fix the magnet and the SPION-Fe_3O_4 nanoparticles attached to the retina to allow the pathologist to identify the retina area that was treated. The biological analysis using optical macroscopy

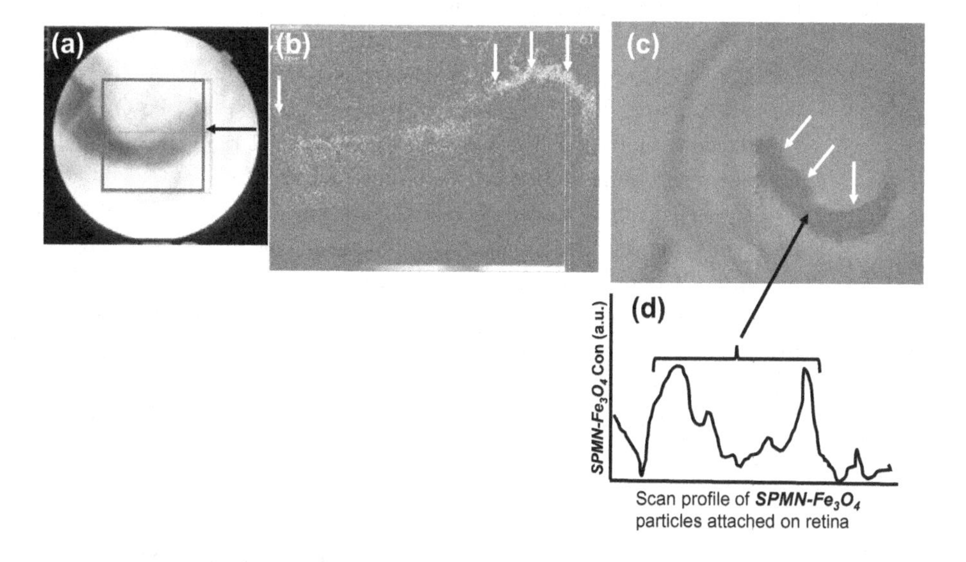

Figure 4.7 (a,b) Images from OCT analysis in a rabbit eye showing the region where SPION-Fe_3O_4 nanoparticles were attracted to the retina's surface (arrows in (a) and (b)). (c) Retinography showing how SPION-Fe_3O_4 nanoparticles agglomerate around the magnet (not shown) d) curve showing counts of SPION-Fe3O4 nanoparticles from imaged area in (c)..

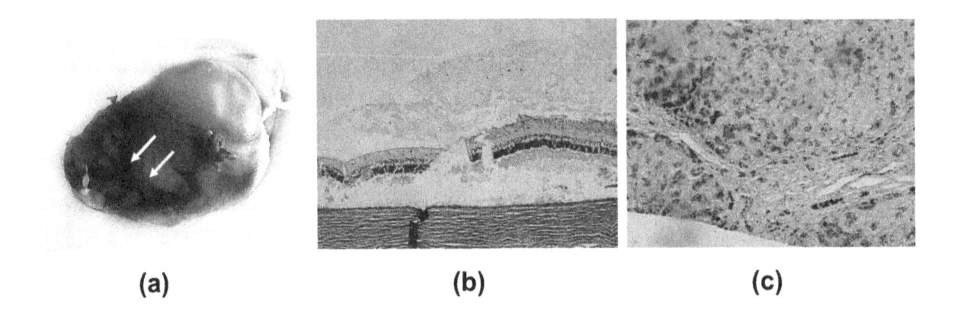

Figure 4.8 (a) Optical macroscopic image of the retina with the magnet and attached SPION-Fe_3O_4 nanoparticles. (b) SEM microscopic image of the retina-attached SPION-Fe_3O_4, which showed no blue color, and thus no SPION-Fe_3O_4 nanoparticle insertion. (c) Microscopic image of a patient's liver, which showed blue color indicative of Fe particle insertion.

(Figure 4.8a) and scanning electron microscopy (SEM) (Figure 4.8b) revealed minor inflammation mainly due to the surgical procedure and not to the magnet or SPION-Fe_3O_4 nanoparticles contacting the eye tissue. Figure 4.8a shows the semicircular magnet over the sclera, indicated by the white arrows, and no evidence of inflammation on the area that was in contact with the magnet or the SPION-Fe_3O_4 nanoparticles. In order to study more accurately the effect of SPION-Fe_3O_4 nanoparticles in contact with the retina, those areas were exposed to coloration sensible to Fe (Figure 4.8b). In order to ensure that there was no false information, samples of livers of patients affected by siderosis (lung disease caused by breathing in fine particles of iron or other metallic dust) were inserted in the same container with the retina samples with SPION-Fe_3O_4 nanoparticles. The analysis of tissue affected by insertion of Fe particles should reveal a blue color. The histological analysis revealed no coloration of the retina tissue, indicating total inertness of the SPION-Fe_3O_4 nanoparticles in contact with the retina, because there was no penetration, while the liver control samples with inserted Fe revealed a strong blue color [51].

The *in-vivo* animal studies on retina reattachment via attraction of injected SPION-Fe_3O_4 onto the detached retina, induced by the magnetic field generated by the magnet on the eye's outer wall, showed that retina reattachment using SPION-Fe_3O_4 nanoparticles would overcome some of the issues associated with the current retina reattachment techniques using silicone oil or gas bubbles. In addition, and very important in comparison with the bubbles-based procedure, the SPION-Fe_3O_4 process is suitable for performing retina reattachment of the inferior quadrant. This procedure is projected for demonstration in the near future.

4.2.3 *In-Vivo* Human Clinical Trials for SPION-Fe_3O_4-Based Retina Reattachment

Clinical trials to test SPION-Fe_3O_4 nanoparticles-based retina reattachment in humans have been initiated using non-functionalized SPION-Fe_3O_4 nanoparticles and a gold-coated magnet positioned in the episcleral area on the surface of the eye's outer wall.

Retina reattachment was performed on male and female patients, whose eyes exhibited good retina reattachment. Further trials are projected to complete clinical trials as the next step for addition of the SPION-Fe_3O_4-based retina reattachment process to the market.

4.3 Conclusions

The animal studies and clinical trials in humans using the novel SPION-Fe_3O_4 based retina reattachment procedure revealed that this new process may provide a revolutionary retina reattachment procedure. In addition, the extremely positive results presented in this chapter indicate the possibility of new potential applications using SPION-Fe_3O_4 nanoparticles with surface biofunctionalization to enable molecular specificity, thus providing a new platform for the treatment of retinal detachment and other eye disorders.

Acknowledgments

O. Auciello is grateful to the University of Texas-Dallas for support through his Distinguished Endowed Chair Professor position providing yearly funding for research. M. J. Saravia, A. Berra, E. Lima, and R. D. Zysler acknowledge support from CONACIT-Argentina. All authors acknowledge great contributions to R&D related to the science and technology described in this chapter by Dr. Pablo Gurman (MD-Argentina).

References

[1] Encyclopedia Britannica, "Sensory reception: human vision: structure and function of the human eye," in *Encyclopedia Britannica* vol. 27, 1987. Edinburgh: Encyclopaedia Britannica.

[2] A. F. Goldberg, O. L. Moritz, and D. S. Williams, "Molecular basis for photoreceptor outer segment architecture," *Prog. Retin. Eye Res.*, vol. 55, p. 52, 2016.

[3] V. Y. Arshavsky and M. E. Burns, "Photoreceptors signaling supporting vision across a wide range of light intensities," *J. Biol. Chem.*, vol. 287 (3), p. 1620, 2012.

[4] J. R. Sparrow, D. Hicks, and C. P. Hamel, "The retinal pigment epithelium in health and disease," *Curr. Mol. Med.*, vol. 10 (9): p. 802, 2010.

[5] J. Letelier, P. Bovolenta, and J. B. Martínez-Morales, "The pigmented epithelium, a bright partner against photoreceptor degeneration," *J. Neurogenet.*, vol. 31 (4), p. 203, 2017.

[6] G. Shepherd, *The Synaptic Organization of the Brain*. New York: Oxford University Press, 2004.

[7] D. Johnson and H. Hollands, "Acute-onset flatters and flashes," *Can. Med. Assoc. J.*, vol. 184 (4), p. 431, 2011.

[8] D. Cline, H. W. Hofstetter, and J. R. Griffin, *Dictionary of Visual Science*, 4th ed. Boston, MA: Butterworth-Heinemann, 1997.

[9] C. D. Gelston, "Common eye emergencies," *Am. Fam. Physic.*, vol. 88 (8), p. 515, 2013.

[10] D. Mitry, D. G. Charteris, B. W. Fleck, H. Campbell, and J. Singh, "The epidemiology of hematogenous retinal detachment: geographical variation and clinical associations," *Br. J. Ophthalmol.*, vol. 94 (6), p. 678, 2010.

[11] N. Byer, "Natural history of posterior vitreous detachment with early management as the premier line of defense against retinal detachment," *Ophthalmology*, vol. 101 (9), p. 1503, 1994.

[12] R. D. Dickerman, G. H. Smith, L. Langham-Roof, et al., "Intra-ocular pressure changes during maximal isometric contraction: does this reflect intra-cranial pressure or retinal venous pressure?," *Neurol. Res.*, vol. 21 (3), p. 243, 1999.

[13] J. A. Shields and C. L. Shields, "Treatment of retinoblastoma with cryotherapy," *Trans. Pa. Academi Ophthalmol. Otolaryngol.*, vol. 42, p. 977, 1990.

[14] J. J. Augsburger and C. B. Faulkner, "Indirect ophthalmoscope argon laser treatment of retinoblastoma," *Ophthalmic Surg.*, vol. 23 (9), p. 591, 1992.

[15] P. J. Banerjee, A. Chandra, P. Petrou, and D. G. Charteris, "Silicone oil versus gas tamponade for giant retinal tear-associated fovea-sparing retinal detachment: a comparison of outcome," *Eye (Lond.)*, vol. 31 (9), p. 1302, 2017.

[16] H. M. Moharram, A. S. Abdelhalim, M. A. Hamid, and M. F. Abdelkader, "Comparison between oil and gas in tamponade giant retina breaks," *Clin. Ophthalmol.*, vol. 14, p. 127, 2020.

[17] M. Ratanasukon, A. Kittantong, S. Visaetsilpanonta, and S. Somboonthanakij, "Pars plana vitrectomy with silicone oil or gas endotamponade in HIV-related rhegmatogenous retinal detachments in Thai patients," *J. Med. Assoc. Thai*, vol. 90(6), p. 1161, 2007.

[18] Yi-X. J. Wang, S. Xuan, M. Port, and J-M. Idee, "Recent advances in superparamagnetic iron oxide nanoparticles for cellular imaging and targeted therapy research," *Curr. Pharmaceut. Design*, vol. 19, 6575, 2013.

[19] A. Hanini, A. Schmitt, K. Kacem, et al., "Evaluation of iron oxide nanoparticle biocompatibility," *Int. J. Nanomed.*, vol. 6, p. 787, 2011.

[20] M. Kowalczyk, M. Banach, and J. Rysz, "Ferumoxytol: a new era of iron deficiency anemia treatment for patients with chronic kidney disease," *J. Nephrol.*, vol. 24 (6), p. 717, 2011.

[21] T. W. Prow, I. Bhutto, S. Y. Kim, et al., "Ocular nanoparticle toxicity and transfection of the retina and retinal pigment epithelium," *Nanomed. Nanotechnol. Biol. Med.*, vol. 4 (4), p. 340, 2008.

[22] R. Weissleder, D. D. Stark, B. L. Engelstad, et al. "Superparamagnetic iron oxide: pharmacokinetics and toxicity," *Am. J. Roentgenol.*, vol. 152 (1), p. 167, 1989.

[23] E. Stefansson, "Ocular oxygenation and the treatment of diabetic retinopathy," *Surv. Ophthalmol.*, vol. 51, p. 364, 2006.

[24] P.-T. Yeh, C.-M. Yang, C.-H. Yang, and J.-S. Huang, "Cryotherapy of the anterior retina and sclerotomy sites in diabetic vitrectomy to prevent recurrent vitreous hemorrhage," *Am. Acad. Ophthalmol.* , vol. 112 (12), p. 2095, 2005.

[25] M. Triantafyllou, U. E. Studer, F. D. Birkhauser, et al., "Ultrasmall superparamagnetic particles of iron oxide allow for the detection of metastases in normal sized pelvic lymph nodes of patients with bladder and/or prostate cancer," *Eur. J. Cancer*, vol. 49 (3), p. 616, 2013.

[26] M. G. Harisinghani, J. Barentsz, P. F. Hahn, et al., "Noninvasive detection of clinically occult lymph-node metastases in prostate cancer," *New Engl. J. Med.*, vol. 348 (25), p. 2491, 2003.

[27] M. Jeun, J. W. Jeoung, S. Moon, et al., "Engineered superparamagnetic $Mn_{0.5}$ $Zn_{0.5}$ Fe_2O_4 nanoparticles as a heat shock protein induction agent for ocular neuroprotection in glaucoma," *Biomaterials*, vol. 32 (2), p. 387, 2011.

[28] M. Jeun, Y. J. Kim, K. H. Park, S. H. Paek, and S. Bae, "Physical contribution of Néel and Brown relaxation to interpreting intracellular hyperthermia characteristics using superparamagnetic nanofluids," *J. Nanosci. Nanotechnol.*, vol. 13 (8), p. 5719, 2013.

[29] E. Buitrago, M. J. Del Sole, A. Torbidoni, et al., "Ocular and systemic toxicity of intravitreal topotecan in rabbits for potential treatment of retinoblastoma," *Exp. Eye Res.*, vol. 108, p. 103, 2013.

[30] F. Ghassemi, C. L. Shields, H. Ghadimi, A. Khodabandeh, and R. Roohipoor, "Combined intravitreal melphalan and topotecan for refractory or recurrent vitreous seeding from retinoblastoma," *JAMA Ophthalmol.*, vol. 132, p. 936, 2014.

[31] P. Nicolas, M. Saleta, H. Troiani, et al., "Preparation of iron oxide nanoparticles stabilized with biomolecules: experimental and mechanistic issues," *Acta Biomaterialia*, vol. 9 (1), p. 4754, 2013.

[32] M. Raineri, E. L. Winkler, T. E. Torres, et al., "Effects of biological buffer solutions on the peroxidase-like catalytic activity of Fe3O4 nanoparticles," *Nanoscale*, 2019. doi: 10.1039/C9NR05799D.

[33] M. L. Mojica Pisciotti, E. Lima, Jr., M. Vasquez Mansilla, et al., "In vitro and in vivo experiments with iron oxide nanoparticles functionalized with DEXTRAN or polyethylene glycol for medical applications: Magnetic targeting," *J. Biomed. Mater. Res. B,* vol. 102, p. 860, 2014.

[34] R. D. Zysler, E. Lima Jr., M. Vasquez Mansilla, et al., "A new quantitative method to determine the uptake of SPIONs in animal tissue and its application to determine the quantity of nanoparticles in the liver and lung of Balb-c mice exposed to the SPIONs," *J. Biomed. Nanotechnol.*, vol. 9 (1), p. 142, 2013.

[35] J. Harrison, C. A. Bartlett, G. Cowin, et al., "In vivo imaging and biodistribution of multimodal polymeric nanoparticles delivered to the optic nerve," *Small*, vol. 8 (10), p. 1579, 2012.

[36] A. Yanai, U. O. Häfeli, A. L. Metcalfe, et al., "Focused magnetic stem cell targeting to the retina using superparamagnetic iron oxide nanoparticles," *Cell Transplant.*, vol. 21 (6), p. 1137, 2012.

[37] J. Wen, K. C. McKenna, B. C. Barron, H. P. Langston, and J. A. Kapp, "Use of superparamagnetic microbeads in tracking subretinal injections," *Mol. Vis.*, vol. 11, p. 256, 2005.

[38] L. A. C. Fandiño and M. Saravia, "Guiones de Oftalmologia," in *Capitulo: Ediciones Universidad de Valladolid*, J. C. P. Jimeno, Ed. Valladolid: Universidad de Valladolid, p. 320, 1993.

[39] R. Zysler, A. Berra, P. Gurman, O. Auciello, and M. J. Saravia, "Material for medical use comprising nanoparticles with superparamagnetic properties and its utilization in surgery," US Patent #9,427,354, 2016.

[40] R. Zysler, A. Berra, P. Gurman, O. Auciello, and M. J. Saravia, "Material for medical use comprising nanoparticles with superparamagnetic properties and its utilization in surgery," Japanese Patent #5,954,797 (2016).

[41] A. S. Fortuin and J. O. Barentsz, "Comments on Ultrasmall superparamagnetic particles of iron oxide allow for the detection of metastases in normal sized pelvic lymph nodes of patients with bladder and/or prostate cancer," *Eur. J. Cancer*, vol. 49 (7), p. 1789, 2013.

[42] H. Dimaras, K. Kimani, E. A. Dimba, et al., "Retinoblastoma," *Lancet*, vol. 379 (9824), p. 1436, 2012.

[43] A. Z. Wang, V. Bagalkot, C. C. Vasilliou, et al., "Super-paramagnetic iron oxide nanoparticle-aptamer bioconjugates for combined prostate cancer imaging and therapy," *Chem. Med. Chem.*, vol. 3(9), p. 1311, 2008.

[44] M. Kowalczyk, M. Banach, and J. Rysz, "Ferumoxytol: a new era of iron deficiency anemia treatment for patients with chronic kidney disease," *J. Nephrol.*, vol. 24(6), p. 717, 2011.

[45] J. M. Vargas and R. D. Zysler, "Tailoring the size in colloidal iron oxide magnetic nanoparticles," *Nanotechnology*, vol. 16, p. 1474, 2005.

[46] A. C. Moreno Maldonado, E. L. Winkler, M. Raineri, et al. "Free radical formation by the peroxidase-like catalytic activity of MFe_2O_4 (M = Fe, Ni and Mn) nanoparticles," *J. Phys. Chem. C*, vol. 123, p. 20617, 2019.

[47] K. R. Reddy, P. A. Reddy, C. V. Reddy, et al., "Functionalized magnetic nanoparticles/ biopolymer hybrids: synthesis methods, properties and biomedical applications," *Meth. Microbiol*, vol. 46, p. 227, 2019.

[48] S. Liu, B. Yua, S. Wanga, Y. Shena, and H. Conga, "Preparation, surface functionalization and application of Fe_3O_4 magnetic nanoparticles," *Adv. Colloid Interf. Sci.*, vol. 281, p. 102165, 2020.

[49] X. Xiao, J. Wang, J. A. Carlisle, et al., "In vitro and in vivo evaluation of ultrananocrystalline diamond for coating of implantable retinal microchips," *J. Biomed. Mater.*, vol. 77B, p. 273, 2006.

[50] J. D. Weiland, W. Liu, and M. S. Humayun, "Retinal prothesis," *Ann. Rev. Biomed. Eng.*, vol. 7, p. 361, 2005.

[51] O. Auciello, P. Gurman, A. Berra, M. Saravia, and R. Zysler, "Ultrananocrystalline diamond (UNCD) films for ophthalmological applications," in *Diamond-Based Materials for Biomedical Applications*, R. Narayan, Ed. Cambridge: Woodhead Publishing, p. 150, 2013.

5 Science and Technology of Biocompatible Ultrananocrystalline Diamond (UNCD™) Coatings for a New Generation of Implantable Prostheses

First Application to Dental Implants and Artificial Hips

Orlando Auciello, Karam Kang, Daniel G. Olmedo, Debora R. Tasat, and Gilberto López Chávez

5.1 Background Information on Human Teeth, the Need for Natural Teeth Replacement by Dental Implants, and Current Dental Implant Technologies

Teeth are vital human organs that enable the nutrition of the human body for a normal life. The main function of teeth is to break down food to enable good digestion. In addition, teeth play a key social function, including enabling the phonetic expression of humans when talking to each other, and a harmonious expression of the face. A good denture reveals good health and welfare. Humans, from the beginning, were concerned about replacing lost natural teeth using dental prostheses.

A brief history of time on dental implants:

- In ancient China, 4000 years ago, carved bamboo pegs were used to replace missing teeth in the first known society to use dental implants.
- Remains of Egyptians from 3000 years ago show that they used precious metals like copper as prosthetic teeth. The Egyptians are the first recorded culture to use a metal to fix destroyed jawbones.
- A 2300-year-old dental implant was discovered in the mouth of a skeleton in a Celtic grave in France. The implant was held in place by an iron pin, which would have been extremely painful when hammered in. Ancient Romans used a similar approach, but with gold pins.
- About 2000 years ago human teeth were replaced with animal ones. People would also purchase teeth from slaves or poor people. Today, replacement of a human tooth with an animal one is defined as a heteroplastic implant, and replacement with

a tooth of another human is called a homoplastic implant. In most cases, animal or other human replacement teeth are rejected by the host, resulting also in infection.

- In 1931, the archaeologist W. Popenoe and his wife made a fascinating discovery by looking at the lower jaw of a young Mayan woman's remains from around 630 BCE. The scientists found three missing natural incisors replaced with pieces of seashell. Interestingly, they observed bone growth around two of the implants, providing evidence of osteointegration. The implants appeared to have been inserted both as functional and aesthetic prostheses in the mouth.

In the 1960s, in Suecia, Dr. Brånemark and colleagues discovered, accidentally, a mechanism for adhesion of metal to human bone. Brånemark was interested in the microcirculation of bones and healing of wounds. To understand the mechanisms, Brånemark used the well-known vital microscopy technique, based on introducing a camera into the tibia of a rabbit, which enabled observation of circulatory and cellular changes in living tissue. Upon using a camera with an external cover made of titanium, implanted in a rabbit's bone, with a low-trauma technique, a significant outcome happened: on removing the camera, it was found that the bone was very well adhered to the titanium surface, which demonstrated that titanium could have good integration with human bones, providing a key material for different prostheses implantable in human bones, including dental implants. The phenomenon of metal integration with bones was named *osseointegration* and extensively investigated initially for dental implants by Brånemark and colleagues [1–3].

Titanium (Ti) or Ti-based alloys are currently widely used to make different prostheses fabricated to replace degraded or destroyed natural components of the human body (e.g., teeth, hips, knees, and other joints). Ti or Ti alloys are used for implants because early R&D demonstrated that these materials were biocompatible and would be resistant to the chemical and mechanical environment of the body. However, in recent years many studies in academic, industrial, and medical communities have demonstrated that metal-based prostheses undergo many types of failure, in most cases with substantial contribution from the chemical and mechanical environment of the body. In the case of dental implants there is now an extensive literature revealing failure of metal-based dental implants due to the chemical and mechanical environment of the mouth and body, to which selective references are included here due to constrained space [4–14]. When corrosion happens on the surface of Ti-based implants due to electrochemical processes induced by body fluids, titanium dioxide (TiO_2) particles from the oxidized Ti (Ti and its alloys are easily oxidized upon exposure to the atmospheric environment) implant surface are dislodged and insert into the surrounding tissue, inducing inflammation and tissue death (necrosis), as shown for Ti alloy dental implants [7, 8, 10, 11]. In addition, particles may enter the blood stream and insert into organs (e.g., liver, spleen, and abdominal lymph nodes) [15] and body fluids [16, 17], producing serious biological deleterious effects, as shown in recent studies by Auciello et al., who compared the effect of TiO_2 and ultrananocrystalline diamond (UNCD) (3–5 nm dimensions) particles injected into the lung and liver of Wistar rats [18]. TiO_2 caused areas compatible with foci of necrosis in the liver and renal hyaline cylinders, while UNCD particles did not induce any membrane damage (thiobarbituric acid reactive substances [TBARS]) or mobilization of enzymatic antioxidants in either the lung or liver samples [18].

5.2 Physical and Chemical Effects on Dental Implants

5.2.1 Chemical and Biological Processes Destroying Natural Teeth

According to the American Dental Association, tooth decay is caused by acids and carbohydrates, such as sugars and starches, which are contained in food. In this respect, milk, fruit, cookies, and/or candy are the common elements of food that are left on the teeth after eating. These are the foods with which bacteria that live in the mouth react chemically to produce acids. Over a period of time, the acids produced by the bacteria–food interaction may destroy the tooth enamel, resulting in tooth decay [19]. The bacteria-induced breakdown of enamel, which is the hard tissue of the tooth, causes caries on the teeth. The tooth decay process is a result of the chemical reaction of bacteria with sugars, starches, or debris from the food left on the tooth surface, resulting in acid formation, which may destroy tooth enamel. Minerals entering the mouth present another risk factor since they are released from the tooth surface. Diabetes mellitus is a condition that results in reduced saliva, which can be a risk factor, as well as medications such as antihistamines and antidepressants, which also reduce saliva. Finally, caries can also be caused by poorly cleaned teeth and receding gums, which will affect the roots of the teeth [20–22].

Tooth caries are also known as cavities, or decay, produced by bacterial activity on the teeth [23]. Caries result in pain and difficulty eating. The color of caries can range from yellow to black [24, 25]. Caries result in infection, tooth loss, and abscess formation, as well as tissue inflammation around the teeth [24, 26, 27].

5.3 Replacement of Destroyed Natural Teeth

To replace the roots of the teeth, currently metal-based dental implants (mainly Ti-6Al-4V alloys) are placed below the gum line to anchor in the bone, as shown in Figure 5.1. Above the gum line they are normally protected with a ceramic crown. The major reasons why a missing tooth must be replaced with a dental implant is to stop the jawbone from collapsing, reestablishing function, and also for esthetic reasons [28]. The dental implant stimulates the normal jawbone functionality. In addition, for the bone to remain healthy, it needs stimulation [29]. However, most importantly, replacement of a natural tooth with a dental implant immediately after extraction of the tooth will provide protection against bone resorption. Research by several groups has shown that there is less resorption (about 75%) when the missing tooth area is replaced by a dental implant than leaving the empty space in the bone. It has been determined that it is critical to insert a dental implant within a year after losing a tooth, since most of the bone resorption happens during this period [28].

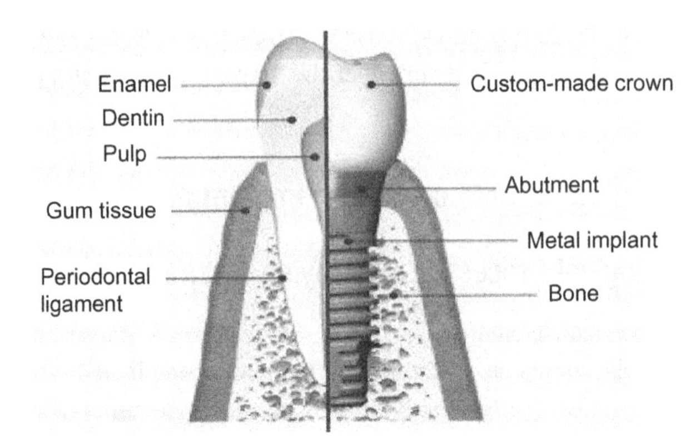

Figure 5.1 Schematic of a natural tooth and a metal-based dental implant inserted in the maxillary bone.

Replacement of teeth in the back region of the denture is more frequently done via insertion of a three-unit fixed partial denture (FPD), also called a fixed bridge, even though it is not the best approach. An alternative way to insert a replacement tooth in the back part of the molar is by positioning the tooth between two teeth with abutment, which are crowned and support a "pontic," as shown in Figure 5.2.

The fixed bridge can function well, looks like a real tooth, and potentially lasts a long time if it is well made, but recent research shows that within seven years about 75% of the three-unit FPD failed [30]. As shown in Figure 5.2, two of the teeth are connected together with the middle tooth, which is the ceramic crown. This is one of the main reasons why there is a high percentage of failure with fixed bridges. Another reason for failure is that there will be extra stress on both teeth that are supporting the crown in the middle and this can cause mechanical breakdown, since the three teeth are connected when people chew. However, even with these problems, the fixed bridge is still the most common treatment choice by patients [30].

Although most dental implant systems perform successfully, it has been reported in the literature that corrosion is a failure mode [1]. Figure 5.3a shows a schematic of a Ti-based dental implant inserted in the maxillary bone and the components of a dental implant. Figure 5.3b shows a picture of a Ti-based dental implant extracted from a patient's mouth, revealing extensive corrosion induced by oral fluids. In addition to dental implant failure due to chemical corrosion, both early- and late-stage dental implant failure can be due to one or a combination of events such as patient related factors, surgical skill, loss of osseointegration due to infection, tissue damage, overloading, and sinus problems [31].

5.4 Challenges with Current Materials in Dental Implants

About 85–90% of dental implants are working, without failures [31, 32], after implantation. However, according to other worldwide statistics, some of the dental

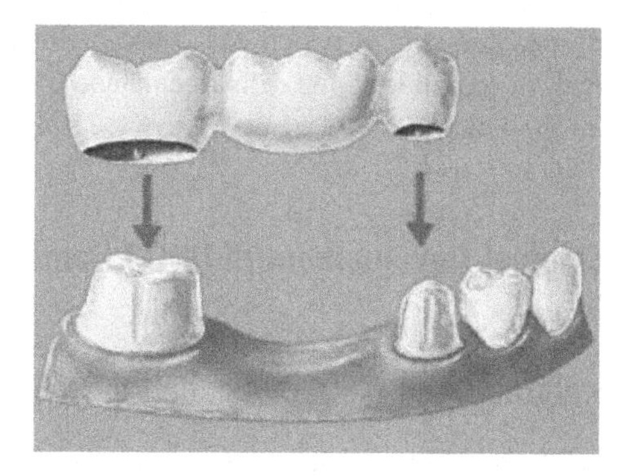

Figure 5.2 Schematic of a bridge with a crown in the middle.

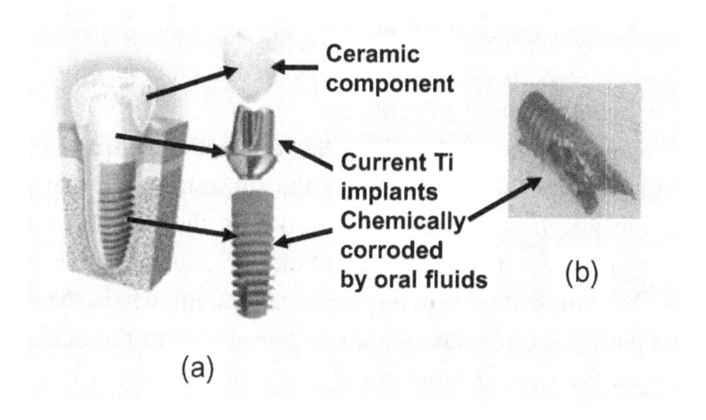

Figure 5.3 (a) Schematic of a Ti dental implant; (b) explanted Ti dental implant corroded by oral fluids.

implants could have problems such as no ossification, bone infection, a broken prosthesis, and infection around the gum after implantation. In addition, smoking and antidepressant use may be factors contributing to dental implant failure [32].

5.5 Background on Biocompatible, Corrosion-Resistant UNCD Coatings for Dental Implants

UNCD coatings exhibit excellent resistance to chemical corrosion by any strong acid [33] and body fluids [34–36], and enables efficient extensive growth and migration of embryonic epidermal stem cells, bone cells, and neural cells on the surface of the film, as demonstrated in pioneering work by Auciello's group [34, 37–39] and confirmed by other groups worldwide [40, 41]. The hypothesis is that the surface of UNCD is formed by carbon atoms (the main element in human DNA and cells), and it has been demonstrated to be extremely biocompatible [34–36]. To develop a better dental implant based on any coating, biocompatibility is the most important issue since it plays a major role in preventing early infection, which can cause premature failure [25], and induces excellent growth of maxillary bone cells to enhance fixation of the UNCD-coated dental implants to the bone.

5.6 Fundamentals and Technological Development of UNCD Coating on Dental Implants

5.6.1 Methodology, Process, Procedures, and Characterization of UNCD Films

The dental implants used for the R&D described in this chapter were machined-thread spiral no-head type test implants (NH1312-05 Ti alloy [Ti-6Al-4V]) manufactured by Abs Anchor Dentos (Figure 5.4).

Ti–6Al–4V because it is the main Ti-based material used for manufacturing current commercial Ti-based dental implants (Table 5.1) and other prostheses (e.g., hips and knees) [42]. In addition, there are several Ti alloys used in other prostheses, for which UNCD coatings are being developed (Table 5.1) [35, 42].

The first step in preparing the test dental implant for coating with a UNCD film is to insert nanodiamond particles on the surface of the substrate, such that the particles behave as seeds upon which the UNCD films grow. In the seeding process, which takes about 1 h, implants are immersed in a solution of methanol with nanocrystalline diamond particles (3–5 nm diameter) in an ultrasonic system, where the sound waves shake the diamond particles, embedding them on the surface of the metal where they act as seeds on which the UNCD films grow, using ether the hot filament chemical vapor deposition (HFCVD) or microwave plasma chemical vapor deposition (MPCVD) techniques (a detailed review of the UNCD growth process can be seen in [33]).

5.6.1.1 Methods for Growing UNCD Coating on Metal Dental Implants

Many different types of diamond thin films, such as microcrystalline diamond (MCD), nanocrystalline diamond (NCD), and UNCD have been synthesized and systematically studied over the years [33]. All of these diamond films revealed different microstructures, surface morphologies, chemistries on external surfaces and grain

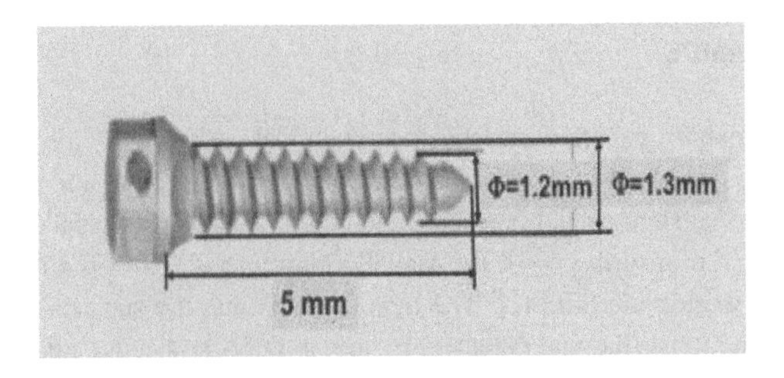

Figure 5.4 Test dental implant NH1312-05 (Ti-6Al-4V alloy).

boundaries, and properties [33]. All polycrystalline diamond films described above have been grown on the surfaces of metals, insulators, and semiconductors, and all of them involved the first step of chemical seeding of the substrate surface with diamond particles. However, for growing UNCD films, the surface of the substrate is seeded with nanocrystalline diamond particles. After seeding, the implant is inserted in a HFCVD system or an MPCVD system, where the UNCD films are grown (see Chapter 1 and [33]).

UNCD Coating on Test Dental Implants via the HFCVD Process

In the HFCVD process, test dental implants were coated by positioning them in a horizontal and a vertical position (Figure 5.5) underneath a set of parallel W filaments, heated to about 2200 °C (see a detailed description of the HFCVD process in Chapter 1 and in [43, 44]), to determine the best position for simultaneously coating a large number of implants with uniform UNCD films, to enable a low-cost fabrication process for commercialization. The parameter window for growing uniform UNCD

Table 5.1 Titanium alloys suitable for medical applications [42].

ASTM	BS/ISO	Alloy designation
F67	Part 2	Unalloyed titanium – CP grades 1–4 (ASTM F1341 specifies wire)
F136	Part 3	Ti-6Al-4V ELI wrought (ASTM F620 specifies ELI forgings)
F1472	Part 3	Ti-6Al-4V standard grades (SG) wrought (F1108 specifies SG castings)
F1295	Part 11	Ti-6Al-7Nb wrought
–	Part 10	Ti-6Al-2.5Fe wrought
F1580	–	CP and Ti-6Al-4V SG powders for coating implants
F1713	–	Ti-13Nb-13Zr wrought
F1813	–	Ti-12Mo-6Zr-2Fe wrought

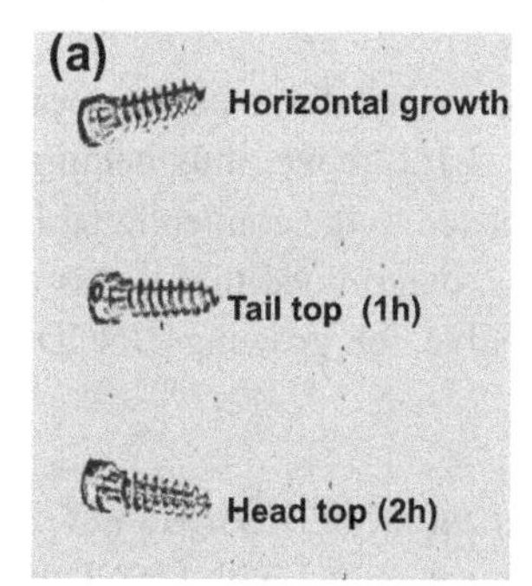

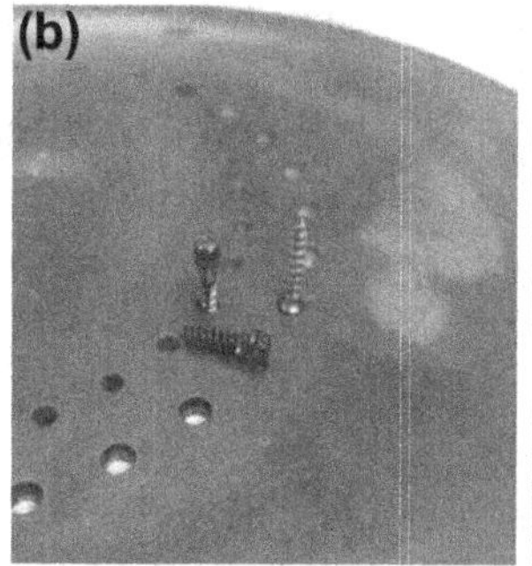

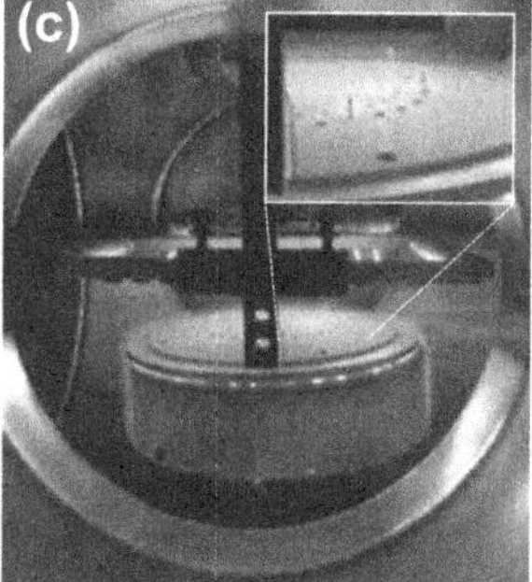

Figure 5.5 (a) Ti-6Al-4V test dental implants coated with UNCD films in a horizontal (top), and vertical 1 h coating with tail on top (middle) and 2 h coating with head on top (bottom) positions. (b) optical picture showing implants in vertical and horizontal positions on the holder, for growth of UNCD coatings on both simultaneously, in the HFCVD system. (c) optical picture during growth of UNCD coating on test dental implants using the bias enhanced nucleation-bias enhanced growth (BEN-BEG) process in the HFCVD system.

films on test dental implants using the HFCVD process is very narrow [43, 44]. That is, the surface temperature of the implant is critical, as well is the substrate-to-filament distance. If the appropriate substrate temperature and the substrate-to-filament distance is not accurately controlled, the films may exhibit a mixture of UNCD and graphite, or be totally made of graphite [43, 44].

When UNCD films are grown by positioning the tests dental implants horizontally on the holder, the coating is developed on one side. Thus, after growing on one side, the sample has to be turned over to coat the other side. However, it is difficult to accurately control the growth all around, since the implant has a circular shape. When the implant is positioned in a vertical position the problem observed is that in the middle part and under the thread of the implant the UNCD coating is too thin or does not grow well. However, an important point is that more implants can be coated by positioning them vertically than horizontally. When the implants are positioned vertically, first with the head up and subsequently with the tail up, the coating is fairly uniform all over, although there are some small areas in which the coating appears thinner. The coating using the HFCVD process requires about 8 h with the head up and another 8 h with the tail up, for a total of 16 h of growth time, which would result in a relatively expensive manufacturing process.

In order to shorten the total UNCD film growth time using HFCVD, a new process was explore, named bias enhanced nucleation-bias enhanced growth (BEN-BEG) [45]. This process involves biasing the substrate holder with a negative voltage to create a plasma between the filaments and the substrate (notice the plasma created around the substrate holder – see a detailed description of the BEN-BEG process in Chapter 1 and in [45]). In the HFCVD BEN-BEG process, positively charged species (Ar^+, CH_x^+, C^+, and H^+), created in the plasma (Figure 5.5c), are accelerated toward the substrate to induce nucleation of the UNCD films via sub-plantation of the C^+-based species on the substrate surface, and subsequent film growth. However, the plasma created in the HFCVD BEN-BEG process is fairly unstable due to discharges produced around the substrate holder.

The BEN-BEG process shown in Figure 5.5c would eliminate the need for the wet chemical seeding process described in Section 3.1.2 above, thus helping reduce the coating process cost. However, the problems related to the conventional HFCVD and HFCVD BEN-BEG processes, described above, determined that it was necessary to investigate the MPCVD process as an alternative for growing UNCD coating on dental implants.

UNCD Coating on Test Dental Implant via the MPCVD Process

In the MPCVD growth process [33] implants were coated with UNCD films only in the vertical position. In the MPCVD process, many crystalline diamond films and specifically UNCD films are grown using a plasma [33], created by coupling microwave power into a mixture of gases, including the main CH_4 molecules, to crack these molecules to grow the diamond films. The process parameter window to grow UNCD films is wider than that for the HFCVD process. For the particular series of first UNCD films grown by MPCVD, a mixture of Ar = 30 sccm, CH_4 = 0.9 sccm, and

$H_2 = 10$ sccm gases was flown into the growth chamber to produce a total pressure of 20 mbar, on which a microwave power of 1900 W was coupled to produce the plasma to grow the UNCD films. To coat the metal dental implants they were first positioned with the head of the implant on the top, and UNCD films were grown for 3 h, following by cooling down in a flow of H_2 gas. The implants were subsequently turned around, positioning them with the tail up and the head down, and coated with UNCD films for another 3 h. The total 6 h coating process produces implants fully encapsulated with the UNCD coating. However, the head needs one more hour of coating to fill little spaces that were not filled, and 1 h less is needed on the tail since it is too thick of a coating. This MPCVD process, without optimization, produced very uniform, dense, hermetic UNCD coatings in a total 6 h process, which is shorter than for the HFCVD growth process described above. In addition, it was determined, using complementary Raman and scanning electron microscopy (SEM) analysis, that the MPCVD process produces a much more uniform UNCD film all along the implant surface. Also, it was demonstrated that the MPCVD process produced uniform UNCD coatings on seven screws on which the films were grown simultaneously. This R&D showed the pathway for an industrial low-cost coating process. Calculations, taking into account the size of the implants (~8 mm long by 3–4 mm in diameter) and the size of the substrate holder of the MPCVD system (200 mm in diameter), indicate that up to 300 implants can be coated simultaneously, with a cost per implant of $\leq$\$3 dollars, which is a very good cost for a high-value product. Work is proceeding to optimize the UNCD film growth process by MPCVD to try to minimize the coating growth time, optimize the implant substrate holder geometry to accept the largest number of implants, and explore using BEN-BEG to eliminate the wet chemical seeding process and minimize growth time, while enhancing film adhesion on the surface of the Ti alloy metal-based implant. Figure 5.6a shows a picture taken inside the MPCVD system growing UNCD coating on 7 dental implants simultaneously.

A more recent MPCVD process with increased gas flows and total pressure (patent under preparation) demonstrated that the total growth time to produce a dense,

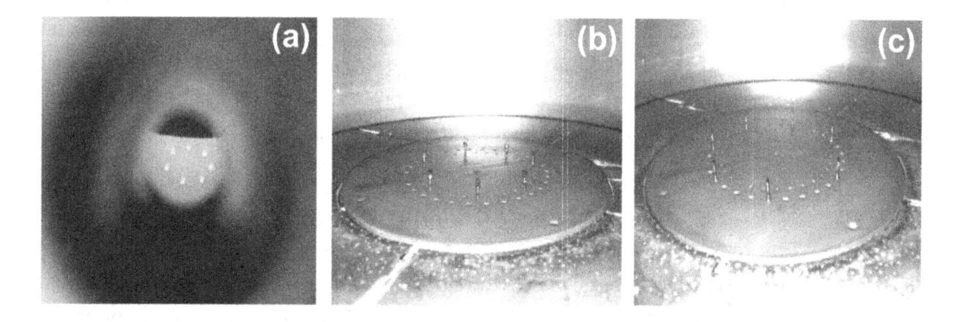

Figure 5.6 (a) view of the MPCVD chamber during plasma-assisted growth of UNCD coatings simultaneously on seven vertically positioned Ti-6Al-4V implants (the bright spots are the implants with the uniform plasma around them). (b) UNCD-coated implants after deposition (head on top) and (c) after deposition (tail on top).

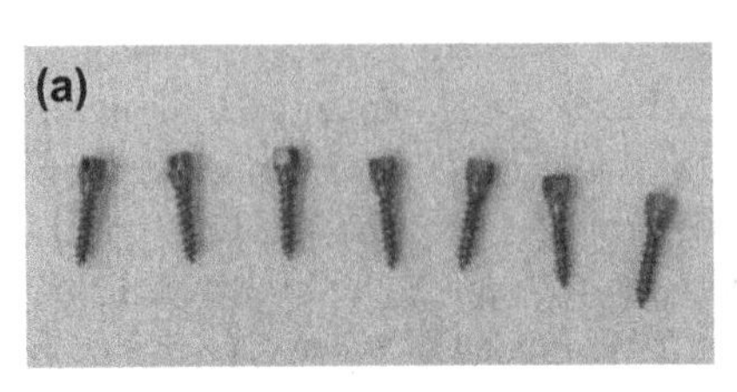

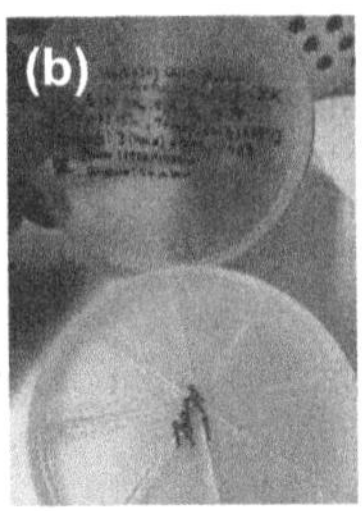

Figure 5.7 (a,b) Seven implants coated with UNCD films after being extracted from the growth chamber, showing a fairly uniform color of the UNCD coating all along the implants.

hermetic UNCD coating on the dental implants can be reduced to a total of 2 h, including the growth with tail and head tops, which can reduce even more substantially the cost for the growth process via MPCVD.

5.6.1.2 Characterization of UNCD-Coated Metal Dental Implants

UNCD-coated implants were characterized via Raman spectroscopy using a Thermo Scientific DXR Raman Spectroscopy system to determine the chemical bonds of C atoms in the UNCD films, SEM using a Zeiss Supra 40 SEM to determine surface morphology of the UNCD films, and X-ray photoelectron spectroscopy (XPS) using a PHI Versa Probe II scanning XPS microprobe to determine surface chemistry of the UNCD films. These measurements provide valuable information on surface molecule identity, surface morphology, and film thickness and density.

Figure 5.7 shows seven test dental implants coated with UNCD films in the simultaneous MPCVD growth process described above, after they were extracted from the MPCVD system. The picture shows a uniform dark color corresponding to the UNCD coating, as opposed to the original bright metallic color before coating.

The data presented below show the results from the characterization of UNCD films grown on Ti-6Al-4V implants using both HFCVD and MPCVD processes. Results and discussion are presented, which provide valuable information on the quality of the UNCD coatings from the chemical, surface morphological, and thickness points of view to optimize the process to achieve the uniform coated dental implant with moderate thickness (0.5~1.0 μm).

Raman Analysis of UNCD Coatings Grown by the HFCVD Process

Figure 5.8 shows Raman spectra from analysis of UNCD-coated dental implants using the HFCVD process. The Raman spectra shown in Figure 5.8 was obtained via analysis from the head to the tail of the implant and head section, which is subdivided into four sections: head center top, head edge, head middle, and head bottom.

On the threaded portion of the UNCD-coated implant, the Raman analysis was done on six sections which in Figure 5.8 are labeled first thread top and bottom, fourth thread top and bottom, and eighth thread top and bottom. The Raman spectra clearly

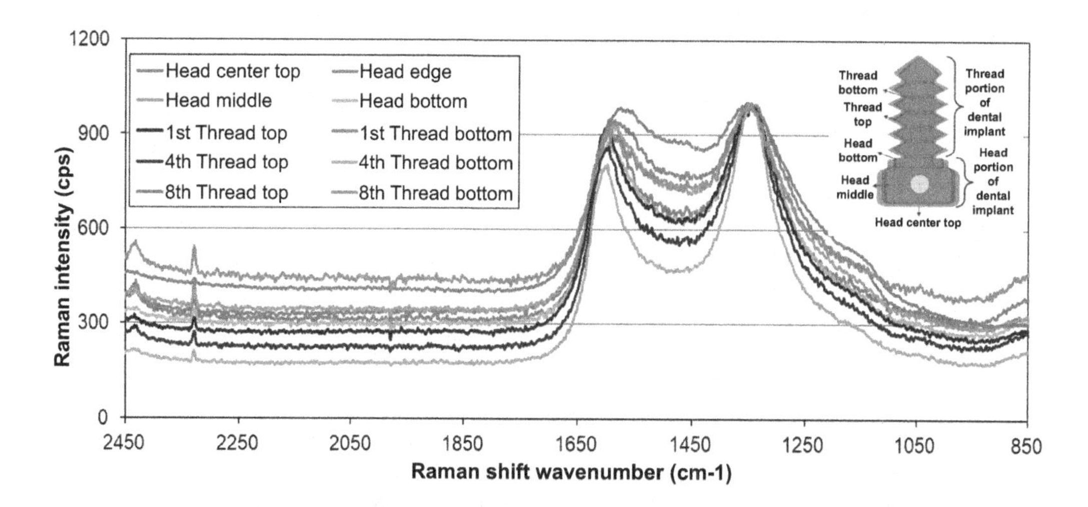

Figure 5.8 Raman spectra obtained from a HFCVD UNCD-coated dental implant.

shows that UNCD coating on the top of the upper head and bottom portion of threaded section is excellent. When compared with prior data from Auciello's group, the Raman spectra in Figure 5.8 shows the characteristic D band at 1341.7 cm^{-1} and G band at 1588.5 cm^{-1} peaks [34, 35]. However, the middle section starting from middle of the head to the middle of the threaded section is a combination of UNCD and graphite, which is revealed by the sharper structure of the two main peaks shown in Figure 5.8 [43, 44]. The different chemical structures of the UNCD coatings along the implant length may be due to temperature differences between implant positions when the head or tail are in the top position. The surface temperature at the top of the implant when the implant is positioned with the top up is close to 650 °C since the top receives substantial radiation from the filaments, adding to the thermal transmission from the heated substrate. However, the middle section may be at lower temperature than the top surface due to the lower conductivity of titanium. Another effect that should be considered is the shape of the implant, which in the threaded portion may perform as an umbrella, or holes in the middle of the head.

Raman Analysis of UNCD Coatings Grown by the MPCVD Process

Figure 5.9 shows spectra from Raman analysis of two samples out of a total of seven metal-based Ti-6Al-4V dental implants coated simultaneously with UNCD films grown by the MPCVD process. The UNCD-coated implants were analyzed from head to the tail of the mini-implant and head section, subdivided into two sections represented by head center, and head bottom. On the threaded portion the Raman spectra is subdivided into nine sections labeled first thread top and bottom, third thread top and bottom, fifth thread top and bottom, seventh thread top and bottom, and tail tip. Following the analysis on one implant, six more implants were analyzed in three different sections represented by head, middle, and tail. The data revealed that all

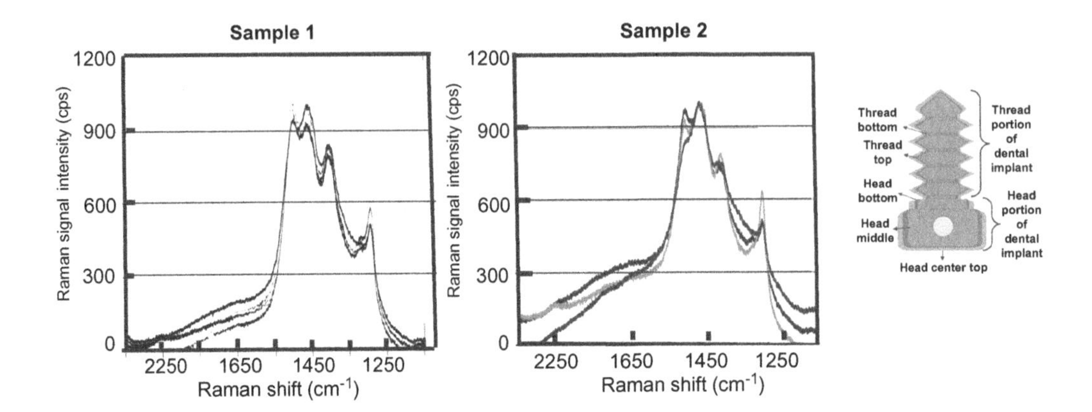

Figure 5.9 Raman spectra from analysis of two out of seven UNCD-coated implants using the MPCVD process. The Raman spectra of UNCD-coated dental implant samples 1 and 2 coated simultaneously in a single process show extreme uniformity of the UNCD coatings grown simultaneously on seven implants. The information presented in this figure is supported by SEM images of UNCD coatings (see Figure 4.13b).

UNCD coatings are very uniform as they were grown simultaneously on seven implants using the MPCVD process. The Raman spectra shown in Figure 5.9 reveal good uniformity in the UNCD films on the head, tail, and even middle thread section, with uniform thickness around all surfaces. Since the MPCVD growth process involves using plasma to grow the UNCD films, the data show that the plasma process induced a uniform coating at the bottom thread that could be hidden by the top thread and so was hard to coat using the HFCVD process. Also, the hole in the middle of the head was uniformly coated, as shown in SEM images (see Figure 5.13). An important issue with UNCD coating of implants is whether uniform UNCD coating can be grown simultaneously and reproducibly on many implants.

XPS Studies of UNCD Coatings Grown on Implants Using the MPCVD Process

An excellent complementary analysis technique to study the chemistry of the UNCD films, particularly on the surface, which is critical in relation to interactions with the human body, is XPS, which enables detection of elements and their chemical bonds on the surface of materials. This provided valuable complementary information to that obtained by the Raman analyses on the chemistry of the UNCD coating surface covering the dental implant under investigation. Figure 5.10 shows the points analyzed with XPS on a UNCD coating on an implant.

The XPS analysis survey spectra (Figures 5.11a–c) show a very small O peak (at 540 eV, due mainly to exposure of the UNCD coating surface to air) and a C peak (at 290 eV, corresponding to C atoms in the UNCD film).

The XPS analysis was focused on analyzing the surface of the UNCD coating on the Ti alloy implant, since it is not critical to do XPS analysis on Ti as that analysis

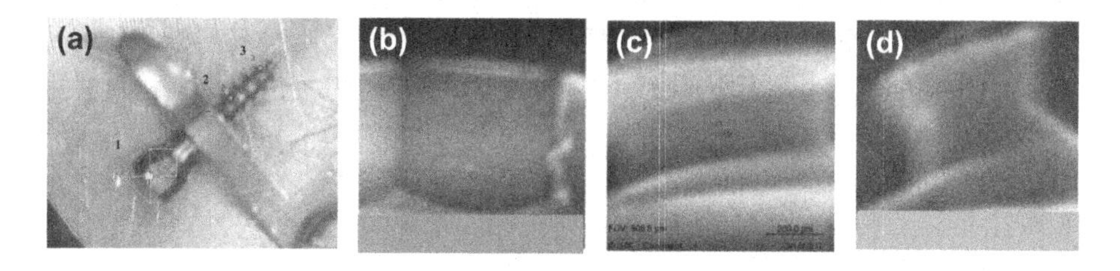

Figure 5.10 Optical image obtained in the XPS instrument to address the points on the surface of the UNCD coating on the implant to perform the XPS analysis: (a) picture of the XPS signal position on the UNCD coating on the implant. (b–d) Magnified picture of the three spots where XPS analyses were done (head, thread, and tip of the implant).

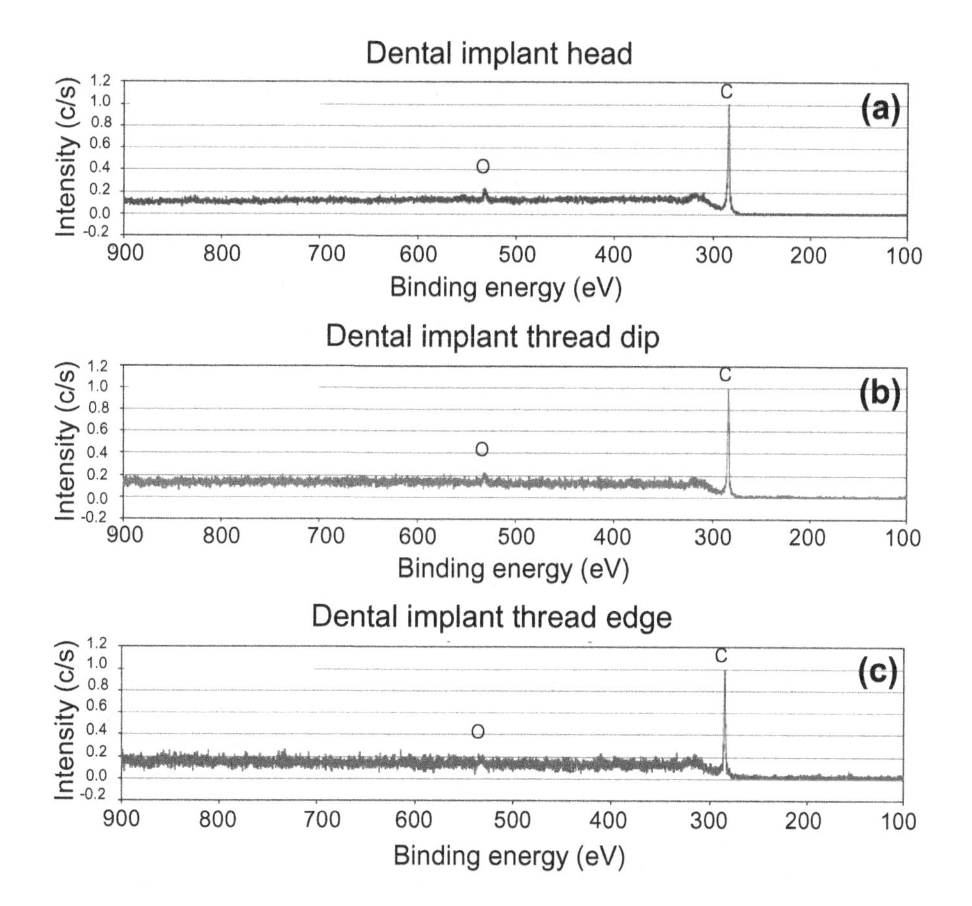

Figure 5.11 XPS analysis survey spectra on the surface of a UNCD coating on a dental implant: (a) on the implant head; (b) on the top of a thread at mid-distance between the head and the tail; and (c) on the tail. The O peak relates to oxygen adsorbed on the surface of the UNCD film exposed to air, and the C peak corresponds to the C atoms forming the UNCD coating.

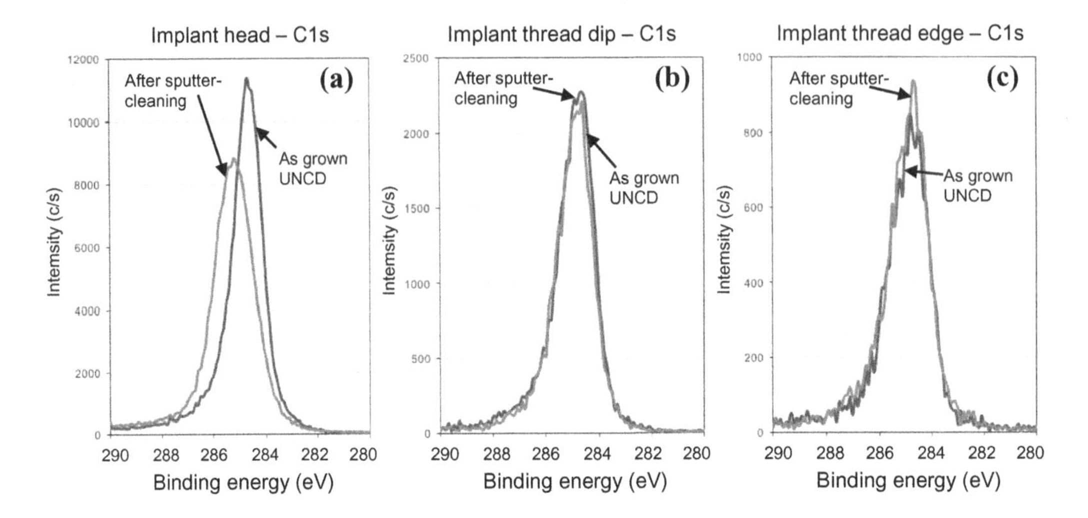

Figure 5.12 XPS spectra showing the C1s peak on the surface of the UNCD coating on a point on the implant head (a), on top of the thread at mid-distance between the head and tail (b), and on a point at the tip of the implant (c). The XPS peak labeled "As" corresponds to XPS analysis of the UNCD coating as inserted in the XPS system. The APS peak labeled after corresponds to the XPS analysis after the 30 s irradiation with the 1 keV Ar+ ion beam to practically eliminate the O atoms adsorbed on the UNCD surface from the atmosphere.

would not add valuable information in relation to characterization of the UNCD coating. The surface of the UNCD coating on the Ti alloy implant was slightly sputtered at grazing incidence, using a 1 kv Ar^+ ion beam for 30 s to remove contaminants from the surface of the UNCD coating when exposed to the atmospheric environment. As shown in Figure 5.12, the C1s peak appears at the energy characteristic of C-sp^3 bonding, which corresponds to diamond. A comprehensive study of UNCD films using XPS analysis has been performed by Auciello's group in the last year, which enabled determining with great accuracy the position of the C1s peaks in the XPS spectra to characterize graphite vs. diamond structures [46].

SEM Studies of UNCD Coatings Grown by the MPCVD Process

The Raman and XPS studies of UNCD coatings grown by the MPCVD process on Ti alloy implants, shown in Figures 5.9 and 5.12, respectively, provided strong evidence of excellent chemical bonding uniformity on the UNCD films grown simultaneously on several implants. In order to complement those studies, SEM imaging was performed to obtain information about the surface morphology and roughness of the UNCD coatings.

As described above, UNCD coatings can be grown using the HFCVD and MPCVD processes, as demonstrated by Raman analysis showing good chemistry for both types of UNCD films. SEM imaging can provide complementary information to check the differences between HFCVD and MPCVD in relation to surface morphology and roughness of the UNCD films grown by the two different methods. The surface of the

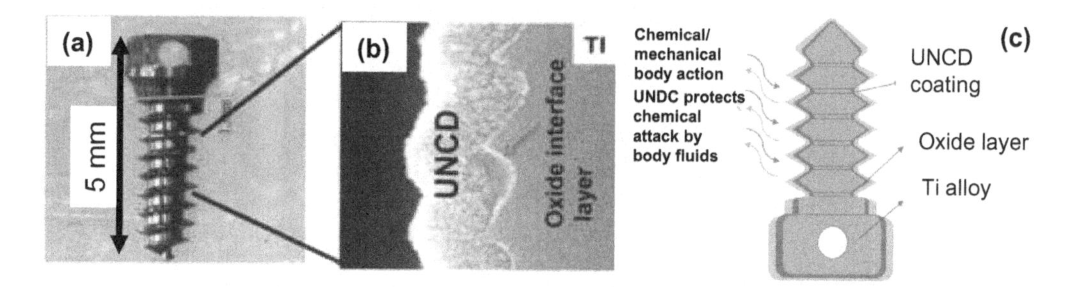

Figure 5.13 (a) Optical picture of a Ti alloy implant coated with a UNCD film produced by MPCVD. (b) A cross-section SEM image of the UNCD conformal coating shown in (a) on the threaded implant, with the film obtained on all seven UNCD-coated implants using the MPCVD process. (c) A schematic of the conformal UNCD coating on the implant.

UNCD coating on the implant head produced by HFCVD and MPCVD shows very uniform and well-grown UNCD films produced by both processes, even though there is a deposition time difference. However, the threaded section of the implant shows less growth on the bottom part of the thread for UNCD films grown by HFCVD, while those grown by MPCVD exhibit vey uniform dense/hermetic surface UNCD films (Figure 5.13b).

Figure 5.13a shows a Ti-6Al-4V alloy implant coated with an extremely uniform and dense UNCD film, as confirmed by the high-resolution cross-section SEM image of Figure 5.13b, which shows that the UNCD coating is very conformal around the threaded geometry of the implant, as shown by the schematic in Figure 5.13c.

Figure 5.14 shows SEM images of the UNCD coating on different parts of a Ti alloy implant for UNCD film produced with the MPCVD process (Figure 5.14a,b) and specific sections (Figure 5.14c–g). The SEM images show that the UNCD surface morphology and roughness provide the tailored surface that induces the excellent biological performance demonstrated in animal studies, as discussed in Section 5.8. In general, the surfaces of the UNCD films produced by the MPCVD process are very smooth and dense, and the films cover the convoluted screw regions of the implant and the inner surface of the hole in the head better than when the UNCD films are produced by the HFCVD processes.

SEM Studies of Surface Morphology, UNCD–Ti Interface, and Nanostructure of UNCD Coatings Grown on Ti Alloy Substrates by the MPCVD Process

Considering that the MPCVD processes appears to be superior to the HFCVD process, additional studies of surface morphology, the UNCD–Ti interface, and the nanostructure of the UNCD coatings grown by the MPCVD process on Ti-based material were performed using atomic force microscopy (AFM) surface imaging at the nanoscale, cross-section SEM of the UNCD–Ti interface, and high-resolution transmission electron microscopy (HRTEM) of the UNCD nanostructure. The AFM study revealed that the surface roughness of the MPCVD UNCD coating is in the range 5–10 nm (Figure 5.15a), which is very important information since surface roughness of the

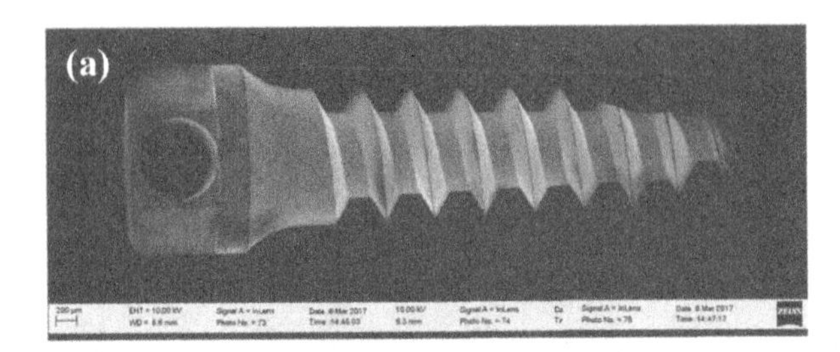

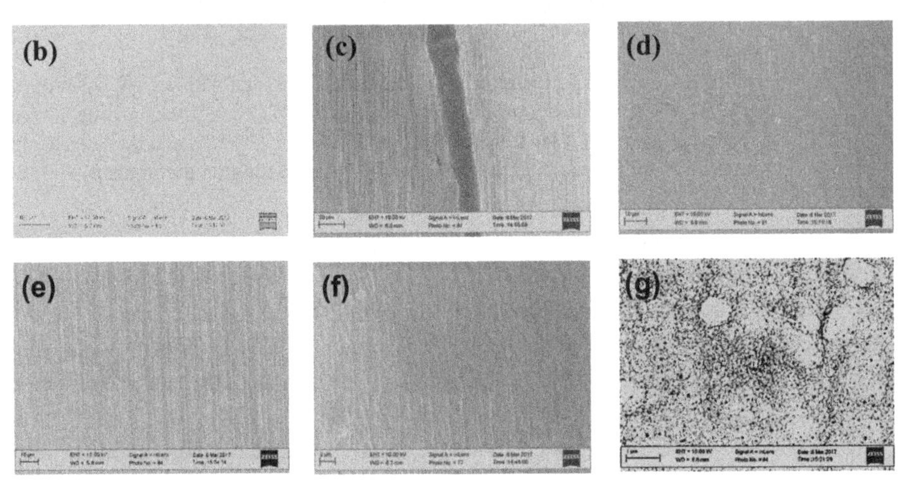

Figure 5.14 SEM images of an implant coated with UNCD by MPCVD. (a) Head section (100 μm resolution); (b) thread section (100 μm resolution); (c) middle top of the thread section (10 μm resolution); (d) tip of the threaded section (10 μm resolution); (e) higher magnification image of edge of the head (10 μm resolution); (f) highest magnification that can be obtained with SEM on the side section of the head (2 μm resolution); (g) highest magnification that can be obtained with SEM on the implant (1 μm resolution).

UNCD coating on Ti may play a critical role in the UNCD-coated dental implant performance. Cross-section SEM imaging of the UNCD–Ti interface revealed a sharp interface and dense UNCD film (Figure 5.15b), as required to provide strong protection against oral fluid penetration, thus protecting the Ti-based dental implant from chemical corrosion by oral fluid. Finally, the HRTEM study of the UNCD film demonstrated the nanostructure (3–7 nm) characteristic of UNCD (Figure 5.15c) [33].

SEM Studies of UNCD Coatings Grown on Dental Implants Using the HFCVD Process

Figure 5.16 shows high-resolution SEM images of the UNCD coating on a Ti alloy implant coated with a UNCD film produced with the HFCVD process (Figure 5.16 a, b) and specific sections (Figure 5.16c–f). In general, the uniformity across the UNCD coating and the surface morphology of the UNCD films produced by the HFCVD process are not as good as those produced by the MPCVD process. In addition, the work performed until now indicates that the MPCVD process will provide a most

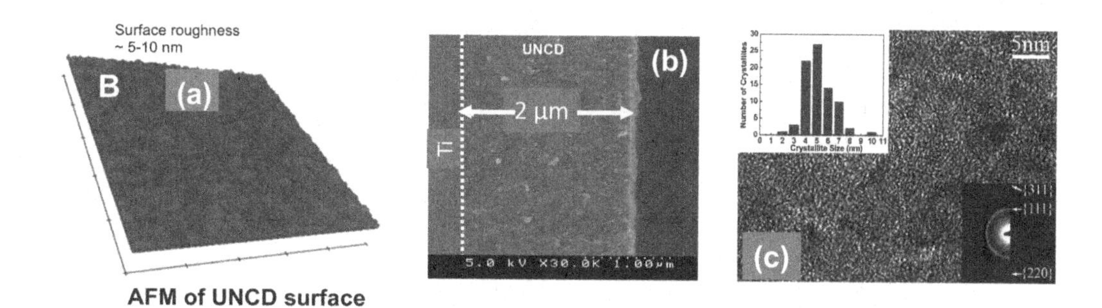

Figure 5.15 (a) Nanoscale-resolution AFM image of the surface of the UNCD coating grown on Ti metal by MPCVD; (b) Cross-section SEM image of the UNCD coating and UNCD–Ti interface, showing the high-density UNCD coating and hermetic nanoscale UNCD–Ti interface. (c) HRTEM image of the UNCD coating grown on Ti, showing the characteristic nanostructure of UNCD.

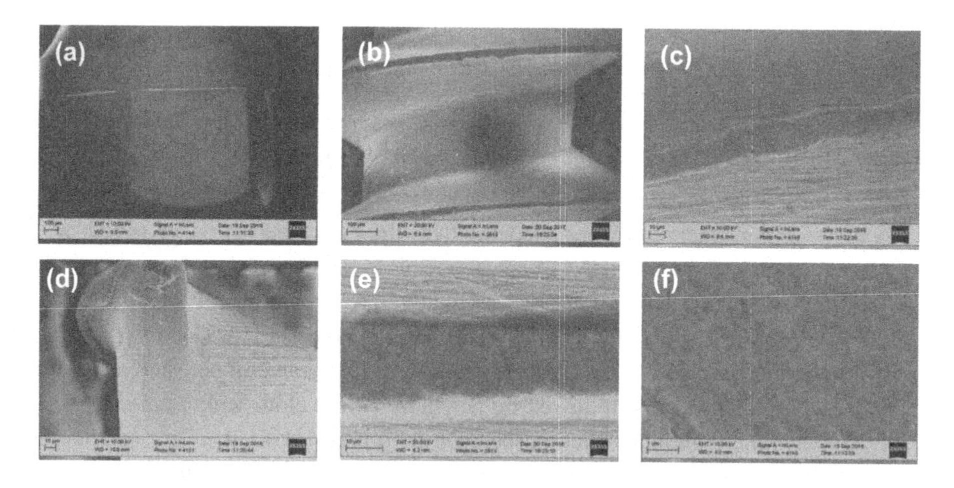

Figure 5.16 SEM images of a micro-implant coated with UNCD by HFCVD. (a) Head section (100 µm resolution); (b) thread section (100 µm resolution); (c) middle top of the thread section (10 µm resolution); (d) tip of the threaded section (10 µm resolution); (e) higher magnification image of the edge of the head (10 µm resolution); (f) highest magnification that can be obtained with SEM on the side section of the head (1 µm resolution).

appropriate scale-up method to coat hundreds of dental implants positioned vertically in a circular pattern.

5.7 *In-Vitro* Studies of Corrosion of Ti-6Al-4V and UNCD-Coated Ti-6Al-4V Dental Implants in Saliva

Commercially pure Ti (Cp-Ti, identified as Ti-2) and Ti-6Al-4V alloy (the main alloy used in commercial dental implants, identified as Ti-5) discs with 15 mm diameter and

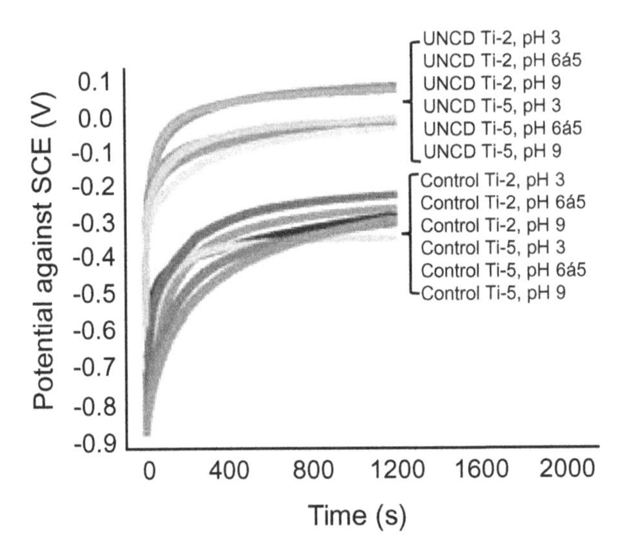

Figure 5.17 OCP curves show the evolution of the potential of both control Ti-based samples and UNCD-coated Ti-based samples. The data show superior performance for the UNCD-coated Ti-based samples (reprinted with permission from *Surf. Innov.*, vol. 5 (2), p. 106, 2017 (Fig. 2) in [47] with permission from ICE Publisher).

2 mm thickness were used to study the corrosive action of saliva on the exposed metals and on the UNCD-coated metals to compare the performance in simulated oral conditions. The samples were divided into four groups (Ti-2, Ti-5, UNCD Ti-2, and UNCD Ti-5), tested with three different pH levels (3, 6.5, and 9) of artificial saliva as an electrolyte. The Ti sample disks were mechanically polished using silicon carbide grinding papers, followed by polishing with a cloth with diamond paste, and finally with a colloidal silica suspension to produce a final mirror-smooth finish on the uncoated Ti surfaces. The samples were finally subjected to sonication in 70% isopropanol for 10 min followed by deionized water to remove surface impurities. A more detailed description of the experimental conditions and instrumentation used in these studies can be found in a recent publication [47].

An artificial saliva was prepared based on the widely used Fusayama Meyer's solution components [47]. The electrochemical test involved a starting step with the open-circuit potential (OCP) being monitored to stabilize the system and to evaluate the evolution of potential during the electrochemical analysis. A potential was applied to two electrodes, and the current measured during electrochemical impedance spectroscopy (EIS) analysis (see the instruments in Figure 1 in [47]) to understand the corrosion kinetics.

The electrochemical behavior of UNCD-coated (Ti-2) and (Ti-5) Ti-based samples were compared with control (uncoated) Ti-based samples via immersion in artificial saliva with different pH levels. The OCP measurements in pH 3 artificial saliva showed that UNCD-coated Ti-2 and UNCD-coated Ti-5 exhibit a higher potential (0.05 V) than at the other two pH conditions (–0.1 V). All the control Ti-based samples, without coating, exhibited potentials in the range of –0.4 to –0.3 V. The data show that UNCD-coated titanium samples exhibit practically no electrochemical corrosion compared with the uncoated titanium samples.

Complementary EIS measurements provide Nyquist and Bode plots, which determine corrosion kinetics from the resulting variation in impedance as a function of frequency during corrosion. Nyquist plots measure the electrochemical resistances on the surfaces of the Ti and UNCD-coated Ti samples. Figure 4(a) in [47] shows that the uncoated Ti-2 and Ti-5 have poor electrochemical resistance compared to both UNCD-coated Ti samples (Ti-2 and Ti-5) (Figure 4(b) in [47]), which exhibit the highest corrosion resistance for artificial saliva with pH 6.5. Bode plots, related to measurements of impedance and frequency along with phase angle (figure 5 in [47]), showed increased impedance values and a much wider phase angle for UNCD-coated Ti samples compared to control Ti samples without coating. UNCD-coated Ti-5 samples showed slightly increased impedance with respect to UNCD-coated Ti-2 samples, which is relevant since Ti-5 corresponds to the alloys (Ti-6Al-4V) currently used for commercial dental implants.

Another key characterization of the surface of the uncoated and UNCD-coated Ti-based samples relates to the measurement of surface roughness before and after exposure to the artificial saliva. This parameter is important because roughening of the dental implant surface can result in enhanced bacterial growth. Figure 5.18 shows measurements of surface roughness via high-resolution cross-section SEM. It is clear that the UNCD-coated Ti samples generally exhibit less surface roughening after exposure to the artificial saliva than the uncoated Ti-based samples, which indicates that the UNCD-coated Ti-based dental implants will perform much better from the point of view of surface roughness change.

In conclusion, from the *in-vitro* studies described above, exposing uncoated and UNCD-coated Ti-based samples to artificial saliva, it is clear that UNCD-coated Ti-based samples exhibit far higher resistance to chemical corrosion than the uncoated Ti-based dental implant materials, and thus UNCD-coated Ti-based dental implants may provide the next generation of superior performance with orders of magnitude longer-life dental implants.

5.8 *In-Vivo* Studies in Animals of Biological Performance of UNCD-Coated Ti Alloy Dental Implants

Male Wistar rats ($n = 20$, mean body weight = 90 g) were used to study peri-implant reparative process. Guidelines of the National Institutes of Health (NIH) and the Statement of Ethics Principles of the Faculty of Dentistry, University of Buenos Aires (Res (CD) 352/02 y Res (CD) 694/02) for the use and care of laboratory animals were followed.

The animals were anesthetized intra-peritoneally with a solution of 8 mg of ketamine chlorhydrate (Fort Dodge® – Argentina) and 1.28 mg of xylazine (Bayer Argentina S.A.) per 100 g of body weight. The skin of both tibiae was shaved prior to

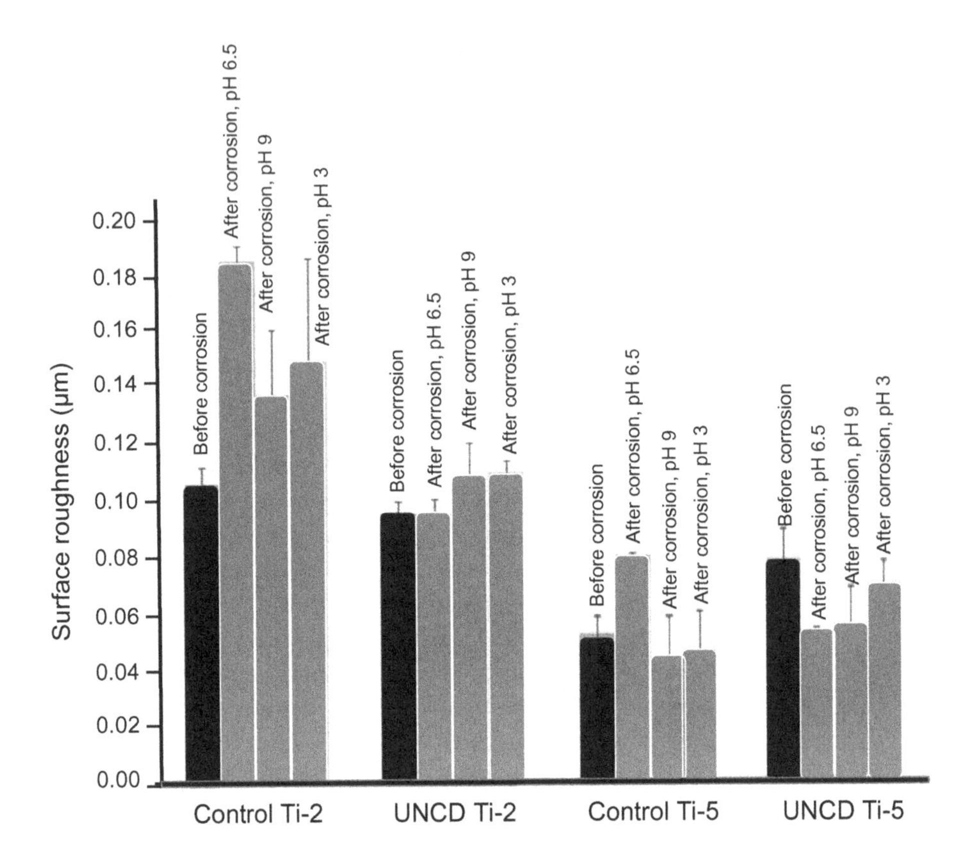

Figure 5.18 Change in roughness values before and after corrosion studies at pH 3, 6.5, and 9 on uncoated and UNCD-coated Ti-based samples, showing the much better performance of the UNCD-coated Ti-based samples (reprinted from *Surf. Innov.*, vol. 5 (2), p. 106, 2017 (Fig. 7) in [47] with permission from ICE Publisher).

performing a 1.5 cm incision along the tibial crest. The subcutaneous tissue, muscles, and ligaments were dissected to expose the lateral external surface of the diaphyseal bone. An end-cutting bur was used to drill a hole of 1.5 mm in diameter with manual rotating movements to avoid overheating and necrosis of the bone tissue. UNCD-coated Ti implants (6.0 × 1.0 × 0.1 mm) were placed in the hematopoietic bone marrow compartments of the rats' right tibias. Bare Ti implants were placed in the left tibia and used as controls. A separate-stitch suture was performed. No antibiotic therapy was administered. At 30 days post-implantation, the animals were euthanized by intraperitoneal overdose of ketamine chlorhydrate (Holliday-Scott S.A., Buenos Aires) and the tibiae were resected and fixed in 10% buffered formalin and radiographed.

Studies were performed to determine osseointegration of the uncoated and UNCD-coated Ti implants using the rat diaphyseal tibia model, which is a good animal model for these studies, as demonstrated in prior R&D performed by researchers at the University of Buenos Aires, Argentina on metal implants [48, 49]. Histological

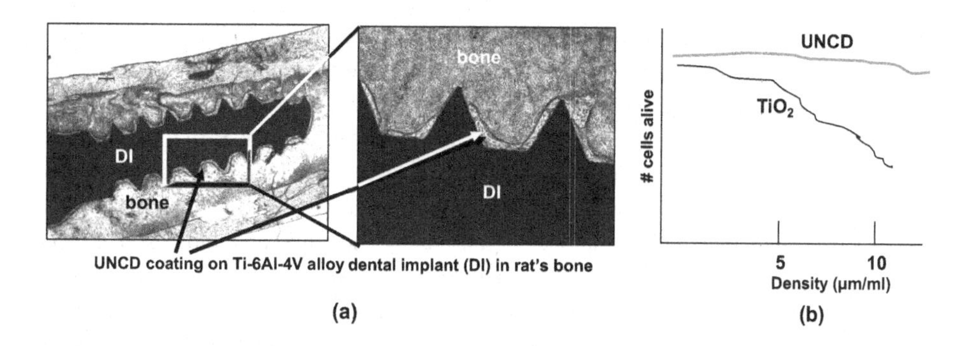

Figure 5.19 (a) Optical pictures showing excellent osseointegration of UNCD-coated Ti-6Al-4V dental implant in a rat's bone. (b) Study of the biological effect of TiO_2 and UNCD particles in contact with live cells, showing that TiO_2 particles induce substantial death of cells, while UNCD particles do not have any effect.

analysis demonstrated that UNCD-coated Ti implants provide excellent osseointegration and do not elicit inflammatory reactions. The osseointegration demonstrated in this work is supported by prior work in Auciello's group, which demonstrated extensive growth and proliferation of embryonic epidermal stem cells, bone cells, and neural cells on the surface of the UNCD films [37–39] (see Figures 9.8–9.13), confirmed by other groups worldwide [40, 41].

An additional process that can enhance osseointegration of dental implants in maxillary bone is to induce surface roughening of the implant via mechanical, chemical, synergistic mechanical/chemical treatment of the surface. It has been demonstrated that roughening the surface of Ti-based implants induces enhanced bone cell growth [50–52]. In this sense, since UNCD coatings have been demonstrated to grow very conformally on roughened surfaces of Ti dental implants (Figure 5.13b), the osseointegration on UNCD-coated Ti dental implants will be enhanced by inducing the micro-topographies on the UNCD surface. Osseointegration enhancement on UNCD-coated roughened Ti implant surfaces will be superior to that of the uncoated Ti implant due to the fact that cells grow and proliferate much more efficiently on the UNCD surface. Figure 5.19a shows a picture of a UNCD-coated Ti plate several months after implantation in a rat's bone. The picture shows excellent osseointegration. On the other hand, Figure 5.19b shows a study to determine the biocompatibility of TiO_2 particles, which are known to be etched from the surface of dental, hip, knee, and many other implants based on Ti [53–55], which oxidize when exposed to air, before implantation in the human body. In relation to this issue, it is well known now that electrochemical corrosion releases metal-oxide particles (mainly TiO_2) from the oxidized Ti surface into the tissue adjacent to the implant, which induces potential long-term systemic toxicity [56]. The metallic implant acts as an *in-vivo* electrode leading to electrochemical degradation of the implant's surface, weakening the implant–bone interface and thus the attachment to the bone [56]. Figure 5.19b shows that TiO_2 particles induce substantial death of live cells. On the other hand, UNCD particles do not induce any death of live cells.

In addition to the effect of TiO_2 particles on live cells, as shown in Figure 5.19, TiO_2 particles released from the implant can migrate to distant sites in the biological system through blood circulation, producing additional biological deleterious effects in several organs, as recently demonstrated by an R&D collaboration between researchers at the University of Texas-Dallas and University of Buenos Ares [53].

In relation to the critical effects that particles released from implants may have when interacting with biological organs, the biokinetics and biological effects of UNCD particles, compared to TiO_2 nanoparticles, were evaluated *in vivo* using Wistar rats ($n = 30$) intraperitoneally injected with TiO_2 and UNCD nanoparticles, or saline solution as control. After six months, blood, lung, liver, and kidney samples from the rats were histologically analyzed. Oxidative damage by membrane lipidperoxidation (TBARS), generation of reactive oxygen species (O_2) and antioxidant enzymes (SOD, CAT) were evaluated in lung and liver. Histological observation showed agglomerates of TiO_2 or UNCD in the parenchyma of the studied organs, though there were fewer UNCD than TiO_2 particle deposits. In addition, TiO_2 particles caused areas compatible with foci of necrosis in the liver and renal hyaline cylinders (Figure 5.20).

Regarding the effects of UNCD particles, no membrane damage (TBARS) or mobilization of enzymatic antioxidants was observed either in lung or liver samples [53–55, 57], and no variations in O_2^- generation were observed in the lungs (35.1 ± 4.02 vs. UNCD: 48 ± 9.1, $p > 0.05$). On the contrary, exposure to TiO_2 nanoparticles caused production of O_2^- in alveolar macrophages and consumption of catalase ($p < 0.05$) (Figure 5.21) [53].

The biological studies described above indicate that UNCD particles caused neither biochemical nor histological alterations, and therefore behave as an extremely good biocompatible material to enable a new generation of superior biomedical implants,

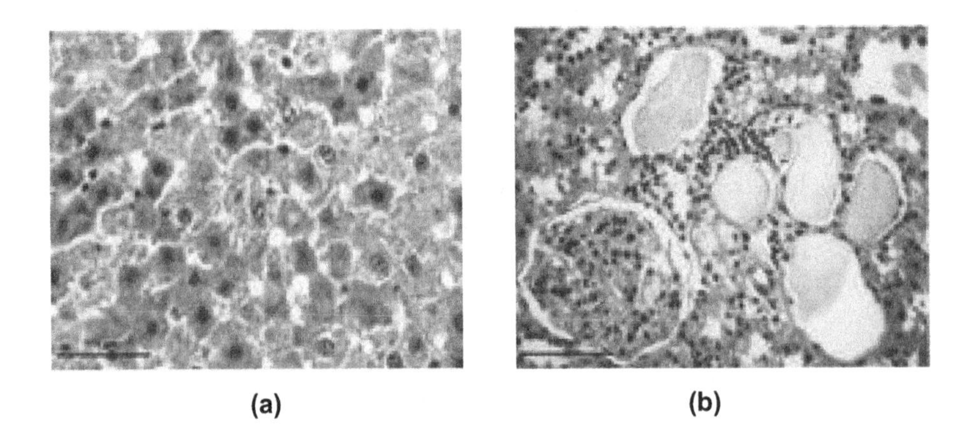

Figure 5.20 Optical pictures of TiO_2 particles in the liver (a) and kidney (b) of a Wistar rat. An area of necrosis can be observed (H&E stain, orig. mag. 400×), produced by the TiO_2 particles in the liver (a). On the other hand, kidney tubules containing hyaline cylinders can be seen clearly (reprinted from *J. Biomed. Mater.* Rest Part B, vol. 105, p. 2408, 2016 Fig. 4 in [53] with permission from Wiley Publisher).

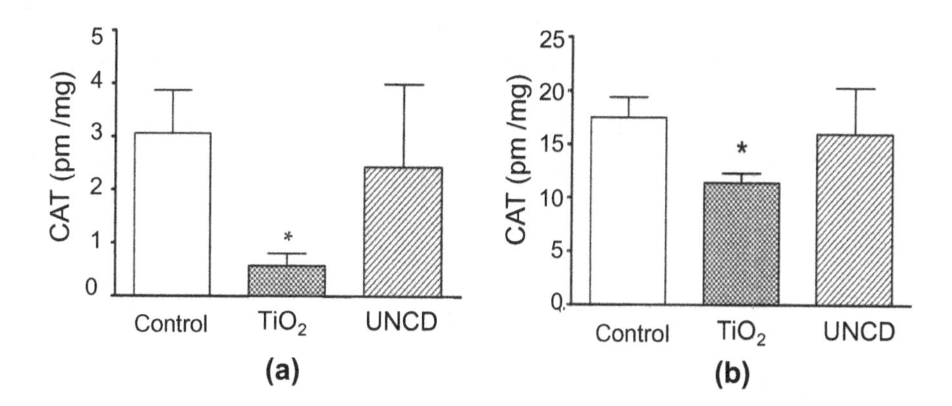

Figure 5.21 Determination of enzymatic antioxidants showing lung (a) and liver (b) catalase (CAT) levels in control, TiO_2- and UNCD-treated Wistar rat. Only TiO_2^- nanoparticles caused significant changes in CAT levels in lung and liver homogenates. The histograms show the mean ± SD ($n \geq 6$), $*p < 0.05$, compared to the control group (reprinted from *J. Biomed. Mater. Res. B Appl. Mater.*, p. 008, 2016 (Fig. 7) in [53] with permission from Wiley publisher Publisher).

since, in addition, it has been demonstrated in prior work by Auciello's group that UNCD coatings are extremely resistant to chemical attack by body fluids such as eye humor [56].

In 2018, López and Auciello and colleagues performed scaled-up animal studies with respect to the initial studies in rats' bones, implanting UNCD-coated Ti alloy dental implants in the maxillary bones of 15 sheep for three months. Subsequently, the implants were extracted and subjected to histopathological studies that demonstrated a strong ankylosis, as described by Brånemark in his publications. These studies provided the basis for the new concept in terms of osteointegration, as supported by three key processes observed in the studies described here:

1. macro-retention: mechanical retention generated by the Ti alloy or stainless steel screw and inserted into the human bone;
2. micro-retention: generated by the treatment of the implant surface, additive–subtractive or mixed process (acid etching, sandblasting, hydroxyapatite coating, or other process);
3. nano-retention: retention generated by the surface of the unique biocompatible UNCD coating, made of C atoms, which induces superior growth of bone cells because C atoms generate an order of magnitude more efficient biological reaction with the bone cells than the metals.

5.9 Clinical Trials of UNCD-Coated Dental Implants in Bioingeniería Humana Avanzada

Following the extensive animal studies described in the previous section, clinical studies have been performed on 20 patients via implantation of UNCD-coated

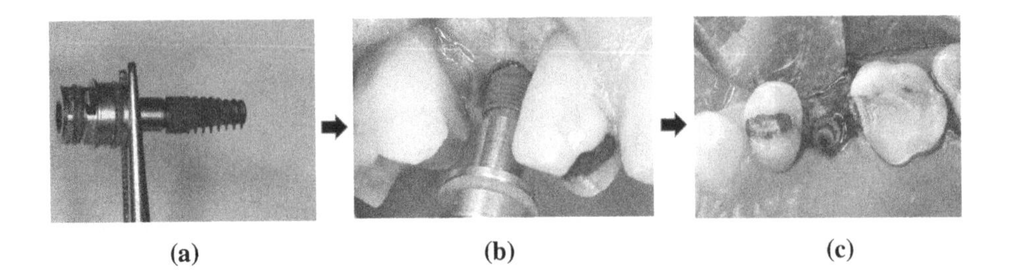

Figure 5.22 Optical pictures of UNCD-coated Ti-6Al-4V dental implant: (a) before implantation, (b) implantation of the first UNCD-coated dental implant in a human maxillary bone worldwide, and (c) surgical boarding for implantation in the patient's maxillary bone using a biological countersink.

commercial Ti-6Al-4V alloys dental implants since May 2019. These clinical studies, conducted in the world-class clinic of Dr. Gilberto López Chávez in Querétaro-México, demonstrated superior osteointegration (100% success for UNCD-coated dental implants inserted in the maxillary and jaw bones of the patients). The UNCD-coated dental implant exhibit an exceptional attachment to the human bones due to excellent bone cell growth on the C atom-terminated surface of the UNCD coating, as demonstrated in prior research [37, 38]. In addition, the UNCD coating exhibits excellent chemical biocompatible interactions with the soft tissue (gum) and adjacent roots of natural teeth. Figure 5.22 shows the initial procedure for implantation of UNCD-coated Ti-6Al-4V commercial dental implant in the maxillary bone of a patient.

Following the implantation of the UNCD-coated dental implant in a patient, as shown in Figure 5.22, a three-month period was allowed for the osteointegration process to develop (Figure 5.23). During that period, clinical observations were pursued, including X-ray imaging (Figure 5.23e), such that the clinical observations demonstrated no infection, no pain, and no tumefaction in the implant area. Also observed was excellent sealing of the gingival tissue on the surface of the UNCD coating all around the implant.

About six months following implantation of a UNCD-coated Ti alloy dental implant in a patient, tests of implant osseointegration in the maxillary bone were performed by applying the strongest possible torque with a top-of-the-line torquemeter (Figure 5.24). The test demonstrated superior osseointegration of the UNCD-coated dental implant over uncoated Ti alloy implants.

5.10 Development of an Industrial Process for Commercial-Scale Production of UNCD-Coated Ti Alloy Dental Implants

A process has been demonstrated, using the large-area industrial MPCVD system (IPLAS, Germany) (Figure 5.25a), coating seven commercial Ti-6Al-4V dental

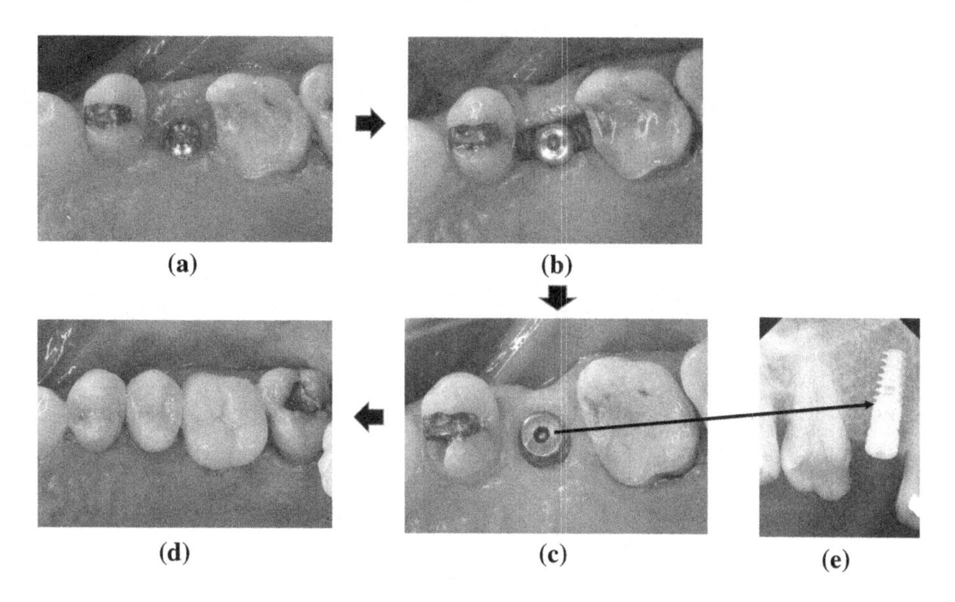

Figure 5.23 Optical pictures of UNCD-coated Ti-6Al-4V dental implant: (a) before implantation, (b) during implantation of the first UNCD-coated dental implant in a human worldwide, and (c) immediately after implantation in the patient's maxillary bone. (d) Ceramic crown insertion on top of UNCD-coated Ti–6A–4V dental implant, after five months; (e) X-ray image of implanted UNCD-coated dental implant, showing excellent osseointegration.

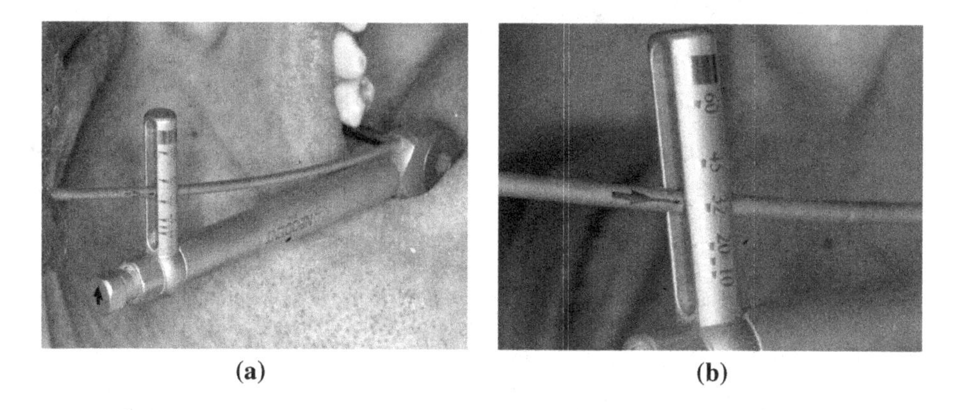

Figure 5.24 (a) Torquemeter used in performing measurement of attachment of the UNCD-coated Ti-6Al-4V dental implant in the maxillary bone of a patient. (b) Torque applied to the implant demonstrated superior osseointegration.

implants simultaneously (Figure 5.25b). Industrial IPLAS systems, available in Auciello's lab in UTD and in Original Biomedical Implants-México, have the capability to coat up to 200–300 commercial dental implants simultaneously, distributed vertically on a large-area plate (Figure 5.25c).

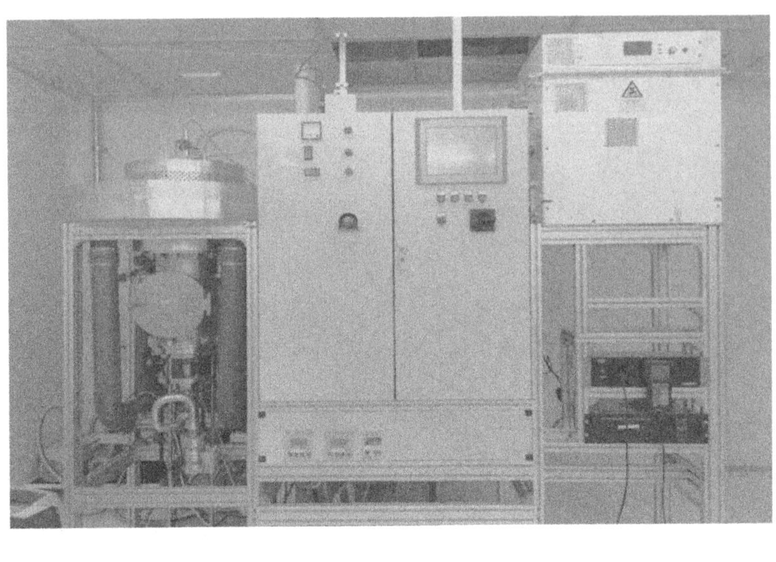

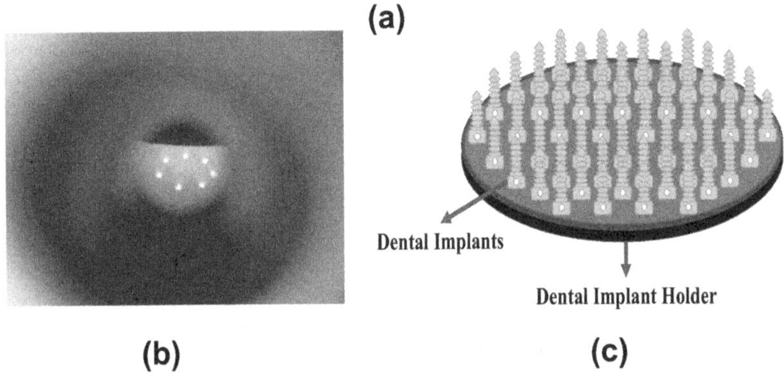

Figure 5.25 (a) Industrial MPCVD system (IPLAS, Germany) used to implement the commercial-scale coating of metallic dental implants with the unique UNCD coating. (b) Seven implants simultaneously coated with UNCD films demonstrated excellent uniformity, as characterized by Raman spectroscopy (see Figure 5.8) and SEM (see Figure 5.13) analysis. (c) Large-area platform, under development, capable of holding 200–300 commercial dental implants (~5-8 mm long, 2 mm diameter) to implement a low-cost industrial process to produce UNCD-coated commercial metal implants.

5.11 Application of Biocompatible, Low-Friction, Body Fluid Corrosion-Resistant UNCD Coatings for a New Generation of Superior Prostheses

Natural human joints are remarkable pieces of the human body, moving with very little friction between the components due to natural lubrication by synovial fluid and helped

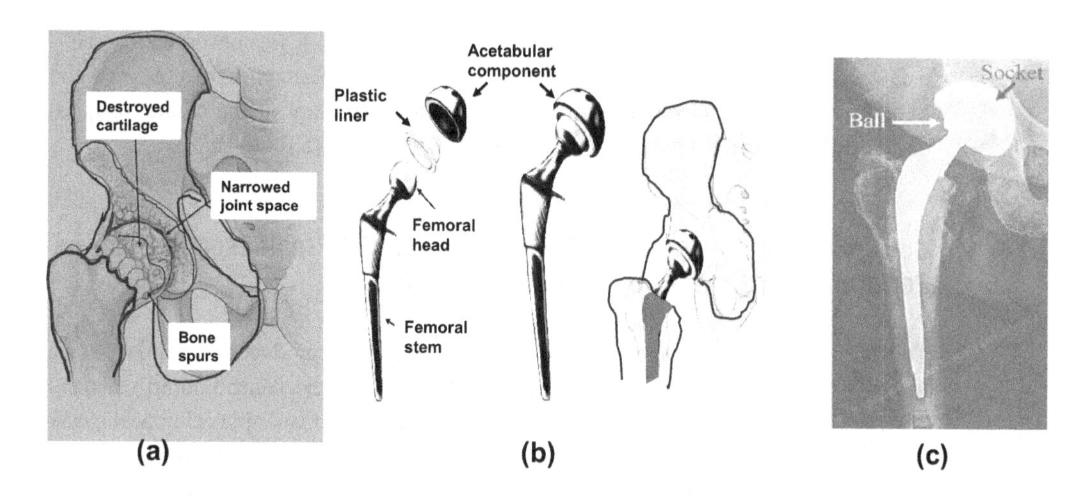

Figure 5.26 (a) Schematic showing cartilage wear and tear in a natural hip, which leads to the need for artificial hip implantation. (b) Schematic of an artificial hip's components (left), assembled artificial hip (center), and implanted hip (right). (c) An X-ray picture of an artificial hip implanted in a human body.

by healthy cartilage. The knee connects the femur (thigh bone), tibia (shin bone), fibula (outer shin bone), and patella (kneecap). A mixed network of muscles, ligaments, and tendons holds the knee joint together, enabling motion. The hip is the other key joint enabling motion, which is composed of two main components: a ball and a socket made of natural bone. The ball rotates inside the socket during walking. These key natural joints are affected by a condition known as osteoarthritis (OA), the most common arthritis worldwide, affecting about 80% of people over 75 years old. This number is expected to rise as life expectancy increases. Typically, symptoms first begin after age 40. Through wear and tear, the cartilage wears out in certain joints – primarily the hip (Figure 5.26a), knee, lower back, neck, and hands, leading to stiffness, pain, and eventually almost total immobility. In many cases the natural components of the joints need to be replaced by artificial prostheses. The components of a hip replacement prosthesis include a round ball on a stem that is inserted inside the thigh bone and a smooth, lined socket that is attached to the pelvic bone (Figure 5.26b). Figure 5.26c shows an X-ray of an artificial metal-based hip implanted in a human body.

In relation to orthopedic prostheses, recent R&D on Ti-based implants in patients showed osseointegration problems and corrosion induced by chemical attack by body fluids [58]. One problem that was not initially recognized is that Ti and Ti alloys oxidize readily when exposed to normal atmospheric conditions. This effect has been identified as a deleterious condition, because body fluids induce chemical attachment of the TiO_2 surface layer, which results in TiO_2 micro- and nanoparticles dislodging from the surface of the implant, with the latter having a greater surface-to-volume ratio than the former, therefore being more biologically reactive and potentially more harmful to human health, as demonstrated in recent studies [53, 54]. Recent histological studies of lung, liver, and kidney tissue samples showed the presence of TiO_2

or UNCD particles in the parenchyma of the studied organs. It was observed that TiO_2 nanoparticles cause substantial deleterious morphological changes in the liver and kidneys,' while UNCD particles caused no alterations in any of the studied organs, demonstrating the outstanding biocompatibility of UNCD [53].

5.12 Current Performance of UNCD-Coated Industrial Prostheses Predict the Performance of Future UNCD-Coated Prostheses

The outstanding performance of UNCD-coated industrial products, currently on the market by Advanced Diamond Technologies [59], such as mechanical pump seals and bearings, which involve outstanding resistance to friction and wear plus chemical attack of corrosive fluids, provide a strong basis that shows that UNCD coatings have great potential to solve all the problems described above in relation to failure of Ti-based prostheses (e.g., hips, knees, elbows, and more) due to synergistic mechanical (high friction) and body fluid chemical attack degradation. Ti and Ti-6Al-4V (the main alloy used in fabrication of approved implantable prostheses) exhibit relatively high friction coefficients (~0.3–0.5, as measured with polytetrafluoroethylene (PTFE) and stainless steel sliding balls, respectively) [58], resulting in the wear and performance problems characteristic of currently approved metal-based prostheses implanted in people today. In addition, the chemical attack by body fluids strongly contributes to the mechanical-induced degradation due to a synergistic combined mechanical–chemical process.

UNCD coatings are currently used in commercial mechanical pumps seals and bearings, commercialized by ADT [59], the company co-founded by Auciello and colleagues in 2003 [59], where the UNCD-coated surfaces exhibit friction coefficients in the range 0.02–0.04 [33, 61, 62], one order of magnitude lower than the friction coefficient of any metal or ceramic used today in any approved prostheses, and practically no wear for UNCD-coated SiC mechanical pump seals (Figure 5.27b) [61, 62], while there is extensive wear for uncoated SiC seals (Figure 5.27a) [61, 62], which slide upon each other at ~6000 rpm, with forces orders of magnitude higher than those present in artificial prostheses. The performance of UNCD-coated SiC seals indicates that UNCD-coated metal and ceramic prostheses used today will exhibit performance and lifetime improved by orders of magnitude over uncoated prostheses. R&D is underway to confirm and prove this hypothesis.

Work is currently in progress to start performing animal tests for UNCD-coated Ti alloy artificial hips. An important fact is that Ti-based prostheses can be coated with excellent dense UNCD films, as demonstrated for the optimized process developed to coat dental implants (see Figures 5.13 and 5.14).

5.13 Conclusions

Systematic R&D was performed focused on investigating the conditions needed to grow biocompatible and corrosion-resistant UNCD coatings on Ti alloy dental

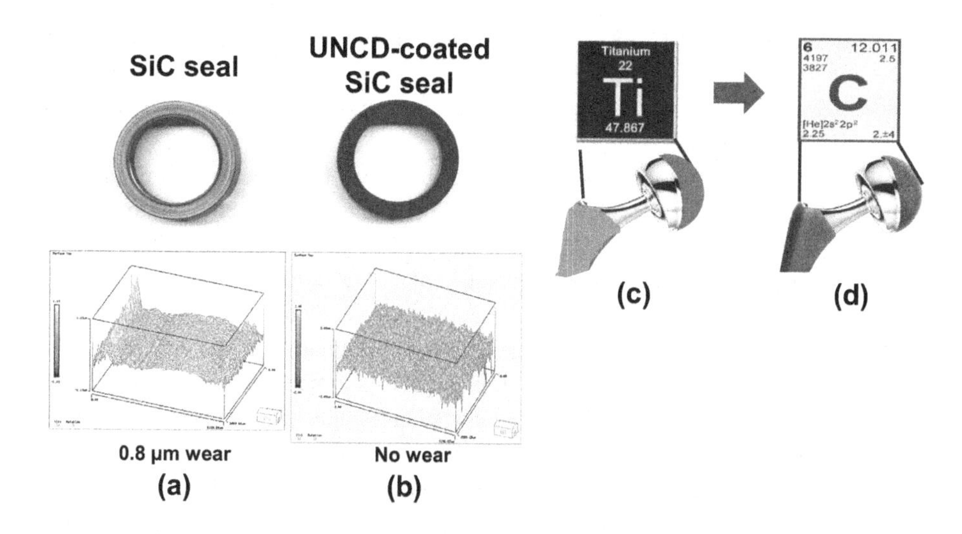

Figure 5.27 (a) A commercial SiC seal with the wear measured by AFM after one year of continuous operation in a mechanical pump at 6000 rpm. (b) A UNCD-coated SiC seal (currently on the market [59]) with no wear measured by AFM after one year of continuous operation in the same mechanical pump for which the SiC shows the wear indicated in (a). (c) Schematic of an uncoated Ti alloy–based artificial hip coated with the natural TiO_2 surface layer. (d) Schematic of a UNCD-coated Ti alloy–based artificial hip.

implants. The purpose of coating commercial Ti alloy dental implants is to protect them from the chemical and mechanical environments of the mouth and body, which currently are the main contributors to failure of commercial Ti alloy dental implants. Two processes were investigated to grow UNCD coatings on dental implants: HFCVD and MPCVD, in order to determine which is the best process to grow UNCD films on a large number of dental implants to develop a low-cost industrial growth process. The fully optimized MPCVD growth process produced dense/pin-hole-free UNCD films. The key features of the new generation of dental implants proposed here rely on the unique set of properties of the UNCD coatings, relevant to the protection of metal-based dental implants from the chemical and mechanical environment of the mouth: (1) excellent conformal growth on 3D structures such as dental implants; (2) extreme resistance to chemical attack by strong acids such as hydrofluoric acid, and body fluids, including oral fluids, that are critical to degradation of metal-based dental implants; (3) excellent surface chemistry based on carbon atoms, which has been proven to facilitate extremely efficient growth of biological cells (e.g., embryonic stem cells, neural cells, fibroblast cells), thus also inducing efficient maxillary bone cell growth on the UNCD surface to produce strong osseointegration of UNCD-coated metal dental implants in the maxillary bone; and (4) the lowest coefficient of friction compared to other materials, which can facilitate insertion of the screw-type dental implant into the maxillary bone with minimal mechanical-induced degradation of the bone.

Raman and XPS analysis revealed that the surface of the UNCD coatings exhibit C-sp^3 chemical bonds characteristic of diamond, which provide excellent resistant to chemical attack because there are no chemical bonds open for atoms from corroding fluids to react with.

High-resolution SEM studies revealed that the hitherto optimized MPCVD process produces dense pinhole-free UNCD coatings with extreme conformity on the screw-type structure of test dental implants, which indicates that similar performance can be expected for coating real dental implants in order to provide protection against the chemical and mechanical environment of the mouth.

The demonstration that several dental implants can be coated simultaneously, as shown in Figures 5.6 and 5.7, prompted us to perform calculations to determine how many dental implants could be coated in a single growth process to produce a low-cost industrial manufacturing process for coating dental implants with UNCD films. It was determined that the 200 mm diameter holder in the MPCVD system can hold up to 300 dental implants (~8 mm high by ~3–5 mm diameter) positioned vertically in the industrial MPCVD system. This provides the industrial production capability to grow UNCD films on a large number of dental implants in a single/low-cost coating growth process.

The current MPCVD process to grow UNCD films involves seeding the surface of the substrate with nanocrystalline diamond particles in a wet chemical process that adds about 1 h of extra time to the overall process. Therefore, it is desirable to eliminate the wet chemical seeding process to reduce costs. In this respect, Auciello's group recently demonstrated a new process, BEN-BEG, using the MPCVD and HFCVD methods. In the BEN-BEG process, the substrate is biased with a negative voltage, which enables it to attract positive C and CH_x ($x = 1, 2, 3$) ions generated in a plasma toward the substrate surface, impacting with a relatively high energy (~100–200 eV), which results in sub-implantation of the C and CH_x ions on the surface of the metal dental implant or other prosthesis, generating a carbide template layer that induces the growth of the UNCD film without the need for a wet chemical diamond seeding process. The BEN-BEG process reduces fabrication cost by at least an order of magnitude. The other big advantage of the BEN-BEG process is that because of the energetic insertion of the C-based ionic species on the surface of the implant, the adhesive force of the UNCD coating has the potential to be a magnitude higher than for the conventional growth process, thus eliminating potential generation of cracks or delamination of the coating.

In-vitro studies exposing UNCD-coated Ti metal plates to artificial saliva, with properties similar to natural saliva, demonstrated that UNCD-coated implants are extremely resistant to chemical attack, as opposed to Ti-based metal implants that are substantially corroded by oral and other body fluids.

In-vivo animal studies involving implantation of UNCD-coated Ti-based implants in rats' bones and clinical trials implanting 20 patients, since 2018, with UNCD-coated commercial Ti–6Al–4V dental implants showed excellent biocompatibility and osseointegration, plus outstanding resistance to chemical attack by body fluids, providing the basis for superior next-generation metal-based implants coated with UNCD to make a transformational positive impact on the quality of life of people worldwide.

Based on the information presented in this chapter, the future for implantable medical devices is based on the unique UNCD coating.

Acknowledgments

O. Auciello acknowledges the funding provided by his Distinguished Endowed Chair Professor position at the University of Texas-Dallas, He also thanks B. Patel, A. C. Duran-Martinez, V. Barao, S. Campbell, C. Sukotjo, and Professor M. T. Mathew (head of the group) who conducted the R&D cited in [47]). O. Auciello also acknowledges the work done on XPS analysis of UNCD coatings by J. F. Veyan and E. de Obaldia at UTD.

O. Auciello, D. G. Olmedo, and D. R. Tasat acknowledge the great contributions of M. E. Bruno, M. Domingo, P. Gurman, M. L. Paparella, P. Evelson, and M. B. Guglielmotti (deceased) to the R&D conducted in the collaboration between the University of Texas-Dallas, USA and the University of Buenos Aires, Argentina during the last five years.

References

[1] I. Ericsson, P. O. Glantz, and P. I. Brånemark, "Titanium implants of Brånemark type for oral rehabilitation of partially edentulous patients," *Tandlakartidningen*, vol. 81 (24), p. 1357, 1989.

[2] P. I. Brånemark, G. Zarb, and T. Albrektsson (Eds), *Introduction to Osseointegration Tissue-Integrated Prostheses: Osseointegration in Clinical Dentistry*. Chicago, IL: Quintessence Publishing, Co., Inc., 1985.

[3] R. Adell, B. Eriksson, U. Lekholm, P. I. Brånemark, and T. Jemt, "Long-term follow- up study of osseointegrated implants in the treatment of totally a dentulous jaw," *Int. J. Oral Maxillofac. Implants*, vol. 5, p. 347, 1990.

[4] W. Becker, B. E. Becker, M. G. Newman, and S. Nyman, "Clinical and microbiologic findings that may contribute to dental implant failure," *Int. J. Oral Maxillofac. Implants*, vol. 5(1), p. 31, 1990.

[5] J. J. Jacobs, J. L. Gilbert, and R. M. Urban, "Current concepts review: corrosion of metal orthopaedical implants," *J. Bone Joint Surg.*, vol. 80, p. 268, 1998.

[6] P. K. Moy, D. Medina, V. Shetty, and T. L. Aghaloo, "Dental implant failure rates and associated risk factors," *Int. J. Oral Maxillofac. Implants*, vol. 20 (4), p. 569, 2005.

[7] D. G. Olmedo, R. L. Cabrini, G. Duffó, and M. B. Guglielmotti, "Local effect of titanium corrosion: an experimental study in rats," *Int. J. Oral Maxillofac. Surg.*, vol. 37 (11), p. 1037, 2008.

[8] D. G. Olmedo, D. Tasat, P. Evelson, M. B. Guglielmotti, and R. L. Cabrini, "Biological response of tissues with macrophagic activity to titanium dioxide," *J. Biomed. Mater. Res. Part A*, vol. 84 (4), p. 1087, 2008.

[9] R. A. Gittens, R. Olivares-Navarrete, R. Tannenbaum, B. D. Boyan, and Z. Schwartz, "Electrical implications of corrosion for osseointegration of titanium implants," *J. Dent. Res.*, vol. 90 (12), p. 1389, 2011.
[10] D. G. Olmedo, M. L. Paparella, M. Spielberg, et al., "Oral mucosa tissue response to titanium cover screws," *J. Periodontol.*, vol. 83, p. 973, 2012.
[11] J. Mouhyi, D. M. D. Ehrenfest, and T. Albrektsson, "The peri-implantitis: implant surfaces, microstructure, and physicochemical aspects," *Clin. Implant. Dent. Relat. Res.*, vol. 14, p. 170, 2012.
[12] D. C. Rodriguez, P. Valderrama, T. G. Wilson, Jr., et al., "Titanium corrosion mechanism in the oral environment: a retrieval study," *Materials*, vol. 6, p. 5258, 2013.
[13] M. Abdel-Hady Gepreel and M. Niinomi, "Biocompatibility of Ti-alloys for l-term implantation," *J. Mech. Behav. Biomed. Mater.*, vol. 20, p. 407, 2013.
[14] T. P. Chaturvedi, "An overview of the corrosion aspect of dental implants (titanium and its alloys)," *Indian J. Dent. Res.*, vol. 20, p. 91, 2009.
[15] R. Urban, J. Jacobs, M. Tomlinson, et al. "Dissemination of wear particles to the liver, spleen, and abdominal lymph nodes of patients with hip or knee replacement," *J. Bone Joint Surg.*, vol. 82(A), p. 457, 2000.
[16] A. Hartmann, F. Hannemann, J. Lützner, et al., "Metal ion concentrations in body fluids after implantation of hip replacements with metal-on-metal bearing: systematic review of clinical and epidemiological studies," *PLoS One*, vol. 8 (8), p. e70359, 2013.
[17] D. G. Olmedo, D. Tasat, M. B. Guglielmotti, and R. L. Cabrin, "Titanium transport through the blood stream: an experimental study on rats," *J. Mater. Sci.*, vol. 14 (12), p. 1099, 2003.
[18] D. R. Tasat, M. E. Bruno, M. Domingo, et al., "Biokinetics and tissue response to ultrananocrystalline diamond nanoparticles employed as coating for biomedical devices," *J. Biomed. Mater. Appl. Biomater.*, vol. 8, p. 1, 2016.
[19] Dr. Axe, "How to reverse cavities naturally & heal tooth decay," 2017. https://draxe.com/naturally-reverse-cavities-heal-tooth-decay/.
[20] R. Nagel, "Living with phatic acid," 2010. www.westonaprice.org/ health-topics/living-with-phytic-acid.
[21] NIH, "The tooth decay process: how to reverse it and avoid a cavity." www.nidcr.nih.gov/health-info/tooth-decay/more-info/tooth-decay-process.
[22] Cure Tooth Decay, "How to heal a tooth abscess naturally." www.curetoothdecay.com/Tooth.abscess.htm.
[23] H. Silk, "Diseases of the mouth," *Primary Care*, vol. 41 (1), p. 75, 2014.
[24] J. M. Laudenbach and Z. Simon, "Common dental and periodontal diseases: evaluation and management," *Med. Clin. North Am.*, vol.98 (6), p. 1239, 2014.
[25] WHO, "Oral health fact sheet N°318," 2012.
[26] Taber's, *Taber's Cyclopedic Medical Dictionary*, 22nd ed. Philadelphia, PA: F.A. Davis Co., 2013.
[27] Wikipedia, "Tooth decay." https://en.wikipedia.org/wiki/Tooth_decay#cite_ref-Silk2014.
[28] D. N. Uditsky, "A new, improved solution for loss of teeth in the lower jaw," 2012. www.schaumburgdds.com/blog/post/a-new-improved-solution-for-loss-of-teeth-in-the-lower-jaw.html.
[29] Complete Dental Implant Cost Guide, "Dental implant cost guide." www.dentalimplantcostguide.com

[30] A. Appleton, "What's the best way to replacing missing tooth or teeth," 2012. http://hubpages.com/health/Whats-the-best-way-to-replace-a-missing-tooth#.
[31] E. Mijiritsky, Z. Mazor, A. Lorean, and L. Levin, "Implant diameter and length influence on survival: interim results during the first 2 years of function of implants by a single manufacturer," *Implant Dent.*, vol. 22, p. 394, 2013.
[32] B. R. Chrcanovic, J. Kisch, T. Albrektsoon, and A. Wennerberg, "Factors influencing early dental implant failure," *J. Dental Res.*, vol. 95, p. 995, 2016.
[33] O. Auciello and A. V. Sumant, "Status review of the science and technology of ultrananocrystalline diamond (UNCD) films and application to multifunctional devices," *Diam. Relat. Mater.*, vol. 19 (7–9), p. 699, 2010.
[34] O. Auciello, "Novel biocompatible ultrananocrystalline diamond coating technology for a new generation of medical implants, devices, and scaffolds for developmental biology," *Biomater Med. Appl. J.*, vol. 1 (1), p. 1000103, 2017.
[35] O. Auciello, P. Gurman, M. B. Guglielmotti, et al., "Biocompatible ultrananocrystalline diamond coatings for implantable medical devices," *MRS Bulletin*, vol. 39, p. 621, 2014.
[36] O. Auciello, P. Gurman, A. Berra, M. J. Saravia, and R. Zysler, "Ultrananocrystalline diamond (UNCD) films for ophthalmological applications," in *Diamond Based Materials for Biomedical Applications*, R. Narayan, Ed. Cambridge: Woodhead Publishing, p. 151, 2013.
[37] B. Shi, Q. Jin, L. Chen, et al., "Cell growth on different types of ultrananocrystalline diamond thin films: special issue coating deposition and surface functionalization of implants for biomedical applications," *J. Funct. Biomater.*, vol. 3 (3), p. 588, 2012.
[38] B. Shi, Q. Jin, L. Chen, and O. Auciello, "Fundamentals of ultrananocrystalline diamond (UNCD) thin films as biomaterials for developmental biology: embryonic fibroblasts growth on the surface of (UNCD) films," *Diam. Relat. Mater.*, vol. 18 (2–3), p. 596, 2009.
[39] P. Bajaj, D. Akin, A. Gupta, et al., "Ultrananocrystalline diamond film as an optimal cell interface for biomedical applications," *Biomed. Microdevices*, vol. 9 (6), p. 787, 2007.
[40] J. Miksovsky, A. Voss, R. Kozarova, et al., "Cell adhesion and growth on ultrananocrystalline diamond and diamond-like carbon films after different surface modifications," *Appl. Surf. Sci.*, vol. 297, p. 95, 2014.
[41] A. Voss, S. R. Staleva, J. P. Reithmaier, M. D. Apostolova, and C. Popov, "Patterning of the surface termination of ultrananocrystalline diamond films for guided cell attachment and growth," *Surf. Coat. Technol.*, vol. 321, 2017.
[42] AZO Materials, "Titanium alloys in medical applications," 2003. www.azom.com/article.aspx?ArticleID=1794.
[43] J. Alcantar-Peña, J. Montes, M. J. Arellano-Jimenez, et al., "Low temperature hot filament chemical vapor deposition of ultrananocrystalline diamond films with tunable sheet resistance for electronic power devices," *Diam. Relat. Mater.*, vol. 69, p. 207, 2016.
[44] E. M. A. Fuentes-Fernandez, J. J. Alcantar-Peña, G. Lee, et al., "Synthesis and characterization of microcrystalline diamond to ultrananocrystalline diamond films via hot filament chemical vapor deposition for scaling to large area applications," *Thin Solid Films*, vol. 603, p. 62, 2016.
[45] J. J. Alcantar-Peña, E. de Obaldia J. Montes-Gutierrez, et al., "Fundamentals towards large area synthesis of multifunctional ultranano-crystalline diamond films via large area hot filament chemical vapor deposition bias enhanced nucleation/bias enhanced growth for fabrication of broad range of multifunctional devices," *Diam. Relat. Mater.*, vol. 78, p. 1, 2017.

[46] J.-F. Veyan, E. de Obaldia, J. J. Alcantar-Peña, et al., "Argon atoms insertion in diamond: new insights in the identification of carbon C1s peak in X-ray photoelectron spectroscopy analysis," *Carbon*, vol. 134, p. 29, 2018.
[47] B. Patel, A. C. Duran-Martinez, P. Gurman, et al., "Ultrananocrystalline diamond coatings for the dental implant: electro-chemical nature," *Surf. Innov.*, vol. 5 (2), p. 106, 2017.
[48] M. B. Guglielmotti, S. J. Renou, and R. L. Cabrini, "A histomorphometric study of tissue interface by laminar implant test in rats," *Int. J. Oral Maxillofac. Implants*, vol. 14, p. 565, 1999.
[49] M. B. Guglielmotti, S. J. Renou, and R. L. Cabrini, "Evaluation of bone tissue on metallic implants by energy dispersive X-ray analysis (EDX): an experimental study," *Implant Dent.*, vol. 8, p. 303, 1999.
[50] E. Orsini, G. Giavaresi, A. Trirã, V. Ottani, and S. Salgarello, "Dental implant thread pitch and its influence on the osseointegration process: an in vivo comparison study," *Int. J. Oral Maxillofac. Implants*, vol. 27 (2), p. 383, 2012.
[51] S. A. Hacking P. Boyraz, B. M. Powers, et al., "Surface roughness enhances the osseointegration of titanium head-posts in non-human primates," *J. Neurosci. Methods*, vol. 211 (2), p. 237, 2012.
[52] A. F. V. Recum, C. E. Shannon, C. E. Cannon, et al., "Surface roughness, porosity, and texture as modifiers of cellular adhesion," *Tissue Eng.*, vol. 2 (4), p. 241, 1996.
[53] D. R. Tasat, M. E. Bruno, M. Domingo, et al., "Biokinetics and tissue response to ultrananocrystalline diamond nanoparticles employed as coating for biomedical devices," *J. Biomed. Mater. Rest Part B*, vol. 105, p. 2408, 2016.
[54] A. Palmquist, O. M. Omar, M. Esposito, J. Lausmaa, and P. Thomsen, "Titanium oral implants: surface characteristics, interface biology and clinical outcomes," *J. Res. Soc. Interface*, vol. 7, p. S515, 2010.
[55] M. E. Bruno, D. R. Tasat, E. Ramos, et al., "Histo-pathological parameters: impact through time of different sized titanium dioxide particles," *J. Biomed. Mater. Res.*, vol. A102 (5), p. 1439, 2014.
[56] X. Xiao, J. Wang, J. A. Carlisle, et al. "In vitro and in vivo evaluation of ultrananocrystalline diamond for coating of implantable retinal microchips," *J. Biomed. Mater.*, vol. 77B (2), p. 273, 2006.
[57] G. Oberdörster, E. Oberdörster, and J. Oberdörster, "An emerging discipline evolving from studies of ultrafine particles," *J. Nanotoxicol*, vol. 113, p. 823, 2005.
[58] J. Qu, P. J. Blau, T. R. Watkins, O. B. Cavin, and N. S. Kulkarni, "Friction and wear of titanium alloys sliding against metal, polymer ceramic counter-faces," *Wear*, vol. 258, p. 1348, 2005.
[59] Advanced Diamond Technologies. Homepage. www.thindiamond.com.
[60] A. V. Sumant, D. S. Grierson, J. E. Gerbi, et al., "Toward the ultimate tribological interface: surface chemistry and nanotribology of ultrananocrystalline diamond," *Adv. Mater.*, vol. 17, p. 1039, 2005.
[61] A. R. Konicek, D. S. Grierson, P. U. P. A. Gilbert, et al., "Origin of ultralow friction and wear in ultrananocrystalline diamond," *Phys. Rev. Lett.*, vol. 100 (23), p. 235502, 2008.
[62] A. V. Sumant, A. R. Krauss, D. M. Gruen, O. Auciello et al., "Ultrananocrystalline diamond film as a wear resistant and protective coating for mechanical seal applications," *Tribol. Trans.*, vol. 48, p. 24, 2005.

6 Science and Technology of Novel Ultrananocrystalline Diamond (UNCD™) Scaffolds for Stem Cell Growth and Differentiation for Developmental Biology and Biological Treatment of Human Medical Conditions

Bing Shi and Orlando Auciello

6.1 Introduction

Biomaterials are being investigated to produce platforms as scaffolds for cell and tissue growth and regeneration. Investigation of cell-material chemical and biological interactions enable the application of more functional materials in the area of bioengineering, which provides a pathway to novel treatment of patients who suffer from tissue or organ damage and face the limitation of donation sources. Many studies have been done on tissue and organ regeneration. Development of new substrate materials as platforms for cell/tissue regeneration is a key research area. This chapter focuses on describing R&D on the novel ultrananocrystalline diamond (UNCD) film, addressed in this book, as the main biological/biocompatible material for biomedical and biological technological applications, as a unique biomaterial for scaffolds for developmental biology. Recent research showed that cell growth on UNCD-coated culture dishes is similar to cell culture dishes, with little retardation, indicating that UNCD films have no or little inhibition of cell proliferation and are potentially appealing as substrate/scaffold materials. Different types of cell growth on UNCD films and cell growth on different types of UNCD films have been and continue to be studied and compared. The mechanisms of cell adhesion on UNCD surfaces are proposed based on the experimental results. The comparisons of cell cultures on diamond powder–seeded culture dishes and on UNCD-coated dishes with matrix-assisted laser desorption/ionization time-of-flight mass spectroscopy (MALDI-TOF MS) and X-ray photoelectron spectroscopy (XPS) analyses provide valuable data to support the mechanisms proposed to explain the adhesion and proliferation of cells on the surface of the UNCD platform.

Arthritis is a problem that affects over 50 million people in the USA. If the pain persists and exercises, injections, pain killers, and other treatments are not working, an artificial joint needs to be implanted to replace the natural joint. Approximately 77,000 total joint replacement surgeries were performed in 2007 and almost one million surgeries were done in 2011 in the USA [1], and this number is still increasing. Current implantable prostheses made of materials, such as Ti alloys and ultrahigh molecular weight polyethylene for artificial joints exhibit surface degradation, wear, biocompatibility issues, and corrosion, all of which contribute to shortening the lifetime of the device in the human body [2]. Well-designed artificial joints have a lifetime of ~20 years. A second joint replacement is required once the first artificial joint fails. Patients might receive a total of two joint replacements in their lifetime. Introducing materials with better physical, chemical, biological, and mechanical properties will improve the quality of implantable devices.

Tissue engineering is the replacement of living tissues with tissue that is designed and constructed to meet the needs of the individual patient [3, 4]. It integrates the fields of biomaterials, cell biology, biochemistry, and clinical medicine. The goal is to create substitutes that functionally repair damaged or diseased bones, tissues, and organs. Although the traditional entire joint implantation will dominate treatment for a while, partial replacement relying on technologies of tissue engineering will be more beneficial to patients. Indeed, replacement is only needed for portions that are damaged, degenerated, or diseased.

Studies are being performed to achieve regeneration of tissues and organs to repair the diseased or damaged ones in human bodies, for which limited supplies exist from human donors [5–7]. The research involves novel approaches using tissue and organ engineering therapies, an interdisciplinary field that applies the principles of engineering and the life sciences toward the development of biological substitutes that restore, maintain, or improve tissue function, prompting the body to regenerate autonomously by growing cells on supporting biomaterials to engineer tissues/organs. Human embryonic stem cells have an almost unlimited developmental potential of providing a limitless amount of tissue for transplantation therapies to treat a wide range of degenerative diseases. The development of new materials that can promote the growth, division, and differentiation of the remaining healthy cells from the patient or stem cells, together with the discovery of growth factors that foster cell proliferation *in vitro*, can provide the basis for advanced regenerative medicine that can restore functionality to diseased or damaged organs or tissues. It is one of the main areas in the study of regenerative medicine.

Investigation of biomaterials as artificial platforms for cell growth and differentiation is one of the most important areas in regenerative medicine. Results from previous research suggest that it is possible to engineer growing tissue by presenting appropriate growth stimuli from the cell transplantation scaffolds, and it is critical to promote the multiplication of transplanted cells if one is to engineer tissue growth *in vivo*. Large-scale cell culture systems will be important to grow sufficient cells *in vitro* [8, 9]. Therefore, materials used as adhesion substrates are a major area of study. Biomaterials used in tissue engineering need to meet the nutritional and biological needs of the cells. They must have good mechanical characteristics and good geometries so as to maintain the structure during new tissue formation.

Diamond biocompatibility has been known about for some years, since the twentieth century. However, it was not until 1991 that a diamond-like carbon was used as a substrate for *in-vitro* culture of mammalian cells (mouse fibroblasts and macrophages) [10]. On the other hand, during the last two decades, many types of adherent cells have been successfully cultured on microcrystalline diamond surfaces. Cultured cells include mesenchymal stem cells [11, 12], dental stem cells [13], neuronal stem cells [14–17], human induced pluripotent stem cell (IPS)-derived neuronal progenitors [18], various types of neurons [19–26], osteoblasts [27–34], fibroblasts [24, 35, 36], macrophages [30, 37, 38], and epithelial cells [39, 40]. While there is no question about the value of diamond as a culture substrate for adherent cells (in fact there is no cell type grown on any other material that cannot be cultured on diamond as well), some controversies have arisen regarding the optimal culture conditions on diamond. In relation to this scientific and technological topic, this chapter is focused on a discussion of key issues related to the various processes related to the growth of biological cells on a novel diamond surface developed by O. Auciello and colleagues, based on UNCD films, described in detail in a recent review [41].

A promising diamond platform for implantable biomedical devices is based on UNCD film. The UNCD thin-film technology is extensively patented (17 patents), of which three key patents can be seen in [42–45]. Also, the UNCD-coating technology is on the market in UNCD-coated mechanical pump seals and bearings for applications in the petrochemical and other non-medical industries, and in the pharmaceutical industry in bearings for mixers for preparing medical drugs (marketed by Advanced Diamond Technologies, a company co-founded by O. Auciello and J. A. Carlisle in 2003, profitable in 2014, sold to large company in 2019 [www.thindiamond.com]).

UNCD films are grown using Ar-rich (Ar [99%]/CH_4 [1%]) plasmas using microwave plasma-enhanced chemical vapor deposition (MPCVD). UNCD films exhibit a unique nanostructure characterized by 3–5 nm grains and 0.4 nm wide grain boundaries, which yield a combination of low friction and high wear resistance, good hardness and fracture strength, good electrical transport properties, and high electric field-induced electron emission [46–49]. A detailed description of the UNCD film growth process can be seen in Chapter 1. The UNCD films, which exhibit properties similar to those of single crystal diamond, are expected to be the top candidates as coatings for implantable biomedical devices that increase the devices' service times, as well as for scaffolds to proliferate cell growth and differentiation for developmental biology.

A review of the pioneering studies on UNCD thin films as the appropriate substrate materials with tailored surfaces to provide the stimulus needed for growth of most mammalian cells is presented in this chapter. Cell culture was performed to investigate cell adhesion and proliferation on the UNCD surface. The results showed that UNCD thin films provide a good platform for cells to attach and grow, and eventually differentiate under appropriate stimulus. Analysis of the UNCD film surfaces provides the information to determine the possible mechanisms of cell adhesion. Studies have indicated that UNCD films are excellent coating candidates for implantable medical devices for humans and also could be applied as substrate materials for tissue engineering.

6.2 UNCD Thin-Film Growth

6.2.1 Substrates

For the R&D discussed in this chapter, UNCD films were grown on Si (100) substrates, quartz watch glasses, and quartz dishes. Si substrates provide a suitable platform for the growth of UNCD thin films [41]. However, for the research of cells growth, Si is not an appropriate substrate because it is not transparent, which is require to see the cells though optical microscopy. Therefore, if Si is used as a substrate for growing UNCD films for cell growth, the cell needs to be fixed for studies using reflection microscopy.

In order to perform regular observations during the cell culture process, a transparent substrate – quartz – was used. Transparent UNCD films were grown on the surface of the quartz substrates for this purpose. Quartz dishes were specially designed and customized. The quartz dishes were specially designed to balance the critical requirements for both UNCD thin-film growth and cell culture. Compared to the cell culture dishes, these dishes were designed with shorter edges and angles wider than 90 degrees, allowing growth of the UNCD films fully inside the dishes, with a spherical shape imitating the 3D culture procedure.

The scenario of bioengineered cells and tissues would involve implantation of cells or tissue cultured on freestanding UNCD films, possibly adhered to flexible polymer surfaces to compensate for UNCD film brittleness. Therefore, cell cultures on free-standing UNCD film were also investigated.

Freestanding UNCD films were produced by growing UNCD films on Si substrates followed by etching of the Si substrate in a mixed solution of nitric acid (HNO_3), acetic acid (CH_3COOH), and hydrofluoric acids (HF). The etching process was started by oxidizing the Si surface by exposure to nitric acid to form SiO_2, which was subsequently removed by hydrofluoric acid. The process is described in the formulas presented below. First, Si was oxidized to form SiO_2 on the surface following the chemical reaction

$$Si + 4HNO_3 + 6HF \rightarrow H_2 + SiF_6 + 4NO_2 + 4H_2O,$$

$$Si + 4HNO_3 \rightarrow SiO_2 + 2H_2O + 4NO_2.$$

Second, the SiO_2 layer was removed by immersion in HF, following the chemical reaction

$$SiO_2 + 6HF \rightarrow H_2 + SiF_6 + 2H_2O.$$

Acetic acid diluted the etching solution since both nitric acid and hydrofluoric acid were concentrated. Acetic acid was added instead of water in the solution to prevent nitric acid dissociation. These two reactions were repeated sequentially until the whole substrate was etched. The freestanding UNCD films obtained after etching were then rinsed with deionized water.

6.2.2 Substrate Surface Seeding with Diamond Nanoparticles to Grow UNCD Films

The substrates used to grow UNCD films were first seeded with ultrananocrystalline diamond powder dissolved in methanol in an ultrasonic bath. The seeding procedure resulted in nanodiamond nanoparticles (seeds) being embedded on the substrate surfaces via sound waves shaking the nanoparticles over the substrate surface. The diamond seeds induce the nucleation and subsequent growth of UNCD films. Polishing seeding plus an ultrasonic seeding procedure introduced a higher density of seeds on the surface compared with using one seeding method alone. Scanning electronic microscopy (SEM) images (Figure 6.1) show a layer of nanodiamond powder on the surface of the Si substrate after seeding. Scratches are observed on the surface of the Si from the seeding procedure.

6.2.3 The MPCVD Process to Grow UNCD Films

UNCD films synthesized by MPCVD were produced via cracking and ionizing methane molecules plus argon-induced collision dissociation of C_2H_2 species with the formation of C_2 dimers in the plasma. Subsequent insertion of C_2 precursors (for

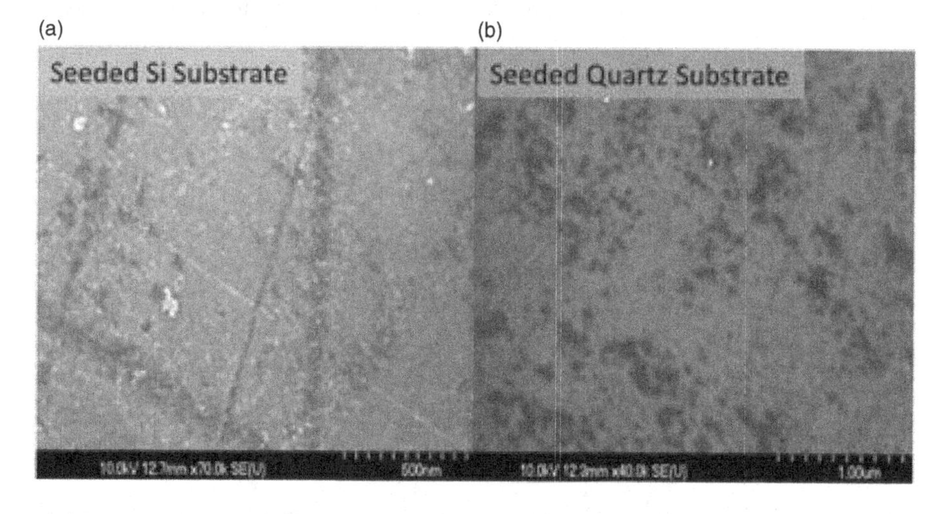

Figure 6.1 SEM images of Si (a) and quartz (b) substrate surfaces seeded with nanodiamond particles via immersion in an ultrasonic bath with a methanol solution of nanodiamond particles (2–5 nm), which are embedded on the substrate surface via shaking by the sound waves (reprinted from *J. Funct. Biomed. Mater.*, vol. 3, p. 588, 2012 (Fig. 1) in [12] with permission from MDPI Publisher).

nucleation) and C_2 and CH_3 radicals into the substrate leads to UNCD film growth via the formation of sp^3 (diamond bonds) carbon atom bonding in the grains and a mixture of sp^2 and sp^3 carbon atom bonding in the grain boundaries. In this plasma chemistry, a small amount of CH_4 (typically 1%) produces carbon dimers, C_2 and CH_x ($x = 1, 2, 3$) radicals, which are derived from methane via molecule cracking [41–48].

The UNCD films are smooth, dense, pinhole free, and phase-pure, with grain size of 2–5 nm [41]. Different UNCD films were synthesized on Si and quartz plates that were heated to 800 °C during the film growth process. Undoped UNCD films were grown using an Ar (99%)/CH_4 (1%) gas mixture; H surface–terminated UNCD films were grown using an Ar (97%)/CH_4 (1%)/H_2 (2%) gas mixture, and electrically conductive nitrogen (N)-grain boundaries-incorporated N-UNCD thin films were grown using an Ar/CH_4/N_2 gas mixture with N_2 (10% and 20%) [41, 44]. Other types of UNCD films used in the research discussed in this chapter include UNCD films exposed to H-plasmas to get H-terminated films: H-plasma-treated UNCD films and H-plasma-treated N-UNCD thin films.

6.3 Characterization of UNCD Films

6.3.1 SEM Imaging

Comparison of SEM images of the top view of UNCD thin films on Si and quartz substrates are shown in Figure 6.2a and b, respectively. A cross-section SEM image of the UNCD film is shown in Figure 6.3. The films exhibit slight thickness variation from the center (thicker) of the 100 mm diameter Si substrate to the edge (thinner) due to the plasma geometry: the shape of the plasma is round, and the higher density is in the center of the plasma.

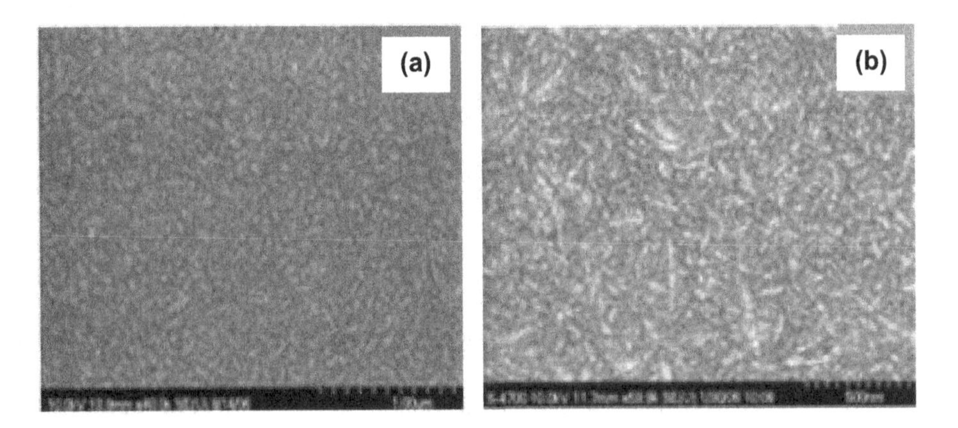

Figure 6.2 SEM images: (a) UNCD film on an Si substrate; (b) UNCD film on a quartz substrate (reprinted from *Diam. Relat. Mater.*, vol. 18, p. 596, 2009 (Figs. 1 (a) and 1 (c) in [50] respectively), with permission from Elsevier Publisher).

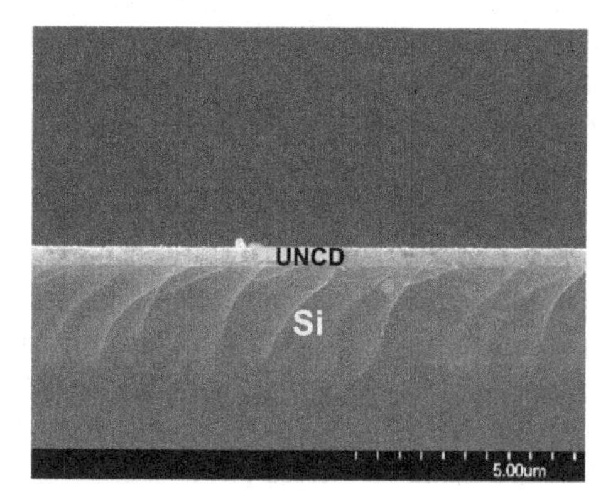

Figure 6.3 Cross-section of UNCD film grown on an Si substrate.

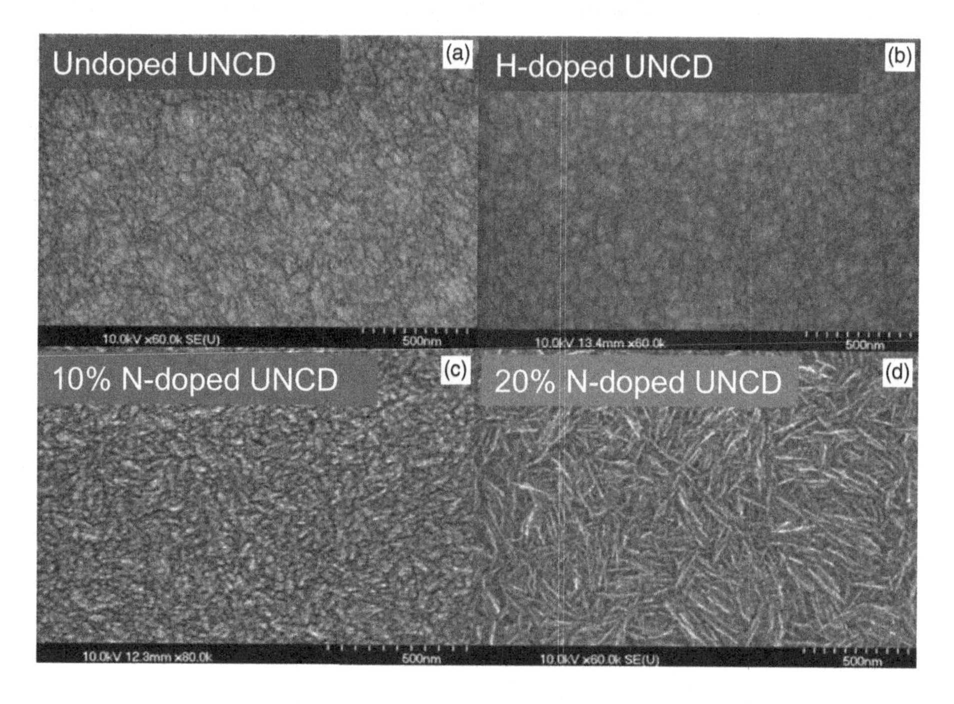

Figure 6.4 SEM images of (a) plain UNCD film, (b) 1% H-UNCD film [50], (c) 10% N_2 and (d) 20% N_2 N-UNCD films (reprinted from *J. Funct. Biomed. Mater.*, vol. 3, p. 588, 2012 (Fig. 2) in [12] and [50] with permission from Elsevier Publisher).

SEM images comparing a plain UNCD film, a H-surface-terminated UNCD film (grown with 1% H_2), an N-UNCD film (grown with 10% N_2), and an N-UNCD 20% film (grown with 20% N_2) are shown in Figure 6.4.

Compared with plain UNCD thin films, 1% H-surface-terminated UNCD thin films are less conductive and showed more charging during the SEM analysis [46]. This effect is because the hydrogen atoms were inserted into the grain boundaries, suppressing the formation of sp^2-bonded carbon, and the hydrogen satisfied the dangling bonds and formed H-terminated bonds. N-UNCD thin films exhibit semi-metallic conductivity [47]. Unlike H-surface-terminated UNCD films, which still have similar surface morphology to plain UNCD films, N-UNCD films showed a more needle-like pattern on the surface (Figure 6.4c,d). The larger the N incorporation percentage in the gas mixture, the longer the needles. However, atomic force microscopy (AFM) and transmission electron microscopy (TEM) studies showed that N-UNCD films have nanograins similar to plain UNCD films. The UNCD and N-UNCD films were exposed to H-plasma to achieve H-terminated UNCD and N-UNCD films. H-plasma treatment of UNCD films resulted in H atoms bonding on the surface to produce a H-terminated diamond surface [48].

6.3.2 Raman Analysis of UNCD Films

Raman spectra from Raman analysis of UNCD films grown on Si substrates (performed with a Renishaw Raman spectrometer) and quartz dish substrates, and from analysis of UNCD films on quartz watch glass (performed with surface-enhanced Raman spectroscopy) are shown in Figure 6.5. The watch glasses have special geometry and the surface-enhanced Raman spectroscope with an He–Ne laser at a wavelength of 632 nm and laser spot size of 6 μm in diameter and a power of 25 mW was used. Raman scattering in the visible range at 632.8 nm is 50–250 times more sensitive to sp^2-bonded carbon than to the sp^3-bonded carbon characteristic of diamond [41]. The results are similar to the characteristics of nanocrystalline diamond films at high temperature grown on Si substrates [41]. The broad D band peak around 1332 cm^{-1} and the G band peak at 1556 cm^{-1} correlate with sp^2-bonded carbon atoms in the grain boundaries. In addition, hidden under the broad peak around 1332 cm^{-1} peak, the 1332 cm^{-1} peak correlates with sp^3 diamond bond of C atoms in the UNCD grans. in addition, the peak at 1140 cm^{-1} is attributed to nanocrystalline diamond [41, 49].

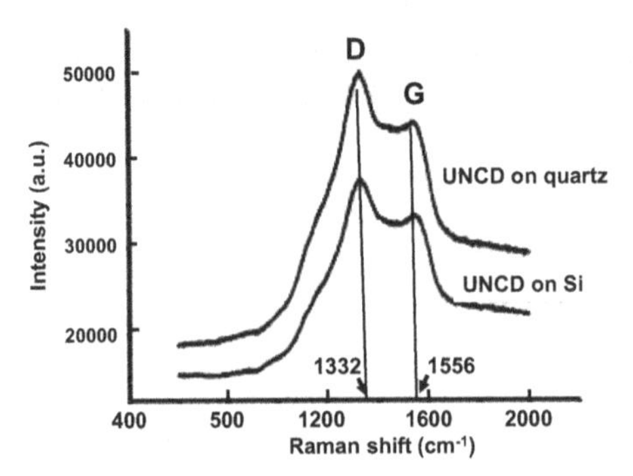

Figure 6.5 Raman spectra of UNCD thin films on Si and quartz substrates (reprinted from *Diam. Relat. Mater.*, vol. 18, p. 596, 2009 (Fig. 2) in [41] with permission from Elsevier Publisher)

6.3.3 Wettability Study of UNCD Films

The wettability of the surface of UNCD thin films was investigated using a contact angle goniometer to measure the water drop contact angle to the surface of the UNCD. The measurements of the contact angle were performed on different UNCD-coated Si wafers with different growth conditions. The film thickness along the wafers' center lines was measured using cross-section SEM. The SEM images showed similar trends for undoped, H-UNCD, and N-UNCD thin films, with the films slightly thicker at the center than at the edges. The contact angles were measured at positions on the center, the edge, and in between. Images of water drop contact angles on UNCD thin films grown for 1 h are shown in Figure 6.6 for the contact angle at the center, middle, and edge. Figure 6.7 shows contact angle values at different positions for plain UNCD, H-UNCD, and N-UNCD thin films. The edge area has a very thin UNCD film layer and the diamond seeds on the Si wafer could still be observed under the optical microscope. Although the contact angles are different for the plain, H-UNCD, and N-UNCD thin films, the contact angles at the edge area of different UNCD films are similar, since the contact angle corresponds mainly to the interaction of the water drop with the seeded Si substrates exposed to the discontinuous UNCD film.

6.4 Cell Culture Studies on UNCD Film Surfaces

Si, quartz dishes, UNCD-coated quartz dishes, and freestanding UNCD films were sterilized with ethanol solution and UV lights in cell culture dishes. After being grown

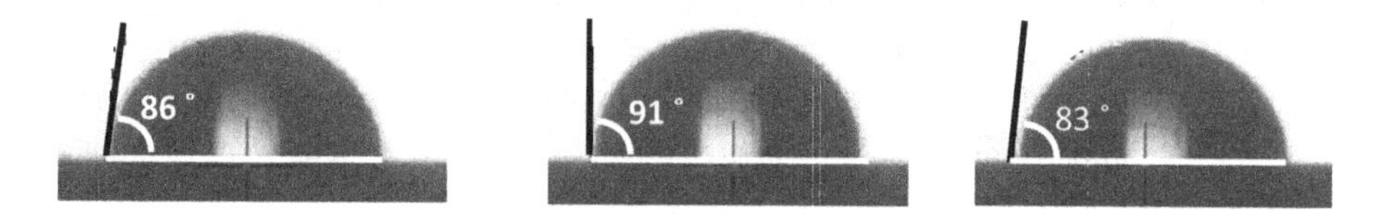

Figure 6.6 Contact angle images in the center area of films for plain UNCD (left), H-UNCD (middle), and N-UNCD (right).

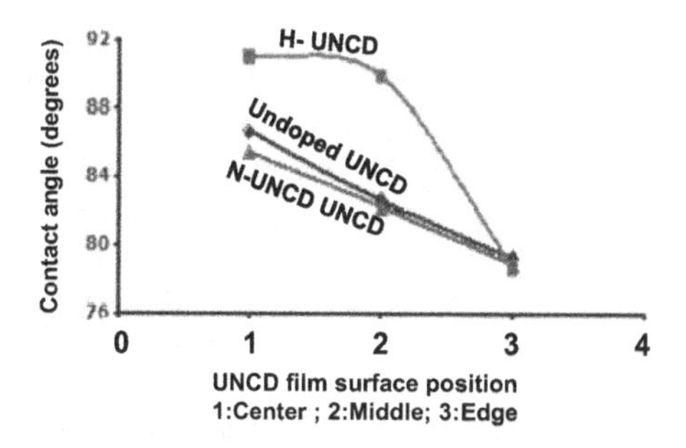

Figure 6.7 Water drop contact angle values for plain UNCD, H-UNCD, and N-UNCD films.

in normal culture flasks, mouse embryonic fibroblasts (MEFs; Chemicon International) were dissociated by trypsin/EDTA and plated onto the cell culture vessels listed above, plus control dishes, at the same starting density of 8.0×10^3 cells/cm^2 and incubated at 37 °C in a humidified atmosphere containing 5% CO_2. Cell growth was monitored daily for four days under optical microscopy for transparent samples. For opaque substrate samples such as the UNCD films grown on Si, the cells were fixed on the substrates at day 4 and observed using an environmental scanning electron microscope (ESEM). The overall cell growth was evaluated by total cell counts at day 4.

6.4.1 Cells Culture on UNCD with Different Substrates

6.4.1.1 Cell Culture on UNCD-Coated Si Substrates

The ESEM technique was used to observe the cell growth on UNCD-coated Si substrates because the substrates are opaque and could not be observed directly using the inverted microscope. Cells were grown for 96 h before they were fixed in 4% paraformaldehyde and preserved in the phosphate buffered saline (PBS) buffer. They were then dried with nitrogen gas and analyzed with ESEM, which showed that cells had grown on the UNCD surface (Figure 6.8). Cells were growing, replicating, dividing, and proliferating while they were being fixed, and nuclear division for cell reproduction was observed. These results are compatible with the cell counts on the non-transparent substrates and on the cell culture dish (control), which showed similar cell densities without significant differences.

6.4.1.2 Cell Culture on UNCD-Coated Quartz Substrates

Cells were plated on UNCD-coated quartz dishes, non-coated quartz dishes, and cell culture dishes at the same density with three replicates. Pictures were taken daily; comparisons of the growth are shown in Figure 6.9. Cells were found attached to the

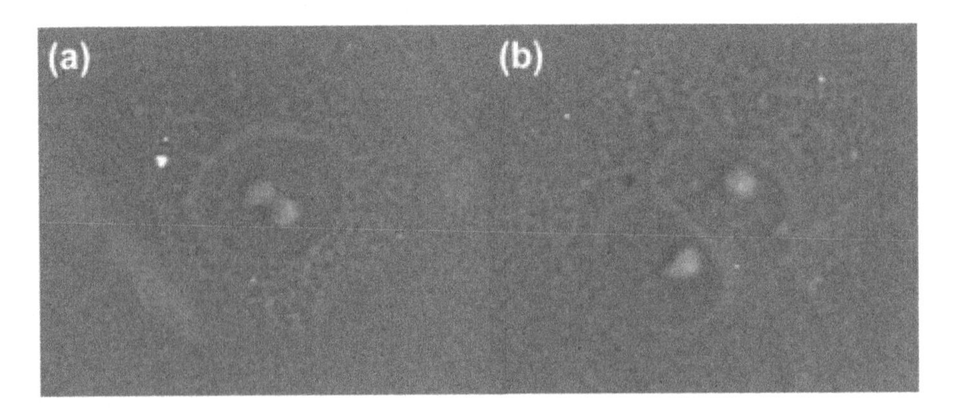

Figure 6.8 ESEM images of MEFs growth for 96 h on UNCD films: (a) replicating nucleus; (b) dividing cells (reprinted from *Diam. Relat. Mater.*, vol. 18, p. 596, 2009 (Fig. 3) in [50] with permission from Elsevier Publisher).

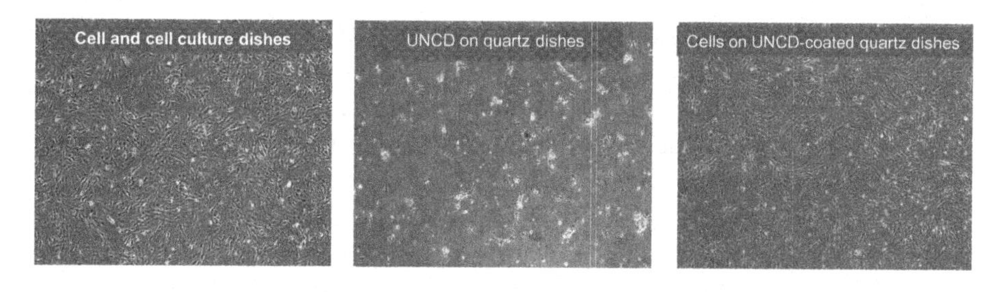

Figure 6.9 Cell culture on culture dish (left), quartz dish (center), and UNCD-coated quartz dish (right) at day 4.

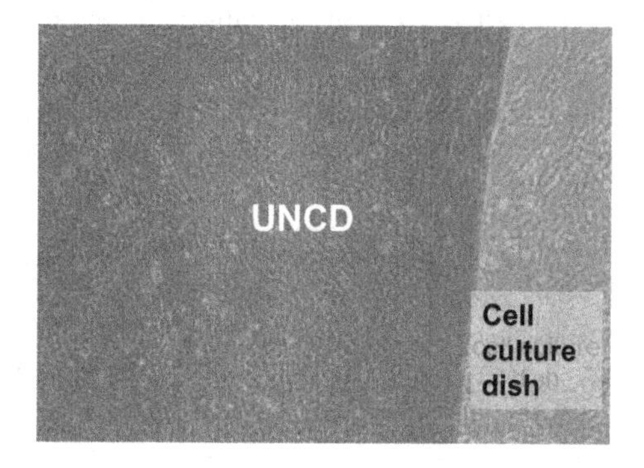

Figure 6.10 MEF growth on freestanding UNCD at 36 h (reprinted from *Diam. Relat. Mater.*, vol. 18, p. 596, 2009 (Fig. 5) in [50] with permission from Elsevier Publisher).

cell culture dishes, the UNCD-coated quartz dishes, and the uncoated quartz dishes at day 1. After day 1, cells on the non-coated quartz dishes started to shrink and aggregate. At day 4, MEFs grown on UNCD-coated quartz dishes reached the same confluence as on the culture dishes, whereas MEFs on non-coated dishes did not show much proliferation. The cells retracted and detached from the dishes and underwent apoptosis. Cell growth densities on UNCD films were on average 2.2×10^4 cells/cm^2, significantly higher than on non-coated substrates (6×10^3 cells/cm^2), and slightly higher than the density on the regular culture dish (2.0×10^4 cells/cm^2). These results indicate that UNCD films were not toxic to the cells.

6.4.1.3 Cell Culture on Freestanding UNCD Films

The MEF growth on freestanding UNCD films was further tested in a culture dish, which allowed for a direct side-by-side comparison. Cell culture on the surface of freestanding UNCD thin films was the closest simulation of the environment of implantable devices. As shown in Figure 6.10 (focused on the cells on the UNCD thin films), the cell growth reached confluence on UNCD as well as on the culture dish. No retardation or reduction of cell growth on UNCD films was observed compared to the culture dish.

6.4.1.4 Cell Culture on Different Types of UNCD Films Grown on Quartz Culture Dishes

Cell growth on UNCD thin-film-coated and uncoated quartz dishes was observed daily. The three repeated samples showed similar results. Except for the quartz dishes that showed cells that were almost totally detached from the dishes, all other substrate surfaces coated with UNCD films showed that cells attached to the surface and proliferated at different levels, but more efficiently than on uncoated dishes.

Cells were attached to the surface of quartz (control) dishes at day 1, but started to shrink and aggregate from day 2. At day 4, most of the cells retracted and detached from the quartz dish surface and underwent apoptosis. On the contrary, quartz dish surfaces coated with nanodiamond powders showed cells attached well to the plate at day 1 (Figure 6.11) and started to shrink and aggregate. The width of the images at 10× is 0.695 mm and the width of the images at 4× is 1.738 mm. At day 4, most of the cells were still attached to the surfaces of quartz dishes coated with nanodiamond powders.

Cell growth on plain UNCD, H-UNCD [51, 52], and N-UNCD (10%), H-plasma-treated N-UNCD film (grown with 10% N_2) films, N-UNCD (20%), and H-plasma-treated N-UNCD film (grown with 20% N_2) coated dishes showed good cell attachment on day 1 and started to proliferate normally. At day 4, cell growth reached confluency and cells were morphologically normal (Figure 6.12).

Cells were dissociated at day 4 and the total numbers of cells from each growth condition were counted. The results are shown in Figure 6.13. Cell culture dishes had

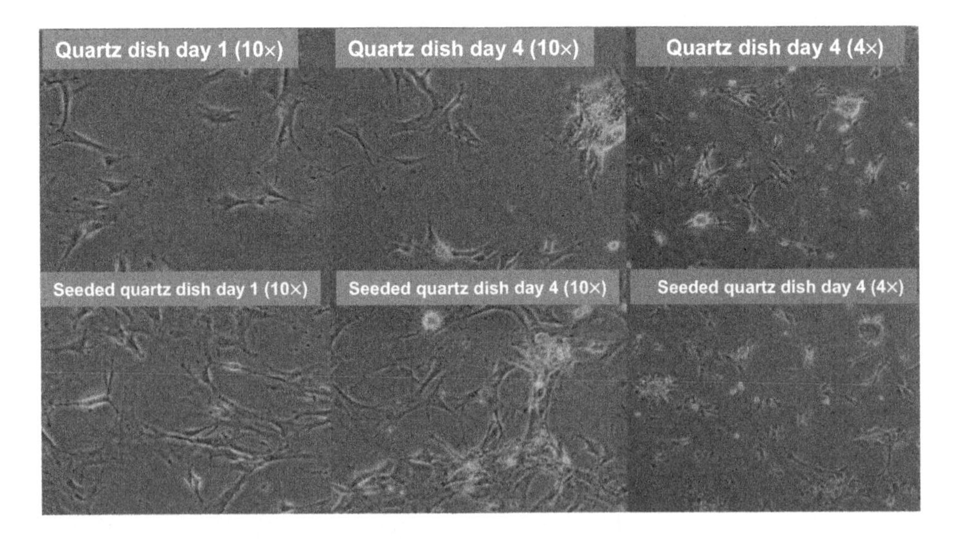

Figure 6.11 MEF growth on quartz dishes and diamond powder–seeded quartz dishes (reprinted from *J. Funct. Biomed. Mater.*, vol. 3, p. 588, 2012 (Fig. 3) in [12] with permission from MDPI Publisher).

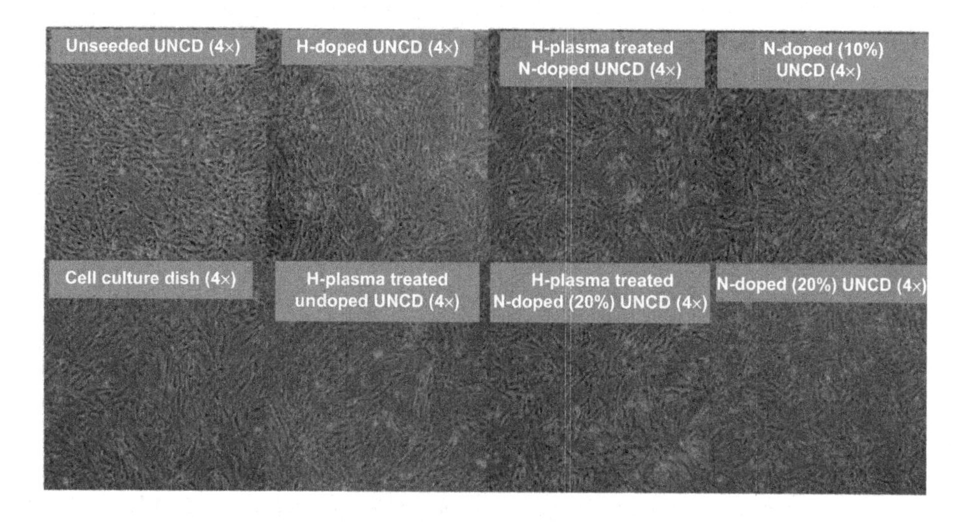

Figure 6.12 Comparison of MEF growths on different UNCD films at day 4 (reprinted from *J. Funct. Biomed. Mater.*, vol. 3, p. 588, 2012 (Fig. 4) in [12] with permission from MDPI Publisher).

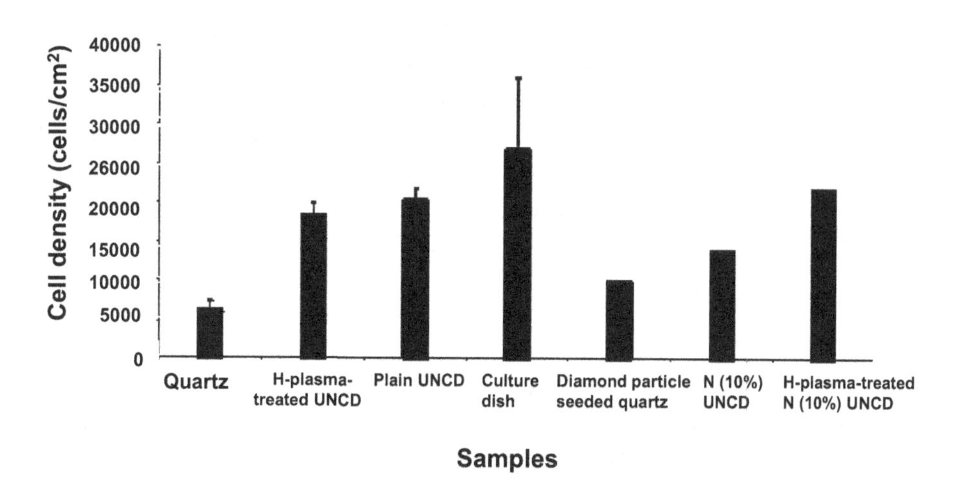

Figure 6.13 Comparison of cell densities on different surfaces at day 4 [12].

the highest densities, yet with the largest variation among three replicates of the cell culture dishes, quartz dishes, undoped UNCD dishes, and the H-plasma-treated undoped UNCD dishes. Quartz dishes had the lowest cell density. Diamond-seeded quartz dishes showed the second lowest densities. The N-UNCD film (grown with 10% and 20% N_2) coated dishes had similar densities of cells. Different types of UNCD-coated dishes had different densities without significant differences [50].

6.5 Studies of Cells' Biological Components by Chemical Analysis Techniques

6.5.1 Matrix-Assisted Laser Desorption/Ionization Time-of-Flight Mass Spectrometry

Matrix-assisted laser desorption/ionization time-of-flight mass spectrometry (MALDI-TOF MS) analysis is one of the best techniques to measure the biological molecules on the surface of materials. Therefore, MALDI-TOF MS analyses were carried out on MEFs and MEFs on a UNCD surface. The results are shown in Figures 6.14 and 6.15. Analysis of the MEFs in Figure 6.14 showed that among the four major phospholipids predominating in the plasma membrane of many mammalian cells – phosphatidyl ethanolamine (PE), phosphatidyl serine (PS), phosphatidyl choline (PC), and sphingomyelin (SM) – PE and PS are negatively charged while PC and SM are positively charged. Among the negatively charged biomolecules on the MEF membrane, phosphatidyl inositol (PI) has the largest intensity compared with PE and PS.

Phosphatidyl inositol is in the outer lipid monolayer of the plasma membrane; PI exposed at the external cell surface attaches only by a covalent linkage via a specific oligosaccharide to phosphatidylinositol. It plays a crucial role in cell signaling. Of the positively charged biomolecules, both PC and SM have a large intensity. There are many other biomolecules that are also positively charged. Figure 6.15 shows the results of MEFs on UNCD. While the spectra of the blank UNCD film and the UNCD film exposed to the buffer solution are fairly similar, the UNCD film with the MEFs shows a series of strong peaks that should be interpreted as lipid ion peaks. These lipids most likely belong to the cell membranes. No high-mass proteins were detected due to the low amount of these species exposed to desorption and ionization by MALDI.

6.5.2 Matrix-Assisted Laser XPS

Similar results from XPS analysis are shown in Figure 6.16. While UNCD showed only carbon peaks, both nitrogen peaks and oxygen peaks were detected after MEFs grew on UNCD. These peaks provide evidence of elements in the membranes of the cells.

6.6 Discussion of Cell Attachment and Proliferation Mechanism on the Surface of UNCD Films

The results described above suggest that the surface of UNCD films can indeed promote cell adhesion, thus showing biocompatibility with embryonic fibroblasts. All of the different types of UNCD thin films investigated could support cell

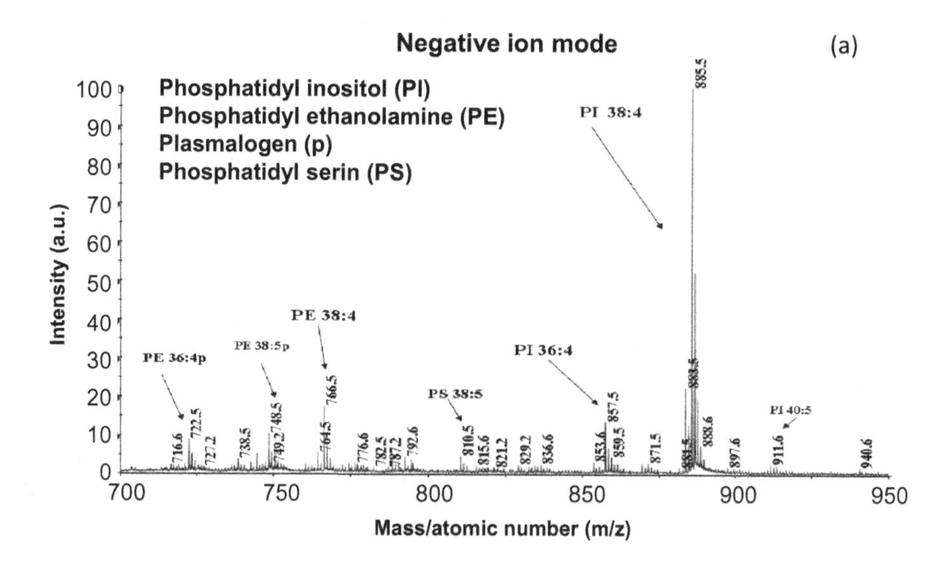

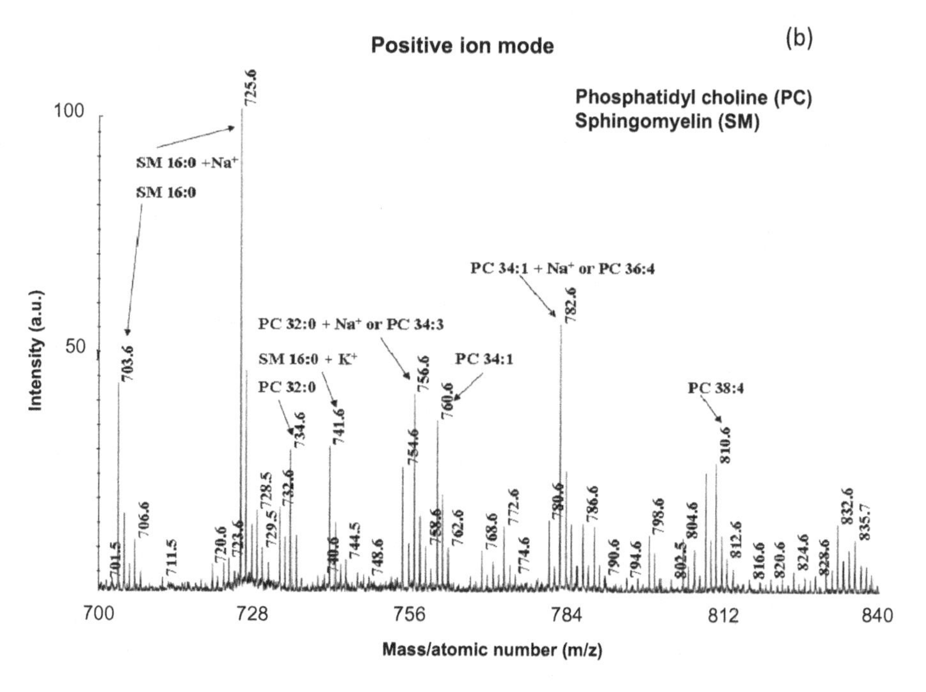

Figure 6.14 (a) MALDI analysis on MEFs – negative ion mode analysis (reprinted from *J. Funct. Biomed. Mater.*, vol. 3, p. 588, 2012 (Fig. 6 (a)) in [12] with permission from MDPI Publisher). (b) MALDI analysis on MEFs – positive ion mode analysis (reprinted from *J. Funct. Biomed. Mater.*, vol. 3, p. 588, 2012 (Fig. 6 (b)) in [12] with permission from MDPI Publisher).

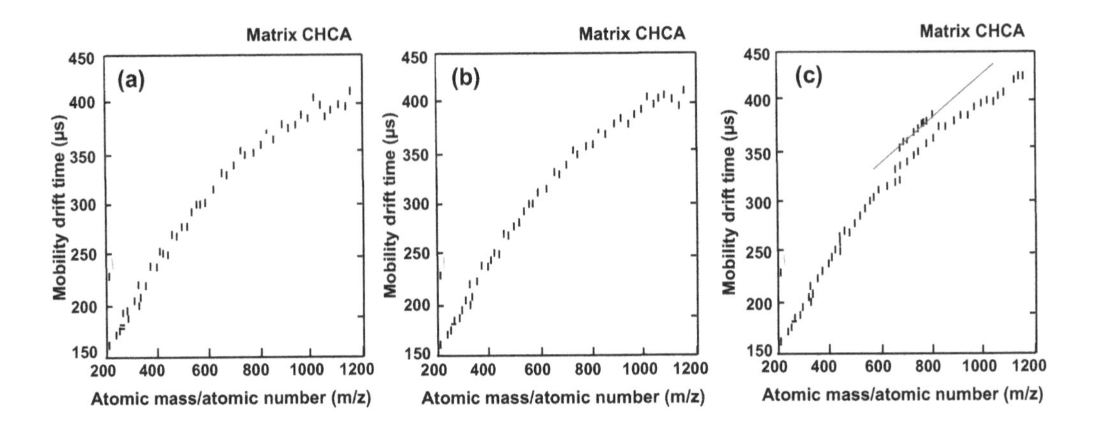

Figure 6.15 MALDI analysis of MEFs on UNCD. (a) UNCD; (b) UNCD in buffer; (c) MEFs on UNCD in buffer (reprinted from *J. Funct. Biomed. Mater.*, vol. 3, p. 588, 2012 (Fig. 7) in [12] with permission from MDPI Publisher).

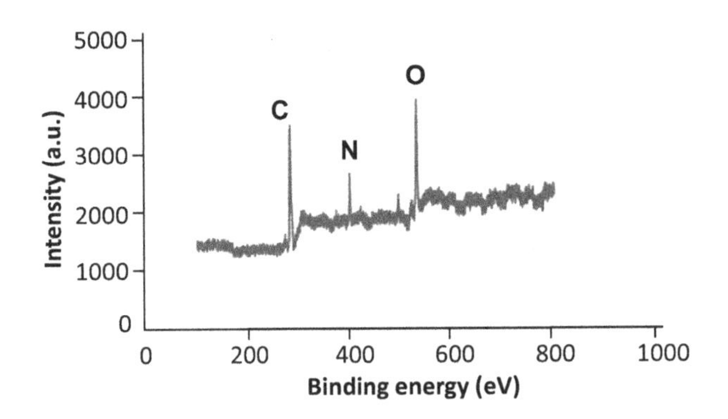

Figure 6.16 XPS analysis of cells on the UNCD film surface (reprinted from *J. Funct. Biomed. Mater.*, vol. 3, p. 588, 2012 (Fig. 8) in [12] with permission from MDPI Publisher).

attachment and proliferation. Raman spectra of UNCD revealed the presence of both sp^2 and sp^3 carbon bonds. The particular nanostructure of UNCD films results in a high density of grain boundaries, which may play a critical role in promoting cell growth. The cell culture dishes made of polystyrene were specially treated to make the surface hydrophilic and negatively charged. Protein-coated culture dishes are also required for some of the cell culture studies. It has been established that adherence between cells and substrate surface is mediated by weak chemical bonds such as hydrogen bonds, polar interactions, and electrostatic interactions between receptor molecules on the cell membrane and the chemical functional groups of the artificial substrates. If cells cannot deposit their own extracellular matrix (ECM) on the substrate surface within a short period of time (24–48 h) they undergo apoptosis. In this respect, it is expected that, similar to the traditional cell culture dish in which the

Pauling's resonance of the benzene rings from polystyrene make the polystyrene electron-negatively charged, the sp^2 bonds along the nanocrystalline boundaries also make the UNCD surface electron-negatively charged, but on a much larger scale (an unknown amount of sp^2 bonds around 2–5 nm UNCD grains vs. 6 sp^2 bonds around 280 pm for benzene).

This hypothesis is further supported by the cell growth on diamond-seeded quartz dishes. Cells attached to the dishes at day 1. Small amounts of cells started to detach and aggregate, which led to the death of these cells. Other cells proliferated on the seeded quartz surface, which suggests the diamond seeds provided the environment that cells could live on. But the cells on seeded quartz dishes did not reach the same confluency as on the other UNCD thin-film-coated dishes. Diamond-seeded quartz dishes could be described as non-continuous UNCD grains. Cells that attached and proliferated on the seeded quartz showed that the grain boundaries did provide the binding sites for cells. Cells partially detached and aggregated on these seeded dishes because the seeds were not as continuous as the UNCD thin films, and therefore the cells were exposed to the quartz surface, which led to the death of these cells.

An additional factor that might contribute to the cell growth on UNCD is the surface chemistry of UNCD, which provides binding sites for cell growth. As-grown UNCD films exhibit dangling bonds at the grain boundaries, which are unsatisfied. Once the chamber is vented to atmosphere, the dangling bonds on the surface of the films can react with oxygen in the air and form C–O bonds. Also, the freshly synthesized UNCD surfaces absorb water molecules in the air, leading to the formation of carboxy groups that may provide the binding sites for the cells. These carboxy groups were proved by XPS analysis on UNCD, which had the O1s peak around 530 eV. The ECM contains a variety of versatile proteins and polysaccharides that could anchor on top of the UNCD. Once the ECM is anchored on the UNCD surface, the cell–cell adhesion on the matrix can change from non-receptor-mediated to receptor-mediated adhesion.

MALDI analysis of the membrane of the MEFs and the analysis of MEFs on UNCD gave further evidence in support of the mechanisms described here. A schematic of the cells' adhesion mechanism on UNCD is presented in Figure 6.17.

The surface chemistry of UNCD films plays a critical role, as demonstrated by the comparison of the cell growth on UNCD film-coated dishes. Among the UNCD film-coated dishes, the N-UNCD showed the lowest cell density. H-plasma-treated N-UNCD-coated dishes showed the largest cell density among the UNCD-coated dishes. As mentioned earlier, H-plasma treatment generated the H-terminated surface, which provided the most inertness on the plain UNCD surface. On the N-UNCD, H-plasma treatment also generated a H-terminated surface, and therefore, besides C–H bonds, other bonds such as N–H, and NH_2 could also be formed on the surface, providing more binding sites for cell attachment and resulting in a higher cell density. Plain untreated UNCD films grown on quartz dishes showed the second largest density. The H-plasma-treated H-UNCD-coated dishes showed a lower density compared with the plain UNCD surface.

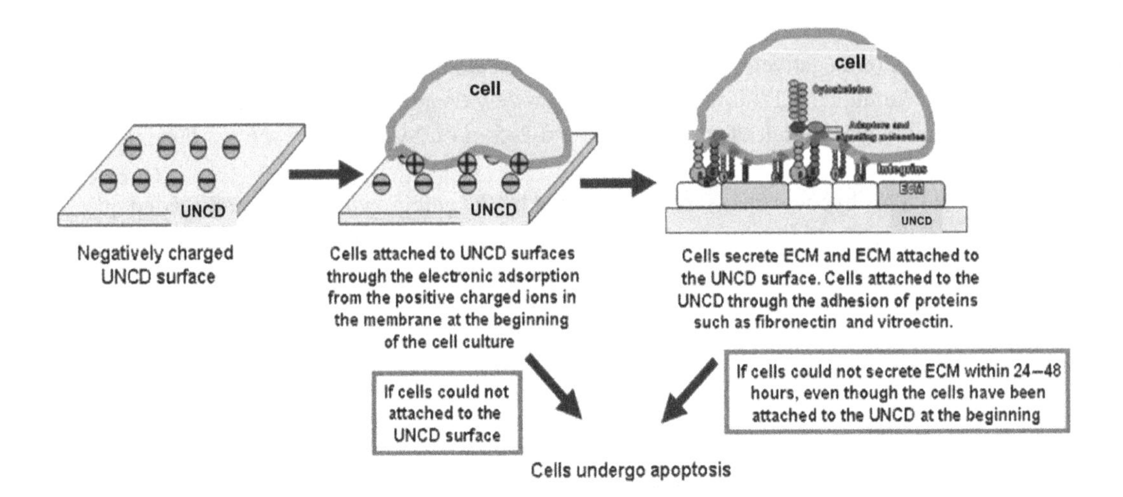

Figure 6.17 Schematic models of cell adhesion mechanisms to UNCD (reprinted from *J. Funct. Biomed. Mater.*, vol. 3, p. 588, 2012 (Fig. 9) in [12] with permission from MDPI Publisher).

6.7 Conclusions

Untreated, H-UNCD, and N-UNCD films were synthesized and H-plasma treatment was introduced to these films. MEF cell growth occurred on these different UNCD-coated dishes, as well as on diamond-seeded quartz dishes. Cell growth on UNCD-coated quartz dishes showed little retardation and reached the same confluence as cell growth on the culture dish. Cell growth on the surface of freestanding UNCD did not show any retardation compared with the cell growth in the cell culture dish. The results showed all these different UNCD film-coated dishes could support cell adhesion and proliferation. The H-plasma-treated N-UNCD-coated dishes showed the largest cell density among all UNCD-coated dishes, because the surface introduced the NH_2 group besides the unique microstructure that could promote cell attachment and growth. The results were analyzed and the mechanisms of cell adhesion on the UNCD surfaces were proposed. The unique nanostructure of UNCD – nano-sized grains and the sp^2/sp^3 mixture bonding along the grain boundaries – and the surface chemistry – remaining dangling bonds and binding of oxygen on the surface – play critical roles in cell adhesion via chemical bonds such as hydrogen bonds, polar interactions, and electrostatic interactions between receptor molecules on the cell membrane and the chemical functional groups of the artificial substrates. All the results in this study showed that UNCD films did not contribute any toxicity to MEF growth, which justified further experiments in cell culture cytotoxicity biocompatibility tests, such as agar diffusion, fluid medium, agar overlay, and flask dilution. With its excellent physical, chemical, and mechanical properties, UNCD can provide an outstanding platform material for a new generation of advanced implantable biomedical devices.

Acknowledgments

The R&D reviewed in this chapter was supported by the US Department of Energy, Office of Science, under contract no. DE-AC02–06CH11357.

O. Auciello acknowledges the support from his Distinguished Endowed Chair Professor grant at the University of Texas-Dallas. He also acknowledges the contribution of A. J. Schultz (Ionwerks, Inc.) related to the MALDI analysis shown in Figures 6.14 and 6.15 and XPS analysis shown in Figure 6.16.

References

[1] OrthoInfo. Homepage. http://orthoinfo.aaos.org.

[2] R. Langer and J. P. Vacanti, "Tissue engineering," *Science*, vol. 260, p. 920, 1993.

[3] J. P. Vacanti and R. Langer, "Tissue engineering: the design and fabrication of living replacement devices for surgical reconstruction and transplantation," *Lancet*, vol. 354 (S1), p. 132, 1999.

[4] S. I. Stupp, "Biomaterials: introduction for annual review of materials research," *Ann. Rev. Mater. Res.*, vol. 31, 2001.

[5] H. Clevers, K. M. Loh, and R. Nusse, "An integral program for tissue renewal and regeneration: with signaling and stem cell control," *Science*, vol. 346, (6205), 2014.

[6] D. N. Kotton and E. E. Morrisey, "Lung regeneration: mechanisms, applications and emerging stem cell populations," *Nat. Med.*, vol. 20, p. 822, 2014.

[7] A. J. Wagers, "The stem cell niche in regenerative medicine," *Cell Stem Cell*, vol. 10 (4), p. 362, 2012.

[8] N. A. Peppas and R. Langer, "New challenges in biomaterials," *Science*, vol. 263, p. 1715, 1994.

[9] M. C. Peters, B. C. Isenberg, J. A. Rowley, and D. J. Mooney, "Release from alginate enhances the biological activity of vascular endothelial growth factor," *J. Biomat. Sci Polymer Edi.*, vol. 9 (12), p. 1267, 1998.

[10] L. A. Thomson, F. C. Law, N. Rushton, and J. Franks, "Biocompatibility of diamond-like carbon coating," *Biomaterials*, vol. 12, p. 37, 1991.

[11] W. C. Clem, S. Chowdhury, S. A. Catledge, et al., "Mesenchymal stem cell interaction with ultra-smooth nanostructured diamond for wear resistant orthopedic implants," *Biomaterials*, vol. 29, p. 3461, 2008.

[12] B. Shi, Q. Jin, L. Chen, et al., "Cell growth on different types of ultrananocrystalline diamond thin films," *J. Funct. Biomed. Mater.*, vol. 3, p. 588, 2012.

[13] S. E. Duailibi, M. T. Duailibi, L. M. Ferreira, et al., "Tooth, tissue engineering: the influence of hydrophilic surface on nanocrystalline diamond films for human dental stem cells," *Tissue Eng.*, vol. A19, p. 2537, 2013.

[14] A. C. Taylor, B. Vagaska, R. Edgington, et al., "Biocompatibility of nanostructured boron doped diamond for the attachment and proliferation of human neural stem cells," *J. Neural. Eng.*, vol. 12, p. 066016, 2015.

[15] Y. C. Chen, D. C. Lee, C. Y. Hsiao, et al., "The effect of ultrananocrystalline diamond films on the proliferation and differentiation of neural stem cells," *Biomaterials*, vol. 30, p. 3428, 2009.

[16] Y. C. Chen, D. C. Lee, T. Y. Tsai, et al., "Induction and regulation of differentiation in neural stem cells on ultra-nanocrystalline diamond films," *Biomaterials*, vol. 31, p. 5575, 2010.

[17] A. Taylor, "Diamond for stem cell biotechnology," PhD thesis, University College London, 2016.

[18] P. A. Nistor, P. W. May, F. Tamagnini, A. D. Randall, and M. A. Caldwell, "Long-term culture of pluripotent stem-cell-derived human neurons on diamond: a substrate for neurodegeneration research and therapy," *Biomaterials*, vol. 61, p. 139, 2015.

[19] C. G. Specht, O. A. Williams, R. B. Jackman, and R. Schoepfer. "Ordered growth of neurons," *Biomaterials*, vol. 25, p. 4073, 2004.

[20] P. W. May, E. M. Regan, A. Taylor, J. Uney, A. D. Dick, and J. McGeehan. "Spatially controlling neuronal adhesion on CVD diamond," *Diam. Relat. Mater.*, vol. 23, p. 100, 2012.

[21] R. J. Edgington, A. Thalhammer, J. O. Welch, et al., "Patterned neuronal networks using nanodiamonds and the effect of varying nanodiamond properties on neuronal adhesion and outgrowth," *J. Neural Eng.*, vol. 10, p. 056022, 2013.

[22] A. Thalhammer, R. J. Edgington, L. A. Cingolani, R. Schoepfer, and R. B. Jackman, "The use of nanodiamond monolayer coatings to promote the formation of functional neuronal networks," *Biomaterials*, vol. 31, p. 2097, 2010.

[23] S. M. Ojovan, M. McDonald, N. Rabieh, et al., "Nanocrystalline diamond surfaces for adhesion and growth of primary neurons, conflicting results and rational explanation," *Front. Neuroeng.*, vol. 7, p. 17, 2014.

[24] W. Tong, P. A. Tran, A. M. Turnley, et al., "The influence of sterilization on nitrogen-included ultrananocrystalline diamond for biomedical applications," *Mater. Sci. Eng. C Mater. Biol. Appl.*, vol. 61, p. 324, 2016.

[25] P. Ariano, O. Budnyk, S. Dalmazzo, et al., "On diamond surface properties and interactions with neurons," *Eur. Phys. J. E.*, vol. 30, p. 149, 2009.

[26] A. Bendali, C. Agnes, S. Meffert, et al., "Distinctive glial and neuronal interfacing on nanocrystalline diamond," *PLoS One*, vol. 9 (3), p. e92562, 2014.

[27] D. Steinmuller-Nethl, F. R. Kloss, M. Najam-Ul-HAq, et al., "Strong binding of bioactive BMP-2 to nanocrystalline diamond by physisorption," *Biomaterials*, vol. 27, p. 4547, 2006.

[28] L. Grausova, A. Kromka, Z. Burdikova, et al., "Enhanced growth and osteogenic differentiation of human osteoblast-like cells on boron-doped nanocrystalline, diamond thin films," *PLoS One*, vol. 6, 2011.

[29] G. Heinrich, T. Grogler, S. M. Rosiwal, and R. F. Singer, "CVD diamond coated titanium alloys for biomedical and aerospace applications," *Surf. Coat. Technol*, vol. 94(5), p. 514, 1997.

[30] M. Parizek, T. E. L. Douglas, K. Novotna, et al., "Nanofibrous poly (lactide-co-glycolide) membranes loaded with diamond nanoparticles as promising substrates for bone tissue engineering," *Int. J. Nanomed.*, vol. 7, p. 5873, 2012.

[31] B. Rezek, L. Michalikova, E. Ukraintsev, A. Kromka, and M. Kalbacova, "Micropattern guided adhesion of osteoblasts on diamond surfaces," *Sensors*, vol. 9, p. 3549, 2009.

[32] L. Grausova, A. Kromka, L. Bacakova, et al., "Bone and vascular endothelial cells in cultures on nanocrystalline diamond films," *Diam. Relat. Mater.*, vol. 17, p. 1405, 2008.

[33] M. Amaral, A. G. Dias, P. S. Gomes, et al., "Nanocrystalline diamond: in vitro biocompatibility assessment by MG63 and human bone marrow cells cultures," *J. Biomed. Mater. Res. A*, vol. 87a, p. 91, 2008.
[34] L. Yang, B. W. Sheldon, and T. J. Webster, "The impact of diamond nanocrystallinity on osteoblast functions," *Biomaterials*, vol. 30, p. 3458, 2009.
[35] K. F. Chong, K. P. Loh, S. R. K. Vedula, et al., "Cell adhesion properties on photochemically functionalized diamond," *Langmuir*, vol. 23, p. 5615, 2007.
[36] M. Amaral, P. S. Gomes, M. A. Lopes, et al., "Cytotoxicity evaluation of nanocrystalline diamond coatings by fibroblast cell cultures," *Acta Biomaterialia*, vol. 5, p. 755, 2009.
[37] H. J. Huang, M. Chen, P. Bruno, et al. "Ultrananocrystalline diamond thin films functionalized with therapeutically active collagen networks," *J. Phys. Chem. B*, vol. 113, p. 2966, 2009.
[38] Y. C. Chen, C. Y. Tsai, C.Y Lee, and I. N. Lin, "In vitro and in vivo evaluation of ultrananocrystalline diamond as an encapsulation layer for implantable microchips," *Acta Biomaterialia*, vol. 10, p. 2187, 2014.
[39] T. Lechleitner, F. Klauser, T. Seppi, et al., "The surface properties of nanocrystalline diamond and nanoparticulate diamond powder and their suitability as cell growth support surfaces," *Biomaterials*, vol. 29, p. 4275, 2008.
[40] B. Rezek, E. Ukraintsev, M. Kratka, et al. "Epithelial cell morphology and adhesion on diamond films deposited and chemically modified by plasma processes," *Biointerphases*, vol. 9, p. 031012, 2014.
[41] O. Auciello, and A. V. Sumant, "Status review of the science and technology of ultrananocrystalline diamond films and application to multifunctional devices," *Diam. Relat. Mater.*, vol. 19, p. 699, 2010.
[42] J. A. Carlisle, D. M. Gruen, O. Auciello, and X. Xiao, "A method to grow pure nanocrystalline diamond films at low temperatures and high deposition rates," US Patent #7,556,982, 2009.
[43] N. Naguib, J. Birrell, J. Elam, J. A. Carlisle, and O. Auciello, "A method to grow carbon thin films consisting entirely of diamond grains 3–5 nm in size and high-energy grain boundaries," US Patent #7,128,8893, 2006.
[44] D. M. Gruen, A. R. Krauss, O. Auciello, and J. A. Carlisle, "N-type doping of NCD films with nitrogen and electrodes made therefrom," US patent #6,793,849 B1, 2004.
[45] A. R. Krauss, D. E. Gruen, M. J. Pellin, and O. Auciello, "Ultrananocrystalline diamond cantilever wide dynamic range acceleration/vibration/pressure sensor," US patent #6,422,077, 2002.
[46] J. E. Gerbi, O. Auciello, J. Birrell, et al., "Electrical contacts to ultrananocrystalline diamond," *App. Phy. Lett.*, vol. 83, p. 2001, 2003.
[47] S. Bhattacharyya, O. Auciello, J. Birrell, et al., "Synthesis and characterization of highly-conducting nitrogen-doped ultrananocrystalline diamond films," *App. Phy. Lett.*, vol. 79, p. 1441, 2001.
[48] M. Hajra, C. E. Hunt, M. Ding, et al., "Effect of gases on the field emission properties of ultrananocrystalline diamond-coated silicon field emitter arrays," *J. App. Phy.*, vol. 94, p. 4079, 2003.
[49] R. J. Nemanich, J. T. Glass, and G. Lucovsky, "Raman scattering characterization of carbon bonding in diamond and diamond like thin films," *J. Vac. Sci. Technol.*, vol. A6, p. 1783, 1988.

[50] B. Shi, Q. Jin, L. Chen, and O. Auciello, "Fundamentals of ultrananocrystalline diamond (UNCD) thin films as biomaterials for developmental biology: embryonic fibroblasts growth on the surface of (UNCD) films," *Diam. Relat. Mater.*, vol. 18, p. 596, 2009.
[51] C. Liu, X. Xiao, J. Wang, et al., "Dielectric properties of hydrogen-incorporated chemical vapor deposited diamond thin films," *J. Appl. Phys.*, vol. 102 (7), p. 074115/1-7, 2007.
[52] B. D. Thomas, M. S. Owens, J. E. Butler, and C. Spiro, "Production and characterization of smooth, hydrogen-terminated diamond C(100)," *Appl. Phy. Lett.*, vol. 65 (23), p. 2957, 1994.

7 New Generation of Li-Ion Batteries with Superior Specific Capacity Lifetime and Safety Performance Based on Novel Ultrananocrystalline Diamond (UNCD™)-Coated Components for a New Generation of Defibrillators/Pacemakers and Other Battery-Powered Medical and High-Tech Devices

Orlando Auciello and Yonhua Tzeng

7.1 Background Information on Li-Ion Batteries and Other Batteries for Implantable Medical Devices and Other Applications

Lithium-ion batteries (LIBs) are the most used energy supplier devices for electric vehicles and portable and fixed electronic and medical devices because of their high-power supply performance. However, they still have shortcomings. They, and other alternative batteries like Li-O_2 and Li-S batteries, still do not meet all the requirements for an energy supplier device with stable energy capacity for charge–discharge cycles in the range $\geq$1000 required for future high-performance electric vehicles [1–3] and portable and fixed electronic or medical devices. LIBs have been the most extensively investigated and developed, exhibiting the fewest technological barriers. They are being used to power electric cars (e.g., GM's Volt), electronic devices (e.g., mobile phones and video cameras) [4], and medical devices (e.g., defibrillators/pacemakers) [4]. It has been shown that several materials used for fabrication of LIB anodes exhibit excellent reversible insertion of Li ions [4–7] via intercalation of layered carbons (e.g., graphite), adsorption in hard carbon surface layers, and binding of hydrogen atoms in carbon containing hydrogen [8]. Mesocarbon microbeads (MCMBs) were the early standard anode material for LIBs due to their high theoretical capacity (372 mAh/g based on the formation of LiC_6 in the natural graphite (NG) material integrated on copper (Cu) base, and the induced low potential profile (0–0.3 V vs. Li/Li^+). Natural graphite emerged as a strong candidate anode material to replace MCMBs because

NG exhibits high Coulombic efficiency and electronic conductivity (0.4–2.5 × 10^4 S cm^{-1} in the basal plane), low volume expansion after full lithiation (~10%), and low cost (≤US$10/kg) [9-11]. However, large irreversible capacity loss and short charging–discharging cycling life are two main problems related to NG/Cu-based anodes in LIBs, although they are used in current commercial LIBs. The chemical reactivity of Li ions inserted in the NG material during the LIB operation with organic electrolytes has to be eliminated to produce efficient NG-based LIB anodes. Reactions of Li ions, inserted in the NG material, with electrolytes forms a local solid–electrolyte interphase (SEI) layer of increasing thickness and electrical resistance, which generates gases (e.g., hydrogen, CO, CO_2, methane, ethylene, and propylene [12, 13]) within interlayers of the NG crystals, resulting in irreversible capacity loss and damage to the NG structure, which reduces the LIB's life.

The effect of the SEI layer on LIBs' performance depends on the electrolyte composition and the microstructures and surface characteristics of the NG material of the anode. The SEI films are formed when the negative-biased anodes are exposed to low potentials in non-aqueous electrolytes containing Li salts, resulting in the reduction of electrolyte species to insoluble salts. The SEI layer blocks electron transfer between the anode and the electrolyte, but still allows conduction of Li ions, as indicated by a model developed to understand the SEI layer performance [14]. However, extensive studies focused on understanding the performance of the SEI passivation layer have shown that flaws are produced in this layer due to the volume change of the NG material of the NG/Cu anode, where the SEI layer forms, exposing the fresh surface of the NG material, which is repaired by surface-electrolyte reactions on the NG material of the anode, resulting in continuous SEI film formation and Li ions deficiency, causing the battery capacity to fade. The process described above becomes worse on anodes where the local charging and discharging current density is higher than average. In addition to the formation of the SEI layer, growth of dendrites [15] based on compounds formed from highly Li-induced reactive NG anode areas result in electrical shorting and hazardous failure of the LIB. The LIB anode problems described above limit increase of LIB market share for electrical vehicles, requiring inexpensive LIBs with NG anodes exhibiting long-life and high-rate charging–discharging cycle performance. Therefore, minimization or, even better, elimination of undesirable Li-induced reactive sites on the NG anode surface and formation of robust SEI films are desirable to produce the next generation of high-performance LIBs [16–20]. Various approaches have been investigated to suppress electrolyte–graphite reactions, including coating NG with carbon layers [21–24], polymer films [25, 26], alkali carbonate films [27], and lithium benzoate films [27]. Research has shown that all of these coatings provide limited improvement in the performance of the commercial NG-based anodes.

Based on the information presented above, it is critical to develop chemically robust, electrically conductive coatings to produce a new generation of high-performance anodes for LIBs. In relation to this statement, nitrogen (N) atoms/grain boundary–incorporated ultrananocrystalline diamond (N-UNCD) films [28, 29] exhibit high-density networks of grain boundaries with extensive sp^2-C open bonds, which provide sites for N atoms to undergo chemical reactions, liberating electrons, which are inserted in the grain boundaries, producing efficient electrical conductivity through the N-UNCD films [28, 29]. In addition, the N-UNCD films exhibit excellent resistance to Li ion–based chemical-/electrochemical-induced corrosion and other chemically reactive environments, high mechanical strength, and one of the lowest coefficients of friction and practically no mechanical wear [29, 30]. Based on the unique combination of properties mentioned above, N-UNCD films were investigated as coatings for commercial NG/Cu anodes in order to suppress undesirable chemical corrosion of the anode material by the extremely reactive Li ions in the LIB electrolyte. Orders of magnitude improvement in the performance of NG/Cu anodes coated with N-UNCD films were demonstrated for LIBs, as reviewed in this chapter. The N-UNCD coatings may also play a critical role in improving the performance of Li–sulfur and Li–air batteries in the future.

7.2 Introduction of Innovative Electrically Conductive N-UNCD and Insulating UNCD-Based Coating Technologies for LIB Components for Superior Performance

This chapter focuses on a review of two unique, novel types of coating: (1) electrically insulating Li-ion corrosion-resistant UNCD coating for covering anode–cathode separation membranes and the inner walls of LIB cases to protect them from Li ion–induced corrosion; and (2) nitrogen grain boundary–incorporated, electrically conductive, Li-ion corrosion-resistant N-UNCD coating for covering the anodes and cathodes of LIBs. The unique UNCD and N-UNCD coating-based technology may enable a new generation of LIBs with orders of magnitude longer stable capacity energy vs. charge–discharge cycles and improved safety in terms of overheating and potential explosion, induced by the continuous formation of the SEI layer, than current LIBs. The new longer-life LIBs may provide a transformational improvement to power a new generation of miniaturized defibrillators/pacemakers and other battery-powered implantable medical devices to improve the quality of life of people receiving them. The UNCD film technology already demonstrated and under development provides new key components for the new LIBs:

1. N-UNCD films encapsulating three type of anodes, under investigation to determine the future optimum LIB anode:

   a. N-UNCD-coated NG/copper composite LIB anode;
   b. N-UNCD-coated W (40–100 nm)/copper composite anode; and
   c. N-UNCD-coated porous Si anode.

   The N-UNCD-coated anodes will provide an order of magnitude superior longer stable capacity energy vs. charge–discharge cycles than for current commercial NG/Cu anodes. In addition, recent research demonstrated that commercial NG/Cu anodes coated with N-UNCD layers exhibit suppressed reactions of NG with the electrolyte and the development of an insulating SEI layer on the anode [31, 32],

which retards anode conductivity and induces stresses, leading to cracks in the NG particles, causing loss of contact between them.
2. UNCD-coated anode–cathode separation membranes with orders of magnitude higher resistance to chemical attack than membranes in current LIBs.
3. UNCD coatings for the inner walls of LIB cases to enable use of less expensive case materials (e.g., Al) than current ones.

7.3 Research and Development on N-UNCD-Coated Commercial NG/Cu Anodes

Several anodes were fabricated by spreading NG powders (30 μm particle size) on Cu foils, followed by coating with pitch and encapsulation with N-UNCD films, as described in the following subsections.

7.3.1 Materials and Methods for Fabrication of NG/Cu Anodes

Powders made of spherical NG particles (~25 μm size) were produced via ball milling of commercial anisotropic NG flakes. The NG powders were subsequently coated with a petroleum-induced pitch layer. The LIB anode (negative electrode) was made by first producing a slurry, made of 92 wt% NG powder and 8 wt% PVDF (KF9130), dispersed in N-methyl-2-pyrrolidone, and then spread on the surface of a Cu foil. The fabricated NG/Cu anode was then dried overnight in a vacuum oven at 75 °C, resulting in the production of an active material with about 10 mg cm^{-2} density. The thickness of the dried NG layer on the Cu foil was about 75 μm, as measured by cross-section scanning electron microscopy (SEM).

7.3.2 Microwave Power Chemical Vapor Deposition for Growth of N-UNCD Coating on NG/Cu Anodes

Electrically conductive N-UNCD coatings were grown on NG/Cu anodes using the microwave power chemical vapor deposition (MPCVD) process, implemented in an industrial-type Innovative Plasma Systems (IPLAS) MPCVD system (GMBH, Germany). The surface of the NG layer on the Cu foil was seeded with diamond nanoparticles (2–5 nm in diameter) via immersing the dried NG/Cu composite matrix anode in a methanol suspension of diamond nanoparticles (2–5 nm in diameter) under ultrasonic agitation, which shakes the diamond nanoparticles and embeds them on the surface of the NG particles. The seeding process was followed by cleaning the surface of the anodes with different alcohol-based fluids.

N-UNCD coatings were grown on two NG/Cu anodes for 3 h and 6 h to investigate the effect of the N-UNCD coating thickness on the electrical performance of the N-UNCD/NG/Cu anode. The N-UNCD coatings were grown using a patented [33, 34] Ar (79 sccm)/CH_4 (2 sccm)/N_2 (20 sccm) gas mixture flown into an air evacuated chamber (~10^{-7} Torr), and coupling microwave power (~2000 W), creating a plasma and producing C_2 dimers and CH_x molecules, which upon landing on the NG/Cu, heated to ~800 °C, induce the nucleation and growth of the N-UNCD film. More details of the nucleation and growth process of the N-UNCD films can be found in [29, 34–36].

7.3.3 Characterization of Chemical and Structural Properties of N-UNCD Coatings on NG/Cu Anodes

7.3.3.1 Raman Analysis of Uncoated and N-UNCD-coated NG/Cu anodes

Figure 7.1 shows Raman spectra obtained with a visible (524 nm wavelength) laser beam for an as-prepared NG/Cu anode and for NG/Cu anodes coated with N-UNCD films for 3 h and 6 h. Even with NG underneath, which has a large resonant Raman scattering intensity, with the N-UNCD coating the Raman spectra shown in Figure 7.1 displays the Raman scattering characteristics of the N-UNCD coating [29, 34–36]. The measured D band at 1332 cm^{-1}, the G band at 1583 cm^{-1}, and a weak transpolyacetylene (TPA) band at 1140 cm^{-1} are comparable with those reported for N-UNCD deposited on a silicon surface [28, 29]. Ultraviolet laser (325 nm wavelength) Raman scattering analysis was also done to further confirm the nature of the N-UNCD coating on the NG electrodes. Figure 7.2 shows the UV Raman spectra measured in randomly selected areas of NG/Cu anodes coated with 3 h and 6 h grown N-UNCD films. Figure 7.2a shows that the NG layer of the NG/Cu anode is not completely covered by 3 h of growth of N-UNCD film. Some areas exhibit only the graphite

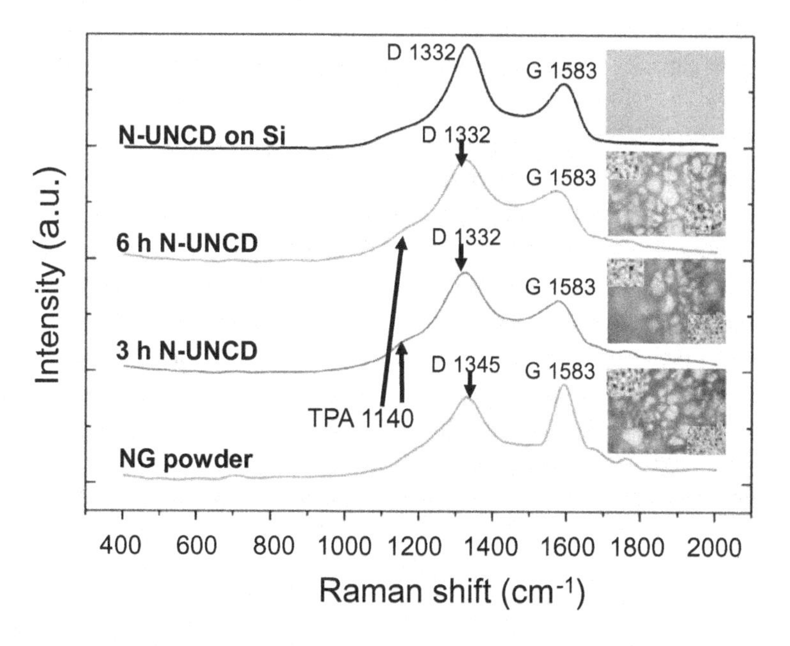

Figure 7.1 Raman spectra using a visible (531 nm wavelength) laser beam on NG powder on 3 h and 6 h grown N-UNCD films on NG/Cu anodes and N-UNCD film on Si, as a template.

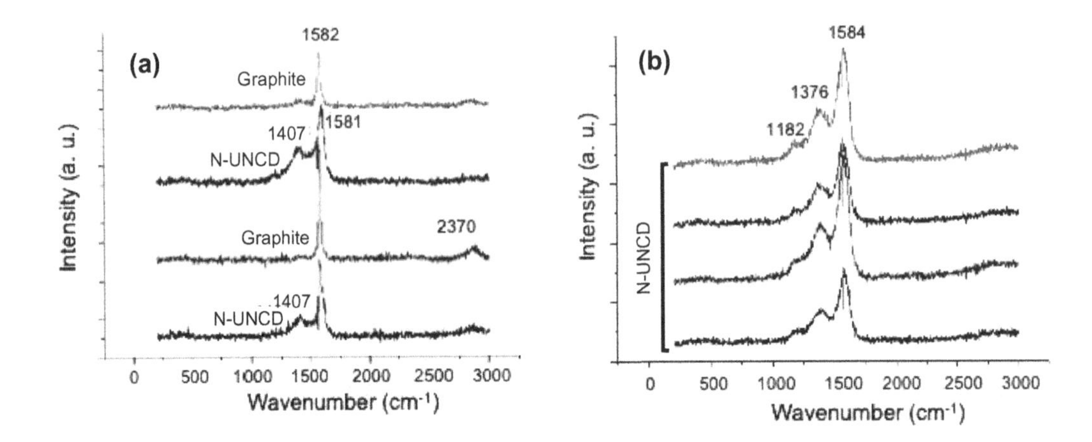

Figure 7.2 Raman spectra using a UV (325 nm wavelength) laser beam on (a) 3 h and (b) 6 h grown N-UNCD films on NG/Cu anodes. Notice that the 3 h grown N-UNCD film exhibits Raman peaks characteristic of graphite mixed with incipient N-UNCD component due to no dense N-UNCD film, while the 6 h grown N-UNCD film shows only the peaks characteristic of N-UNCD, confirming a dense film as shown in the SEM and TEM figures.

Raman peak, indicating that there is no or only very thin N-UNCD coating in that area. On the other hand, the N-UNCD film grown for 6 h provides full coverage of the NG in the NG/CU anode, as revealed by Raman spectra on different areas of the film shown in Figure 7.2b, which confirms that the N-UNCD film grown for 6 h provides a dense encapsulating film for the graphite particles, and this effect correlates with the superior capacity energy vs. charge–discharge cycles (see Figure 7.7a).

7.3.3.2 SEM, MRTEM, and HRTEM Analysis of Uncoated and N-UNCD-coated NG/Cu anodes

SEM images of NG layers coated with N-UNCD for 3 h and 6 h showed that the size of graphite particles is about 25, 31, and 32 μm for the uncoated particles (Figure 7.3a). SEM images also showed that particles coated with N-UNCD for 3 h (Figure 7.3b) and 6 h (Figure 7.3c) exhibit conformal dense coating. The 6 h coating connects the edges of particles, thus resulting in denser encapsulated NG electrodes than those prepared using the 3 h coating growth. The surface morphology of the coated NG grains is very smooth due to the fact that the surface roughness of the N-UNCD films is about 10 nm rms.

Figure 7.4a,b shows SEM top view images of the uncoated NG layer of the NG/Cu anode before charge–discharge and after 100 cycles of charge–discharge measurements, revealing topographical degradation of the NG particles induced by Li-ion corrosive reaction on the NG particles. On the other hand, the N-UNCD-coated NG layer on the NG/Cu anode shows no degradation (Figure 7.4c). In addition HRTEM images of N-UNCD-coated NG (Figure 7.4d,e) confirm no degradation as first observed by SEM analysis.

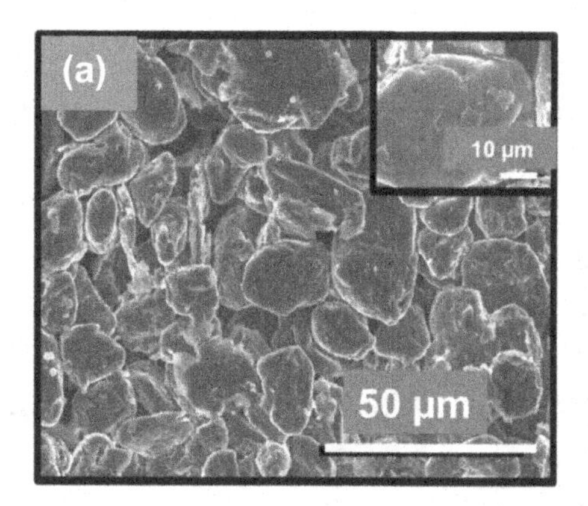

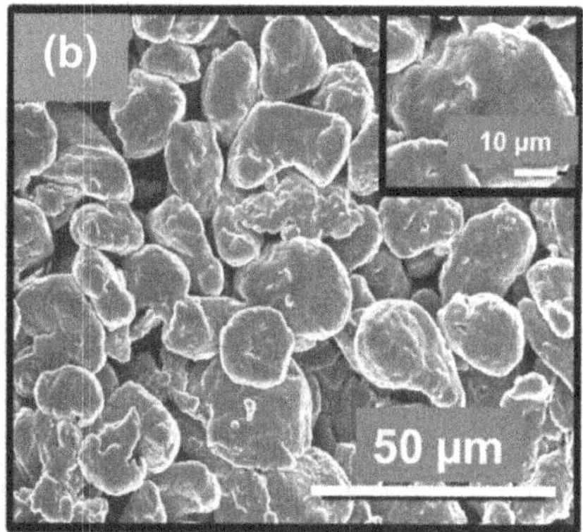

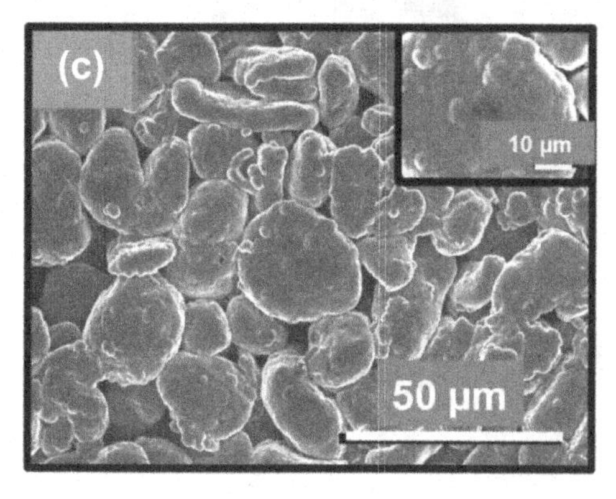

Figure 7.3 SEM images of uncoated NG layer on Cu foils (a) showing NG sizes, 3 h N-UNCD coated NG layer (b), and 6 h N-UNCD-coated NG layer (c).

The N-UNCD coatings were further analyzed by HRTEM to confirm the carbon phases in the coating, growing an N-UNCD film on Si as a well-known template material for UNCD film growth. The N-UNCD films were grown on an Si substrate under the same conditions used to grow the N-UNCD films on the NG/CU anodes. Cross-section TEM samples for HRTEM studies were prepared using a focused ion beam process, in which a high-energy (~80 keV) Ga ion beam is directed perpendicular to the surface of the N-UNCD-coated Si substrate, producing sputtering that results in cutting a rectangular piece of material and exposing the interface between the N-UNCD film and the Si substrate. Figure 7.5a shows a cross-sectional TEM image of an N-UNCD film grown on silicon. Two marked areas in Figure 7.5a are highlighted in Figure 7.5b,c. The HRTEM image of the marked darker region 1 in Figure 7.5a is shown in Figure 7.5b with selected area electron diffraction (SAED) patterns, which indicate the presence of graphitic carbon, while the HRTEM image of the marked lighter region 2 in Figure 7.5a is shown in Figure 7.5c, with SAED patterns that indicate the presence of diamond grains. HRTEM characterization proved that the N-UNCD films are composed of diamond grains with sp^2-bonded carbon atoms in the grain boundaries. The grain boundaries in the N-UNCD encapsulating coating provide

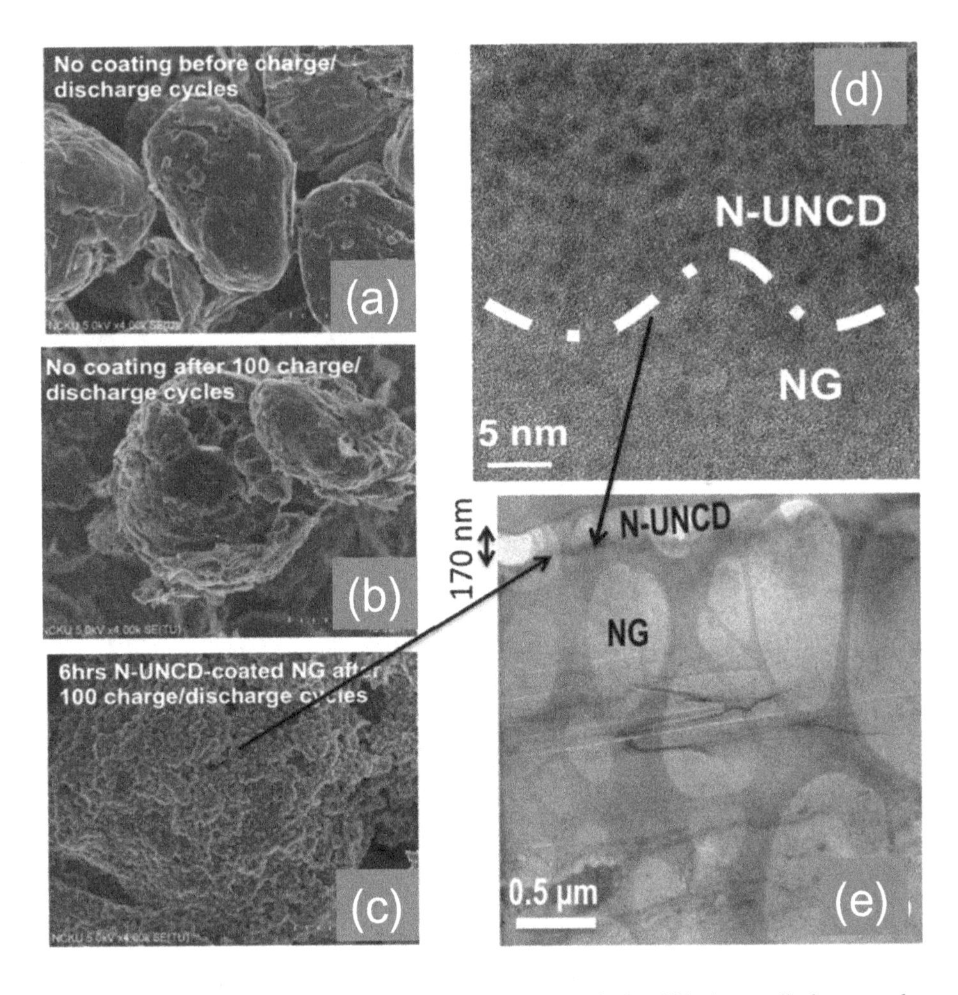

Figure 7.4 SEM images of uncoated NG layer before (a) and after 100 charge–discharge cycles (b), showing NG degradation. (c) SEM image of N-UNCD-coated NG layers on Cu foils after 100 charge–discharge cycles, showing no degradation. (d,e) Cross-section high-resolution transmission electron microscope (HRTEM) images of N-UNCD-coated NG layers on Cu foils after 100 charge–discharge cycles, confirming no degradation as shown in (c) (reprinted from *Adv. Mater.*, vol. 26 (1–5), p. 3724, 2014 (Fig. 3) in [32] with permission from Wiley Publisher).

effective conduction channels for lithium ions between the electrolyte and the NG underlying layer of the NG/Cu anodes, while the uniform distribution of these conduction channels helps distribute Li-ion current more uniformly across the surface of the anode to suppress possible high current density at local areas that may result in dendritic growth of Li-based compound structures.

The materials characterization performed on the uncoated and N-UNCD-coated Ng/Cu anodes strongly support the electrical performance of LIBs with N-UNCD-coated anodes, and provides the basis for extending the N-UNCD coating to cathodes to protect them from Li ion–induced chemical degradation.

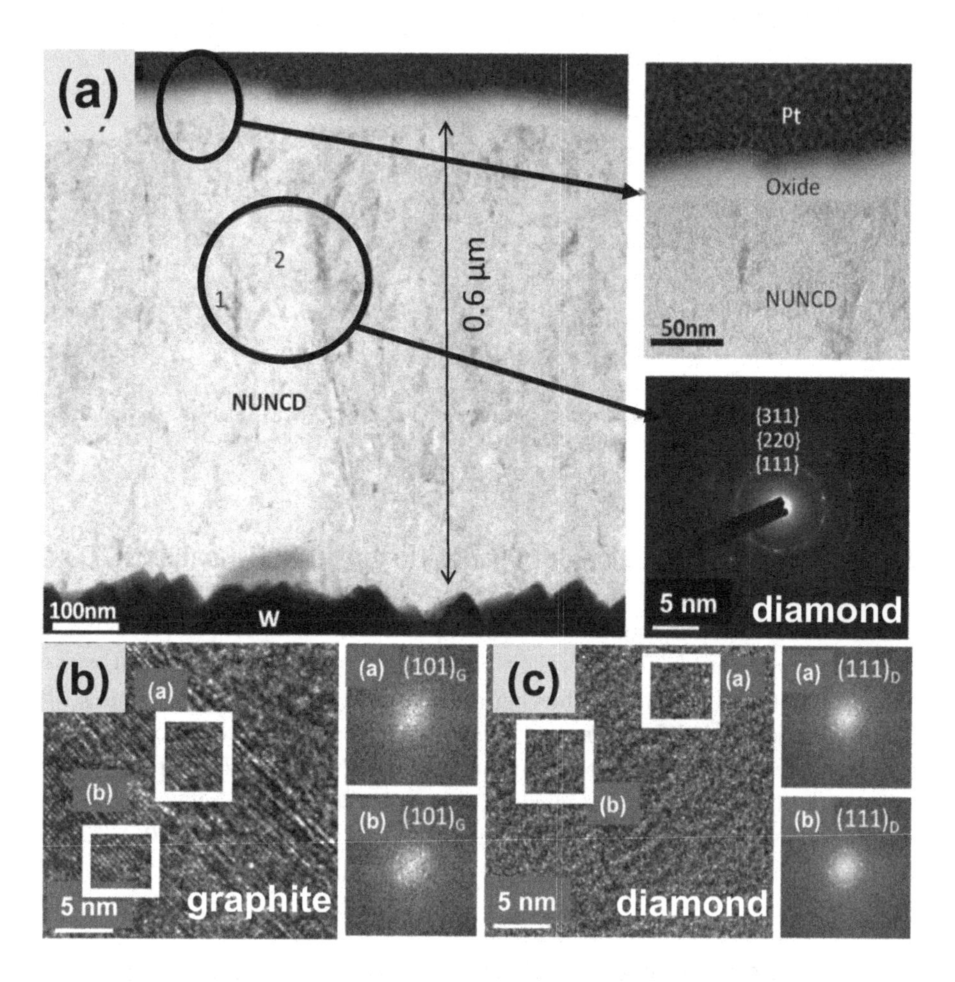

Figure 7.5 (a) Cross-sectional TEM image of an N-UNCD film grown on Si. (b) TEM image of area 1 in (a) with two square areas where electron diffraction (a) and (b) reveal graphite. (c) TEM image of area 2 in (a) with two square areas (a) and (b) where electron diffraction reveal diamond (reprinted from *Adv. Mater.*, vol. 26 (1–5), p. 3724, 2014 (Fig. 1) in [32] with permission from Wiley Publisher).

7.4 Fabrication of LIB Cells

A metal foil made of Li was used as the counter electrode to fabricate an Li/NG/Cu half-cell (Figure 7.6). The NG/Cu/Li cell was formed into a standard coin-shaped battery cell (2032 type). The LIB test cells were fabricated with working NG/Cu anodes, with and without N-UNCD encapsulating coatings, a microporous polypropylene separator (Celgard 2325), a counter electrode (Li), and an appropriate amount of electrolyte. The electrolyte was 1.2M LiPF6 dissolved in mixed solvents of ethylene carbonate (EC) and ethylmethylcarbonate (EMC), with a ratio of 3:7 by weight.

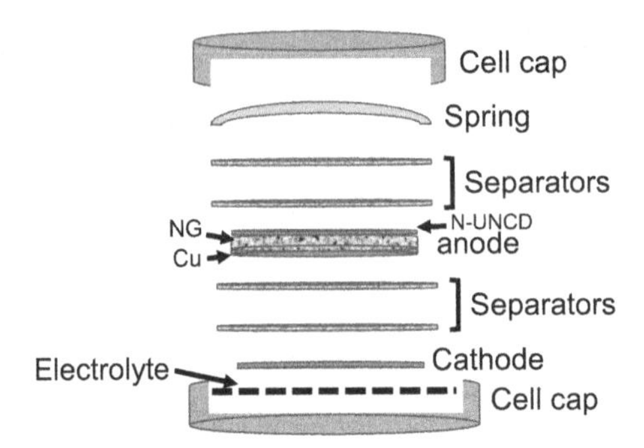

Figure 7.6 Detailed schematic of the components of the LIB cells fabricated to test the capacity energy performance with N-UNCD-coated NG/Cu anodes.

7.5 Test of LIB Cells with Uncoated and N-UNCD-Coated NG/Cu Anodes

The NG/Cu anodes were inserted in several LIB coin-size cells and subjected to systematic tests. The cells were subjected to two charge–discharge cycles at a rate taking 10 hours to achieving full capacity (i.e., C/10, at ~0.35 mA). The cells were then cycled for charge and discharge under tailored conditions: C/10 at room temperature with cell voltage in the range 1 mV to 1.5 V. The capacity energy rate of the lithiated graphite-based anode was evaluated by discharging NG/Cu/Li coin cells to 1 mV by C/10 and then charging them to 1.5 V by C/20, C/10, C/5, C/2, C, 2C, 5C, and 10C rates.

7.5.1 Electrochemical Impedance Spectroscopy Measurements

Electrochemical impedance spectroscopy (EIS) measurements were carried out for the LIB coin-size cells described above under two testing conditions: (1) after three charge–discharge cycles, and (2) after 100 charge–discharge cycles. The test cells were charged using a constant current to a 20%, 40%, 60%, 80%, and 100% state of charge. Subsequently, the test cells were set in a dormant state for 30 minutes before the AC impedance measurements were made, using a frequency response analyzer (Solartron, model 1400). The measurements were done using frequency windows in the range 500 kHz to 0.02 Hz with an applied voltage of 5 mV.

The specific capacity retention of LIBs with uncoated and N-UNCD-coated NG/Cu anodes were measured to determine the difference in performance. Figure 7.7a shows that specific capacity performance of LIBs with N-UNCD-coated NG/Cu electrodes is improved by orders of magnitude with respect to the LIB with uncoated NG/Cu electrodes. In fact, the LIB with uncoated NG/Cu electrodes exhibits a substantial degradation of the specific capacity for 100 charge–discharge cycles, while the LIB with NG/Cu anode with 3 h grown N-UNCD coating exhibits only a minor decrease in specific capacity after 50 charge–discharge cycles, and the LIB with NG/CU anode

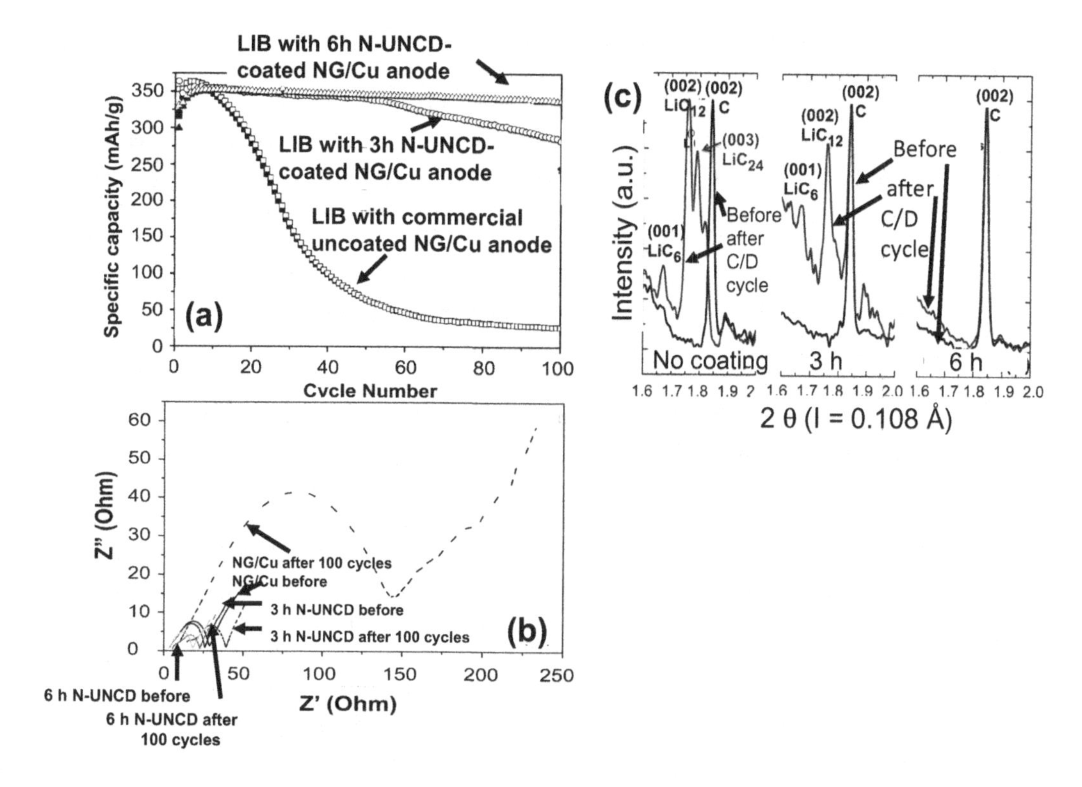

Figure 7.7 (a) Specific capacity vs. charge–discharge cycle performance of NG/Cu anodes without and with 3 h and 6 h N-UNCD coatings at the charging rate of 0.1C. (b) Comparison of alternate-current (0.02 Hz to 500 kHz) electrochemical impedance after the first charge–discharge cycle and after 100 cycles for uncoated NG/Cu anodes and with N-UNCD coatings for 3 h and 6 h. (c) X-ray diffraction (XRD) after the first charge–discharge cycle and after 100 cycles for the uncoated NG/Cu anode and for anodes with 3 h and 6 h grown N-UNCD coatings. The chemical degradation of the NG layer is responsible for the substantial decrease in the specific capacity as a function of charge–discharge cycles (reprinted from *Adv. Mater.*, vol. 26 (1–5), p. 3724, 2014 (Fig. 2) in [32] with permission from Wiley Publisher).

coated with 6 h N-UNCD exhibits practically stable capacity energy up to 100 charge–discharge cycles. On the contrary, a rapid decrease in the discharge–charge capacity of the graphite electrode without N-UNCD is observed after fewer than 10 charge–discharge cycles. The gradual decay of the capacity for the NG/Cu electrode coated with the 3 h N-UNCD and tested at 0.1C can be attributed to the partial exposure of the NG material, where the N-UNCD coating exhibits some pinholes (see Figure 7.5), thus providing pathways for the electrolyte to go through and chemically attack the underlying NG layer of the NG/Cu anode. The polymeric binder between graphite particles is overheated by the substrate temperature during N-UNCD film growth and loses its binding function. NG particles might have thus become loose and eventually separated from each other and the current collector after repetitive expansion and shrinkage during the charging and discharging processes. This effect explains the measured loss of capacity. The 6 h N-UNCD coating provides a superior performance with negligible decay in the specific capacity (Figure 7.7a) because it is fully dense and protects the NG layer from Li ion–induced chemical attack.

The effects of the interfacial layer between the anode and the electrolyte on the electrode performance were investigated via AC impedance measurements after 3 charge–discharge cycles (i.e., formation cycles) and after 100 cycles. The Cole–Cole plots are shown in Figure 7.7b, and reveal that the shifting of the plots from those for the first 3 cycles and those after 100 cycles for the N-UNCD-coated NG/Cu anodes is an order of magnitude smaller than that for the pristine NG/Cu anode. This is consistent with the superior cycle performance of the N-UNCD-coated NG/Cu anodes with respect to the uncoated NG/Cu one, as shown in Figure 7.7a. The high-frequency semicircle is attributed to the SEI layer at the interface between the electrolyte and the NG layer of the NG/Cu anode, with or without N-UNCD coatings. When the SEI passivation film cracks due to the volume change of graphite, a new graphite surface is exposed, which is covered by a new SEI film during the next lithiation process [31]. The comparable or slightly lower impedance for the graphite NG/Cu anode after the first three charge–discharge cycles indicates that the additional coating of N-UNCD on graphite particles does not cause additional impedance between the electrolyte and the anode – that is, N-UNCD allows lithium ions to go through effectively. The additional impedance due to the presence of the N-UNCD coating is more than compensated by the additional effective surface area of the N-UNCD-coated NG/Cu anode. After the binder is exposed to the high-temperature plasma, some graphite particles originally covered by the binder and separated from the electrolyte become N-UNCD-coated graphite particles and are exposed to the electrolyte.

The increase in impedance by more than three times for the pristine NG/Cu anode after 100 charge–discharge cycles with respect to the impedance for the NG/Cu anode coated with N-UNCD (after 3 h deposition) shows that the N-UNCD buffer layer éffectively suppresses the build up of SEI on the anode surface. The undesirable and continuous growth of SEI compounds on the exposed NG layer of the NG/Cu anode surface due to cracks in the graphite causes the depletion of electrolyte as well as the subsequently rise of impedance.

The Warburg diffusion related to the Li diffusion through the graphite lattice and through the N-UNCD is shown in the low-frequency region of the EIS. In the case of the pristine NG/Cu anode, it appears as an inclined line. A comparison of the first three cycles EIS curve for the NG/Cu anode without N-UNCD coating with the first three cycles EIS of the NG/Cu anode coated with N-UNCD shows that the N-UNCD coating does not slow down the diffusion of Li from the electrolyte through N-UNCD into the graphite lattice. Instead, the impedance is slightly lower with the N-UNCD coating than for the pristine NG/Cu anode. The removal of the binder between graphite particles and the replacement by Li ion–conductive N-UNCD on the surface of graphite particles and in the interparticle space makes the contact between the electrolyte and the N-UNCD-coated graphite particles more efficient, thus resulting in lower impedance.

7.5.2 XRD Analysis of Uncoated and N-UNCD-Coated NG/Cu Anodes before and after LIB Charge–Discharge Cycles

Figure 7.7c shows XRD patterns of the NG layer on the NG/Cu anodes before and after charge–discharge cycling tests. The NG/Cu anodes originally show only one peak of C (002). After 100 cycles, the XRD of the NG/Cu anodes without N-UNCD coating shows three different peaks corresponding to LiC_6 (001), LiC_{12} (002), and LiC_{24} (003). This result indicates clearly that part of the Li ion–inserted graphite particle has lost the ability to release Li ions, which indicates either active graphite has peeled off from the anode current collector or unceasing SEI growth interferes with Li-ion diffusion in electrodes. Without the supply of electrons from the current collector through the anode, lithium ions cannot move out of the loose graphite pieces during the discharge. After 100 cycles, the NG/Cu anode with 3 h grown N-UNCD coating shows two peaks corresponding to LiC_6 (001) and LiC_{12} (002). The amount of volume change during the charging and discharging among these compounds is the largest for LiC_6. LiC_{12} is the second largest. LiC_{24} is the smallest among these three measured compounds for the NG/Cu anode without the N-UNCD coating. The NG/Cu anode coated with the 3 h N-UNCD film does not show the LiC_{24} (003) peak. This effect is probably due to the fact that although the 3 h coating of N-UNCD is thin and incomplete in terms of providing full coverage of the NG/Cu anode to withstand the 10% volume change in NG, which occurs during long-term cycling, it is still capable of withstanding the smaller volume expansion due to the formation of LiC_{24}. The graphite particles containing LiC_{24} remain intact and continue to allow the intercalation of lithium. After 100 cycles of charging and discharging, the NG/Cu anode coated with 6 h N-UNCD film shows only the same peak of C (002) as the original NG layer on the NG/Cu anode, indicating that the NG/Cu anode remains intact after repetitive charging and discharging. Figure 7.7 proves that the N-UNCD coating provides an excellent electronic and lithium conductor, which effectively preserves the integrity of the NG layer on the NG/Cu anode and is, therefore, an excellent encapsulation layer for the NG/Cu anode for LIBs. The fact that there is practically no degradation in the specific capacity vs. charging–discharging cycles of the LIB using a 6 h grown N-UNCD-coated NG/Cu anode, and that the XRD shows no chemical change in the anode, is a clear indication of the resistance to chemical attack of the N-UNCD-coated anode. This result confirms prior work [30] in which UNCD- and N-UNCD-based MEMS/NEMS (microelectromechanical/nanoelectromechanical systems) devices were fabricated by etching the SiO_2 sacrificial layer underneath the UNCD layers, after defining cantilevers, using the strong HF acid, which does not attack the UNCD or N-UNCD layers at all.

7.5.3 Measurement of Specific Capacity for LIBs with Uncoated and N-UNCD-Coated NG/Cu Anodes vs. Charge–Discharge Cycles

The results from measurements of specific capacity vs. charge–discharge cycles correlate very well with the SEM and HRTEM images shown in Figure 7.4. In fact,

Figure 7.4a shows SEM images of uncoated NG particles before charge–discharge cycles. Figure 7.4b shows the morphology of NG particles after 100 charge–discharge cycles with uncoated NG/Cu anodes, revealing a substantial destruction of the surface of the NG particles. On the other hand, NG particles coated with the 6 h grown N-UNCD coating show excellent integrity of the surface morphology of the N-UNCD-coated NG layer. The HRTEM images shown in Figure 7.4c,d of the NG/Cu composite electrode coated with a 6 h N-UNCD growth layer show the robust integrity of the heterostructure, without degradation after 100 charge–discharge cycles, mainly due to the protective N-UNCD coating on the NG/Cu composite electrode.

7.6 Conclusions

The R&D described in this chapter has demonstrated that a novel electrically conductive Li-ion corrosion-resistant N-UNCD film provides an excellent chemically robust encapsulation coating for the commercial NG/Cu composite anode currently used in LIBs, thus providing a solution to the problem of LIB materials degradation, as reviewed in the original recent work demonstrating the transformational N-UNCD-coated NG/Cu anodes for a new generation of LIBs [31, 32]. The N-UNCD encapsulating coating allows for good conductivity of both electrons and Li ions while exhibiting outstanding chemical and electrochemical inertness and mechanical strength, resulting in the formation of a robust SEI layer on the anode and preservation of the integrity of the NG layer in the composite NG/CU anodes. NG/Cu anodes encapsulated by N-UNCD films survived 100 charge–discharge cycles with the NG particles remaining intact from the mechanical and chemical points of view. Without the N-UNCD coating, NG/Cu anodes break loose into various compounds of Li ion–inserted carbon, without electrical contacts to the anode and the current collector. The excellent chemical inertness of the N-UNCD encapsulating coating suppresses the co-intercalation of electrolyte in NG. Gas compounds formed inside cracks in graphite particles cause the graphite particles to break into pieces and lose their electrical contact to the current collector and further de-intercalation of Li ions, resulting in the reduction of specific capacity of the LIBs. In addition, the N-UNCD encapsulating coating provides a rigid and isotropic surface for the formation of robust SEI films which effectively eliminates cracking of the SEI film, exposure of a fresh anode surface, and repairing of the SEI film by electrochemical reduction of the electrolyte and the formation of additional SEI compounds, which increases the impedance for Li-ion diffusion and causes the electrolyte to be prematurely depleted. The charge–discharge cycle performance and EIS measurements confirm the orders of magnitude improved durability of NG anodes encapsulated with N-UNCD coatings for LIBs. Further work is underway to optimize the thickness and nanostructure of the N-UNCD coatings to achieve the optimum NG-based anode performance for LIBs. In addition, work in progress in Auciello's laboratory is focused on developing the N-UNCD coating to also encapsulate LIB cathodes, which also exhibit chemical degradation.

Acknowledgments

This work was supported by the National Science Council of Taiwan via grants no. 99-2911-I-006-504, 100-2120-M-006-001, and 100-2221-E-006-169-MY3. N-UNCD films were grown at Argonne National Laboratory (ANL), supported by the US Department of Energy, Office of Science, under contract no. DE-AC02-06CH11357. The National Science Foundation supported part of recent work via grant no. 1343461 to University of Texas-Dallas.

O. Auciello and Y. Tzeng acknowledge the contributions to the R&D performed at ANL from Y.-W. Cheng, C.-K. Lin, Y.-C. Chu, A. Abouimrane, Z. Chen, Y. Ren, and C.-P. Liu.

References

[1] Y.-M. Chiang, "Building a better battery," *Science*, vol. 330, p. 1485, 2010.
[2] M. Armand and J.-M. Tarascon, "Building better batteries," *Nature*, vol. 451, p. 652, 2008.
[3] P. G. Bruce, S. A. Freunberger, L. J. Hardwick, and J.-M. Tarascon, "Li-O2 and Li-S batteries with high energy storage," *Nat. Mater.*, vol. 11, p. 19, 2012.
[4] Nexeon. Homepage. www.nexeon.co.uk.
[5] I. Kovalenko, B. Zdyrko, A. Magasinski, et al., "A major constituent of brown algae for use in high-capacity Li-ion batteries," *Science*, vol. 334, p. 75, 2011.
[6] C. K. Chan, H. Peng, G. Liu, et al., "High-performance lithium battery anodes using silicon nanowires," *Nat. Nanotechnol.*, vol. 3, p. 31, 2008.
[7] A. Magasinski, P. Dixon, B. Hertzberg, et al., "High-performance lithium-ion anodes using a hierarchical bottom-up approach," *Nat. Mater.*, vol. 9, p. 353, 2010.
[8] J. R. Dahn, T. Zheng, Y. Liu, and J. S. Xue, "Mechanisms for lithium insertion in carbonaceous materials," *Science*, vol. 270, p. 590, 1995.
[9] G. Pistoia (Ed.), *Lithium Batteries: New Materials, Developments, and Perspectives*. New York: Elsevier, 1994.
[10] R. R. Hao, Q. L. Zhang, and P.W. Shen (Eds), *Carbon Silicon and Germanium Branch of Inorganic Chemistry Series*. Beijing: Science Press, 1998.
[11] J. Shim and K. A. Striebel, "The dependence of natural graphite anode performance on electrode density," *J. Power Sources*, vol. 130, p. 247, 2004.
[12] C. Jehoulet, P. Biensan, J. M. Bodet, M. Broussely, and C. Tessier-Lescourret (Eds), *Batteries for Portable Applications and Electric Vehicles*, Pennington, NJ: Electrochemical Society, 1997.
[13] A. Ohta, H. Koshina, H. Okuno, and H. Murai, "Lithium bis(oxalato)borate stabilizes graphite anode in propylene carbonate," *J. Power Sources*, vol. 54, p. 6, 1995.
[14] E. Peled and J. P. Gabano (Eds), *Lithium Batteries*. London: Academic Press, p. 43, 1983.
[15] R. Bhattacharyya, B. Key, H. Chen, et al., "In situ NMR observation of the formation of metallic lithium microstructures in lithium batteries," *Nat. Mater.*, vol. 9, p. 504, 2010.
[16] W. A. van Schalkwijk and B. Scrosati (Eds), *Advances in Lithium-Ion Batteries*. New York: Kluwer Academic, 2002.

[17] S.-E. Lee, E. Kim, and J. Cho, "Coating," *Electrochem. Solid State Lett.*, vol. 10, p. A1, 2007.

[18] Y. P. Wu, C. Jiang, C. Wan, and R. Holze, "Influences of surface fluorination and carbon coating with furan resin in natural graphite as anode in lithium-ion batteries," *Solid State Ionics*, vol. 156, p. 283, 2003.

[19] Y. P. Wu, E. Rahm, and R. Holze, "Carbon anode materials for lithium ion batteries," *J. Power Sources*, vol. 114, p. 228, 2003.

[20] M. Yoshio, H. Wang, K. Fukuda, Y. Hara, and Y. Adachi, "Effect of carbon coating on electrochemical performance of treated natural graphite as lithium-ion battery anode material," *J. Electrochem. Soc.*, vol. 147, p. 1245, 2000.

[21] S. Komaba, T. Ozeki, and K. Okushi, "Functional interface of polymer modified graphite anode," *J. Power Sources*. vol. 189, p. 197, 2009.

[22] S. Komaba, M. Watanabe, H. Groult, and N. Kumagai, "Alkali carbonate-coated graphite electrode for lithium-ion batteries," *Carbon*, vol. 46, p. 1184, 2008.

[23] Q. Pan, H. Wang, and Y. Jiang, "Covalent modification of natural graphite with lithium benzoate multilayers via diazonium chemistry and their application in Li-ion batteries," *Electrochem. Commun.*, vol. 9, p. 754, 2007.

[24] X. L. Li, K. Du, J. M. Huang, F. Y. Kang, and W. C. Shen, "Effect of carbon nanotubes on the anode performance of nagural graphite for lithium ion batteries," *J. Phys. Chem. Solids*, vol. 71, p. 457, 2010.

[25] N. Ohta, K. Nagaoka, K. Hoshi, S. Bitoh, and M. Inagaki, "Carbon-coated graphite for anode of lithium ion rechargable batteries: graphite substrates for carbon coating," *J. Power Sources*, vol. 194, p. 985, 2009.

[26] Y.-S. Park, H. J. Bang, S.-M. Oh, Y.-K. Sun, and S.-M. Lee, "Effect of carbon coating on thermal stability of natural graphite spheres used as anode materials in lithium-ion batteries," *J. Power Sources*, vol. 190, p. 553, 2009.

[27] H. P. Zhao, J. G. Ren, X. M. He, et al., "Modification of natural graphite for lithium ion batteries," *Solid State Sci.*, vol. 10, p. 612, 2008.

[28] S. Bhattacharyya, O. Auciello, J. Birrel, et al., "Synthesis and characterization of highly-conducting nitrogen-doped ultrananocrystalline diamond films," *Appl. Phys. Lett.*, vol. 79, p. 1441, 2001.

[29] O. Auciello and A. V. Sumant, "Status review of the science and technology of ultrananocrystalline diamond (UNCD^{TM}) films and application to multifunctional devices," *Diam. Relat. Mater.*, vol. 19, p. 699, 2010.

[30] P. Badziag, W. S. Verwoerd, W. P. Ellis, and N. R. Greiner, "Nanometre-sized diamonds are more stable than graphite," *Nature*, vol. 343, p. 244, 1990.

[31] Y. Tzeng, O Auciello, C.-P. Liu, C.-K. Lin, Y..-W Cheng, "Nanocrystalline-diamond/carbon and nanocrystalline-diamond/silicon composite electrodes for Li-based batteries," US Patent #9,196,905, 2015.

[32] Y.-W. Cheng, C.-K. Lin, Y.-C. Chu, et al., "Electrically conductive ultrananocrystalline diamond-coated natural graphite-copper anode for new long-life lithium-ion battery," *Adv. Mater.*, vol. 26 (1–5), p. 3724, 2014.

[33] D. M. Gruen, A. R. Krauss, O. Auciello, and J. A. Carlisle, "N-type doping of NCD films with nitrogen and electrodes made therefrom," US patent #6,793,849 B1, 2004.

[34] J. Birrell, J. A. Carlisle, O. Auciello, D. M. Gruen, and J. M. Gibson, "Morphology and electronic structure in nitrogen-doped ultrananocrystalline diamond," *Appl. Phys. Lett.*, vol. 81 (12), p. 2235, 2002.

[35] J. Birrell, J. E. Gerbi, J. A. Carlisle, O. Auciello et al., "Bonding structure in nitrogen doped ultrananocrystalline diamond," *J. Appl. Phys.*, vol. 93, p. 5606, 2003.

[36] S. A. Getty, O. Auciello, A. V. Sumant, et al., "Characterization of Nitrogen-Incorporated Ultrananocrystalline Diamond as a Robust Cold Cathode Material," in *Micro-and Nanotechnology Sensors, Systems, and Applications-II*, T. George, S. Islam, and A. Dutta, Ed. Bellingham, WA: SPIE, p. 76791N-1, 2010.

8 Science and Technology of Integrated Nitride Piezoelectric/ Ultrananocrystalline Diamond (UNCD™) Films for a New Generation of Biomedical MEMS Energy Generation, Drug Delivery, and Sensor Devices

Martin Zalazar and Orlando Auciello

8.1 Introduction

A lab-on-a-chip is a miniaturized device that integrates onto a single chip one or several analyses which are usually done in a laboratory; these are analyses such as DNA sequencing or biochemical detection. Research on lab-on-a-chip mainly focuses on human diagnostics and DNA analysis. Miniaturization of biochemical operations normally handled in a laboratory has numerous advantages such as cost efficiency, parallelization, ergonomy, diagnostic speed, and sensitivity [1].

Sensors as a part of a lab-on-a-chip device play a very important role in medical applications. Therefore, those using acoustic waves as the operating principle comprise a very versatile class of sensors: They are highly sensitive to surface mass changes, have many applications as chemical sensors, and also can determine a variety of properties of solid or fluid media in contact with their surfaces, including liquid density, liquid viscosity, polymer modulus, and electrical conductivity [2].

The operation of acoustic wave devices is based on the propagation of bulk-launched acoustic waves (BAW) or surface-launched acoustic waves (SAW) through piezoelectric and other materials. Electroacoustics involves the transformation of acoustic energy into electric energy or vice-versa, and most often this transformation is conducted within a piezoelectric material.

Acoustic waves are easily generated within a piezoelectric material through the transduction of electric fields; such waves propagate within and onward to the boundaries of the material in diverse ways dependent on the field and material geometries [3–6]. The applications of the electroacoustic technology include frequency control, sonar and ultrasound investigations, filter applications, and sensors.

8.2 Medical Applications for Acoustic Wave Devices

When involved in medical applications, the acoustic wave device must have the capability to operate in liquid and gaseous media. In addition, if the acoustic wave device is going to be implanted into the body, it is important to comply with the biocompatibility requirements and the stated norms and regulations.

Once implanted, it is mandatory to avoid an adverse reaction of the body. The foreign body response describes the nonspecific immune response to implanted foreign materials. It is characterized by the infiltration of inflammatory cells to the area to destroy or remove this material, followed by the repair or regeneration of the injured tissue. However, if the foreign material cannot be phagocytosed and removed, the inflammatory response persists until the material becomes encapsulated in a dense layer of fibrotic connective tissue that shields it from the immune system and isolates it from the surrounding tissues.

This response is common to all medical devices or prostheses implanted into living tissue, and ultimately results in fibrosis or fibrous encapsulation, which compromises the efficiency of the device and frequently leads to device failure. This response varies depending on the implant's physicochemical properties (e.g., shape, size, surface chemistry, morphology, and porosity [7]). In order to reduce this unwanted response, having a system capable of sensing and controlling the parameters involved in this process is desirable.

8.2.1 Proposed Solution to the Problems of Current MEMS/NEMS Biosensors

If a material is going to be implanted in the human body, an important requirement is the biocompatibility of the implant. It would be desirable to have a biocompatible piezoelectric material, thus avoiding the need for packaging that represents a significant portion of the cost to the device. In this regard, AlN has emerged as an attractive alternative to other piezoelectric materials, as shown by different groups [8, 9]. AlN is a very attractive piezoelectric material for use in bio-MEMS: it is biocompatible, exhibits high resistivity, high breakdown voltage, high acoustic velocity, and it can be grown by the reactive sputtering technique at relatively low temperature, thus being compatible with complementary metal-oxide semiconductor (CMOS) device technology [10, 11].

It is also becoming important to be able to integrate piezoelectric films with materials used in medical devices. Ultrananocrystalline diamond (UNCD) in thin-film form is a multifunctional material which is extremely bioinert and biocompatible [12–15]; in addition, UNCD is capable of being integrated with CMOS technologies. For example, UNCD has been developed as a hermetic bioinert/biocompatible encapsulating coating for an Si microchip [16] to enable implantation inside the eye on the retina as a critical component of an artificial retina, and as a biocompatible coating for glaucoma valves [17-18] and dental implants (see Chapter 5 of this book). Since both UNCD and AlN films can be processed via photolithography and reactive ion etching processes used in the fabrication of micro-/nanoelectromechanical systems (MEMS/

NEMS), it has been shown that the integration of UNCD and AlN films provides the basis for developing a new generation of biocompatible bio-MEMS/NEMS [19].

Because diamond-based substrates exhibit the highest sound velocity among all materials [20] and AlN exhibits the highest phase velocity among all piezoelectric materials [21], the AlN–diamond heterostructure provides a very promising platform for fabrication of a wide range of devices from SAW to microfluidics. Zalazar et al. [19] reported the integration of films on UNCD layers exhibiting a piezoelectric coefficient of about 5.3 pm/V, one of the highest for AlN film demonstrated today.

The first part of this chapter describes the research performed to develop the integration of AlN films with as-grown UNCD and nitrogen-incorporated (in grain boundaries) UNCD (N-UNCD) films, as well as with boron-doped UNCD (B-UNCD) films. The study was focused on determining which process produced the best AlN film with the highest piezoelectric coefficient integrated on UNCD layers for optimum actuation/sensing of piezoelectric based MEMS for biomedical applications.

The second part is focused on the description of the research done to develop an AlN/UNCD-based thin-film bulk acoustic wave resonator (FBAR) as a sensor/actuator for biomedical applications. In this sense, a piezoelectrically actuated FBAR was fabricated and characterized for its use in a new generation of biosensors and drug delivery devices.

8.3 Integration of AlN and UNCD Films

This section is focused on describing the integration of AlN with as-grown UNCD and N-UNCD films, as well as with B-UNCD films. The study was focused on determining which process produced the best AlN film with the highest piezoelectric coefficient integrated on UNCD layers for optimum actuation of piezoelectric-based MEMS drug delivery devices for biomedical applications. The investigated methods included: (1) direct growth of AlN film on as-grown Pt–Ti–UNCD films with rms surface roughness of ~2–5 nm without any additional chemical mechanical polishing (CMP) or special buffering layer, which would minimize fabrication steps and cost; and (2) develop a method, involving a CMP process, to produce an extremely smooth surface which produce (002)-oriented AlN layer with the highest piezoelectric coefficient due to control of the AlN film orientation. The research was focused on developing biomedical devices, harnessing both the piezoelectric and biocompatibility properties of AlN and the biocompatibility and multifunctionality of UNCD, N-UNCD, and B-UNCD films to create a new generation of drug delivery devices.

8.3.1 Experimental Procedure for the Synthesis of Thin Films and Characterization

8.3.1.1 Synthesis of UNCD and N-UNCD Films

Pt–AlN–Pt heterostructure layers were grown on as-deposited UNCD and N-UNCD layers with a Ti adhesion film on the surface of the UNCD layer. The

Pt–AlN–Pt–Ti–UNCD and N-UNCD multilayer structures were grown in a clean-room facility. N-type Si (100) wafers with surfaces polished to a mirror finish were used as substrates. A nanodiamond seeding layer was produced first via exposure of the Si substrate surface to a solution of nanodiamond particles in methanol in an ultrasonic bath, whereby the diamond nanoparticles are embedded on the substrate surface to provide the seeds upon which the UNCD and N-UNCD films are grown. The N-UNCD films were grown in a microwave plasma enhanced chemical vapor deposition (MPCVD) system (IPLAS, GmbH-Germany) using an Ar-rich/ CH_4 mixture with added N_2. The power was set at 2300 W and the absorbed power ranged between 1900 W and 2000 W, with the pressure set at 50 mbar (Ar flow, 79 sccm; STP N_2 gas flow, 20 sccm; and CH_4 flow, 1 sccm). The UNCD films were grown in the same MPCVD systems used to grow the N-UNCD films, but using 90 mbar pressure with forward power of 1216 W and a mixture of Ar and H_2 (Ar flow, 48.7 sccm; H_2 flow, 0.5 sccm; and CH_4 flow, 0.8 sccm).

8.3.1.2 Synthesis of Metallic Electrode Pt and AlN piezoelectric films

The Pt layer was grown by magnetron sputter deposition on top of a Ti film deposited on the UNCD surface as an adhesion layer. The Pt–Ti heterostructure layer contributes to inducing a highly c-axis (002)-oriented AlN film, which provides the highest piezoelectric coefficient. The Pt layer serves as the bottom and top electrode to apply voltage to excite the piezoelectric effect on the AlN layer for actuation of the MEMS device. Films were grown using the conditions shown in Table 8.1.

AlN (002)-oriented films were grown on the Pt layers at about 500 °C, using reactive sputter deposition using an Ar-plasma to sputter Al atoms from a solid target in an N_2 atmosphere, to provide the nitrogen needed to produce the AlN films. After 5 h of deposition, the thickness of this piezoelectric AlN layer was 417 nm with a deposition rate of 83.4 nm/h. This rate is affected by the low Ar flow used, which reduces the amount of heavy ions impacting the target. A list of deposition parameters can be seen in Table 8.1.

Table 8.1 Sputtering parameters for growth of AlN, Pt, and Ti films.

Parameter	Material	Material	Material
	AlN	Pt	Ti
Base pressure (Torr)	$<1 \times 10^{-7}$	$<1 \times 10^{-7}$	$<1 \times 10^{-7}$
Process pressure (mTorr)	3	5	5
Power (W)	150^a	150^b	150^b
Substrate temperature (°C)	500	25	25
Ar flow (sccm)	3	26	26
N_2 flow (sccm)	20	0	0
Target	Al (99.999%)	Pt (99.99%)	Ti (99.995%)
Target–substrate distance (cm)	10	11	11

[a] DC power; [b] RF power.

8.3.1.3 Characterization of Structures and Properties of UNCD, N-UNCD, and AlN Films

UNCD and N-UNCD were characterized using Raman spectroscopy, involving a 632 nm wavelength laser beam to determine the chemical bonding of C atoms in the films. A typical Raman spectrum from the analysis of the UNCD films is shown in Figure 8.1. Measurements were taken at three different points on the surface of the wafer. The main resonance peaks in the spectrum are 1332 cm^{-1}, characteristic of the diamond sp^3 carbon bonding, and 1532 cm^{-1}, which reveals the existence of disordered sp^2-bonded carbon at the grain boundaries of the film.

The thicknesses of the UNCD films were measured first with a reflectometer tool and then using cross-sectional scanning electron microscopy (SEM) samples. In addition, SEM imaging was used to analyze the microstructure and morphology of the UNCD and AlN films. The crystal structure and the preferred orientation (texture) of the piezoelectric layer were identified by 2θ X-ray diffractometry (XRD). The piezoelectric coefficient of the AlN films was measured using the piezoresponse force microscopy (PFM) technique, which involves using an atomic force microscope (AFM) Pt-coated silicon cantilever to apply AC voltages that excite the piezoelectric activity in the AlN film and monitor the mechanical deformation through the piezomechanical deformation of the film, producing a piezoresponse imaging of the polarization.

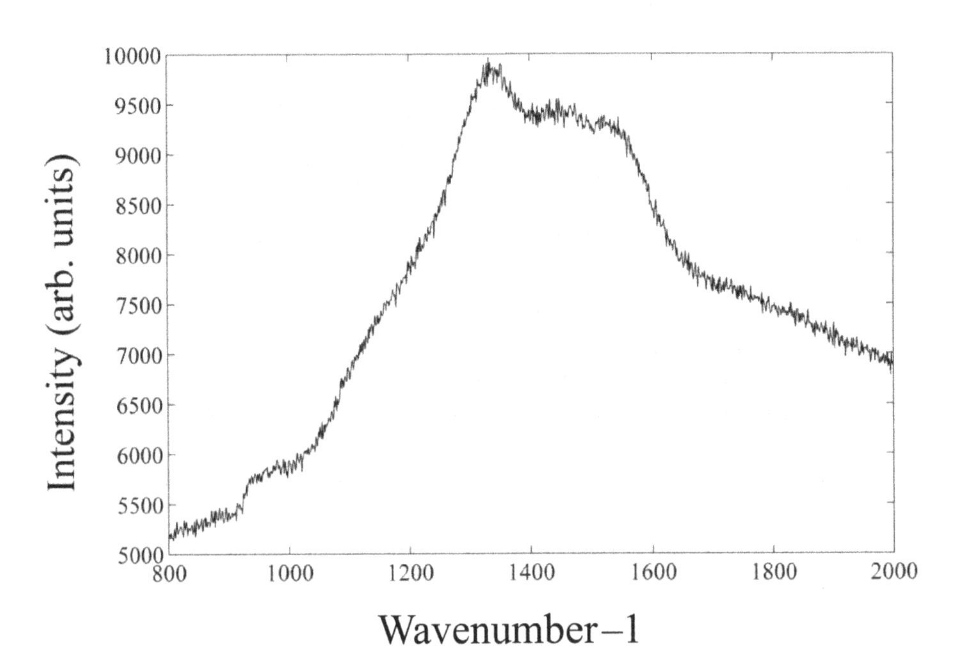

Figure 8.1 Raman spectrum from the Raman analysis of a typical UNCD film showing the 1332 cm^{-1} peak characteristic of diamond sp^3 C atoms bonding and the 1532 cm^{-1} peak, revealing the presence of sp^2-bonded carbon at the grain boundaries of the film.

Influence of Substrate Surface Roughness on AlN Film Synthesis and Properties

AlN layers with thickness in the 260–420 nm range were synthesized on an SiO_2 (100 nm) layer grown on an Si substrate, as-grown UNCD films on an Si substrate, and chemically and mechanically polished B-UNCD films on an Si substrate, respectively, to investigate the effect of the substrate surface nature and rms roughness on the orientation of the AlN film, which has a substantial impact on the corresponding piezoelectric properties. Their thicknesses were obtained first with a reflectometer tool (Filmetric F20 Thin Film Analyzer) and then by via cross-sectional SEM.

SEM Analysis. A SEM micrograph of the top surface of the AlN film and the cross-section of the AlN–Pt–Ti–UNCD heterostructure is shown in Figure 8.2. The thicknesses of the Pt and Ti layers, used as the bottom electrode heterostructure, were 150 nm and 10 nm, respectively, and provided the underlying substrate for growing the piezoelectric AlN layer on the UNCD film grown on an Si substrate. The AlN film exhibits a columnar microstructure which is perpendicular to the UNCD surface with orientation (002), as revealed by the XRD analysis shown in Figure 8.3d, e.

Figure 8.3a, b, c shows cross-section SEM images of AlN films grown on different underlying films with different roughness. The XRD spectra for each of the samples are shown in Figures 8.3d, e, f. These figures show the influence of the roughness of the substrate to produce highly (002)-oriented AlN films.

XRD Analysis. The X-ray wavelength used for XRD analysis was $\lambda = 1.540598$ Å (Kα1, long fine focus, Cu anode radiation). The c-lattice constant for the AlN film was calculated (4.9681 Å) using Bragg's law, which is in good agreement with the c-lattice constant value of AlN published in the literature by other groups.

The XRD $2\theta/\omega$ diffraction pattern from the XRD analysis of AlN–Pt–Ti–UNCD–SiO_2–Si, AlN–Pt–Ti–B-UNCD–SiO_2–Si, and AlN–Pt–Ti–SiO_2–Si multilayers were obtained by scanning between $2\theta = 0°$ to $80°$. Pt and Ti layers, used as the bottom electrode heterostructure, provided the underlying substrate for growing the piezoelectric AlN layer on the UNCD film. Figure 8.3 shows the XRD $2\theta/\omega$ scans of the

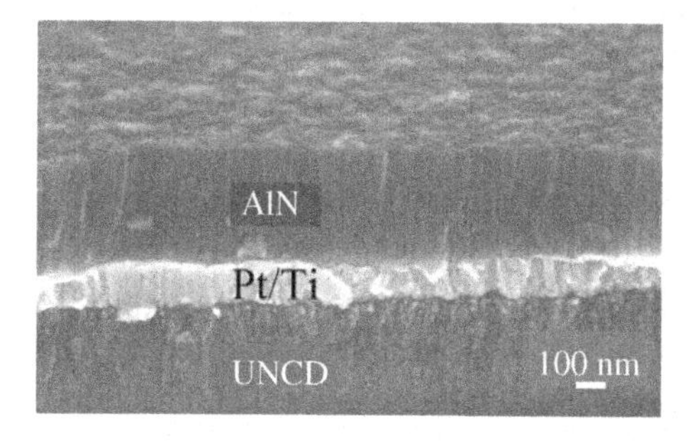

Figure 8.2 SEM micrograph of the surface morphology of the AlN layer and cross-section of the AlN–Pt–Ti–UNCD–SiO_2–Si heterostructure.

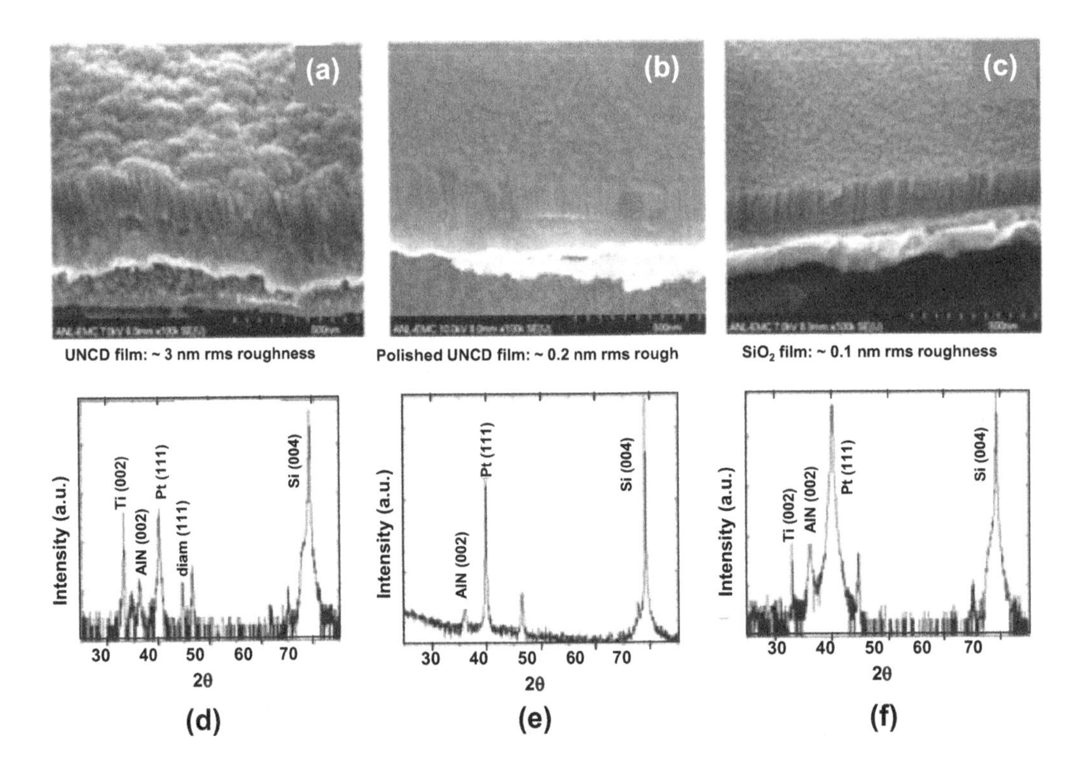

Figure 8.3 Cross-section SEM analysis of AlN films grown on as-grown UNCD films (a), polished UNCD film (b), and on SiO_2 film on Si substrate (c). XRD spectra for AlN films grown on as-grown UNCD films (d), polished UNCD film (e), and SiO_2 film on Si substrate (f) (reprinted from *Appl. Phys. Lett.*, vol. 102, p. 104101, 2013 (Fig. 1) in [19] with permission from AIP Publisher).

multilayer films, showing high c-axis orientation for the AlN film. The presence of the hexagonal AlN (002) diffraction peak at 36.05° is noted, indicating that the AlN layers are highly (002) oriented, which correlates with the high piezoelectric coefficient exhibited by this film. The XRD spectra also show the diffraction peaks of Pt (111) at 40.05° and Ti (002) at 34.45°, corresponding to the electrode, and a peak at 44.7°, which can be attributed to diamond (111) or Ti.

In the XRD spectra also appears a peak at 33.20°, which arises from diffraction in the silicon substrate, and is the result of Si crystalline imperfections. In addition, the very narrow width is an indication that it comes from a single crystal. Figure 8.4 shows the XRD scan of an Si wafer, showing what is known as the extremely sharp and thin ghost Si peak on the left.

While the thick AlN films with >400 nm thickness on different underlying films show similar XRD patterns, the crystallization of thinner AlN films is largely affected by substrate conditions, mostly by their surface roughness. Thinner AlN films with high (002) orientation can be produced on surfaces with rms roughness ≤ 1 nm, as is the case of atomically flat SiO_2 surfaces (Figure 8.3c), unlike the UNCD and

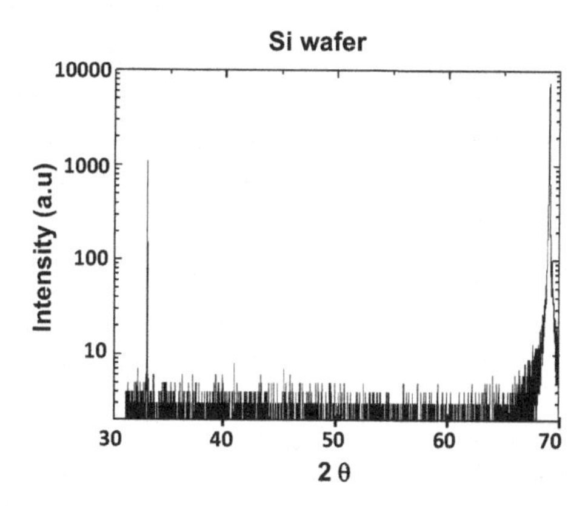

Figure 8.4 XRD scan of an Si wafer showing the ghost peak related to the silicon wafer.

N-UNCD layers that exhibit roughness of 5–7 nm and 7–10 nm, respectively, as previously measured extensively by Auciello et al. [12–16]. AlN films grown on CMP B-UNCD films with a Pt bottom electrode exhibit surface roughness of about 0.2 nm and high (002) orientation (see the XRD spectrum in Figure 8.3e), as for films grown on atomically flat semiconductor surfaces. The results presented here indicate that the smoothness of the substrate surface is critical for achieving highly (002) oriented very thin AlN films.

8.3.2 Measurements of Piezoelectric Response of AlN Films

The piezoelectric coefficient of AlN films grown on UNCD, N-UNCD, and B-UNCD layers was measured using the PFM technique developed by Auciello and Gruverman [22–24]. Briefly, the PFM technique involves application of an AC voltage to the AlN films between the AFM tip, positioned on the top surface of the AlN film, and the bottom electrode layer. The AC voltage induces displacement of positive and negative ions in the lattice of the AlN film, which produces the mechanical displacement of the film, named piezoresponse of AlN films, which appears as a function of the applied AC voltage. Using the average piezoresponse amplitude of AlN films and inverse optical lever sensitivity between the AFM tip and AlN films, the piezoelectric coefficients of AlN films were calculated. These measurements revealed a uniform piezoresponse (Figure 8.5). However, the AlN films grown on as-deposited UNCD and N-UNCD surfaces exhibited piezoelectric coefficients of 1.91 pm/V and 1.97 pm/V, respectively, while the AlN films grown on the CMP B-UNCD surfaces exhibit ~5.3 pm/V, which is the highest among currently reported values for AlN films.

The data shown above indicate that the surface roughness of the substrates is critical to produce highly (002)-oriented AlN films. Growth of highly (002)-oriented AlN films on UNCD layers is critical for achieving the highest possible piezoelectric

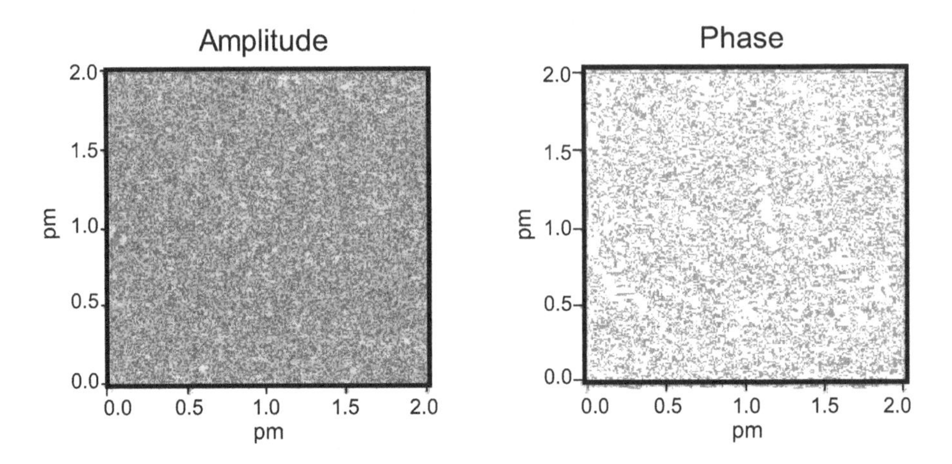

Figure 8.5 PFM imaging of piezoelectric activity in (002)-oriented AlN film (~200 nm thick) grown on CMP B-UNCD film with rms surface roughness of about 0.2 nm. The PFM measurement was used to determine the piezoelectric coefficient of the AlN film, which was 5.3 pm/V, one of the highest demonstrated today for AlN (reprinted from *Appl. Phys. Lett.*, vol. 102, p. 104101, 2013 (Fig. 2) in [19] with permission from AIP Publisher).

coefficient, which is of paramount importance for achieving optimum actuation in piezoelectrically actuated MEMS/NEMS devices. In addition, the Pt (111)-oriented electrode layers used to grow the AlN films contribute to producing oriented AlN layers by inducing local epitaxial correlation between the Pt and AlN lattices.

It has been observed in the past that AlN films grown on smooth Si surfaces exhibit highly c-oriented crystallites, while similar films grown on rough surfaces exhibit a weak (002) orientation. The hypothesis is that a rough surface favors a heterogeneous orientation of the AlN crystallites. This phenomenon is established when the local surface inclination varies on a scale between the diameters of the AlN grain and the nuclei, thus hampering the c-axis (002) orientation; therefore, the high density of oblique crystallographic orientations prevents the growth of highly textured material, as stated by Artieda et al. [25].

The same concept of Si surface roughness effect on AlN films should apply to UNCD surfaces, and indeed, the data presented here show that the atomically smooth B-UNCD surface induced the growth of (002)-oriented AlN film exhibiting one of the highest piezoelectric coefficient seen for AlN films today.

In addition to the effect of the substrate surface roughness, it is important to take into account the effect of the microstructure of Pt electrode layers grown on the substrate, used to apply voltages to induce the piezoelectric effect on the AlN layer to achieve actuation. In this sense, Artieda et al. [25] showed that AlN films grown on Pt (111) electrode layers are highly (002)-oriented, largely independent of the roughness of the Pt film. This can be explained due to a good alignment of Pt (111) planes on top of the Pt grains where the AlN (002) nucleates epitaxially. The Pt roughness can be due to facets at grain boundaries, on which epitaxy of AlN is still compatible with c-planes parallel to the substrate surface. In the case of rough amorphous substrates, the large number of germination sites having various orientations prevents the formation of a dense, well-aligned, c-oriented fiber structure of AlN, inducing a disordered microstructure with pores and, consequently, resulting in low residual tensile stresses [25]. The stress increases gradually from compressive (negative values) to slightly tensile (positive values) with the surface roughness of the substrate. Large compressive stress was observed by Artieda et al. [25] for AlN films grown on smoother SiO_2 surfaces. The large amount of germination sites for AlN growth on rougher substrates leads to smaller grains and a high density of voided grain boundaries [25]. On the other hand, AlN growth on smooth surfaces is facilitated by high mobility of the atoms on the surface, leading to dense grain boundaries and compressive stresses. The impact of roughness on the stress seems to follow quite general rules in as much as a Pt electrode surface leads to a stress that fits the trend observed with Si surfaces.

8.4 New Generation of Biomedical MEMS Based on Integrated AlN–UNCD Films

8.4.1 MEMS Sensors

Sensors produce an output signal in response to some input quantity (Figure 8.6). The output signal is usually electrical, such as an analog voltage or current, a stream of digital voltage pulses, or an oscillatory voltage whose frequency represents the value of the input quantity. The range of input quantities includes physical quantities such as the mechanical properties of thin films and chemical and biological quantities such as the concentrations and identities of unknown species in air or liquid media.

Inside the sensor shown in Figure 8.6, the transduction process is carried out, converting an input event to an electric signal. Also, the sensor could contain the needed circuitry for conditioning the transduction process signal into a more robust and suitable one for use outside the sensor itself.

It is known that the dominant physical quantities may change as a function of length scale. In this context, devices at the microscale show that inertial forces become less affective, whereas van der Waals forces, electrostatic forces, and surface tension forces become dominant. In this sense, it is important to have knowledge about the scaling behavior and other properties of the sensors in order to take a decision on what sensor to use.

8.4.2 Materials Science and Fabrication of Thin Film–Based Bulk Acoustic Wave Resonator

Currently, there are efforts focused on the development of thin film–based acoustic resonators. This technology was born as a direct extension of quartz crystal resonators.

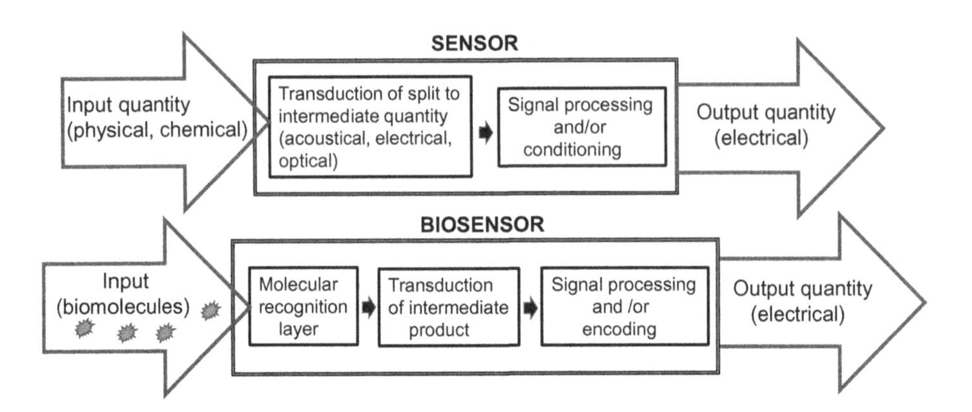

Figure 8.6 Schematic diagram of a sensor producing an electrical output in response to an input quantity (top). Biosensor comprising the generic device shown at top with a molecular recognition layer that has a highly selective response (bottom) [1].

These are BAW-based thickness mode wave resonators, permitting a reduction in the piezoelectric thickness of the resonator, thus conducting to an increased resonance frequency (>1 GHz). Thin film–based resonators [26] provide much higher sensitivity and resolution in mass detection, and in addition they are much more easily integrated with microelectronic drivers.

Film-based bulk acoustic wave resonators could be suspended on a substrate, usually silicon, in order to facilitate its resonance, thus avoiding vibration damping (system loses). As opposed to freestanding or suspended resonators, composite resonators can also be used where the piezoelectric film is deposited on a nonpiezoelectric substrate, such as silicon, with intermediate layers with different acoustic impedances. Composite film resonators can display improved thermal stability due to the property matching that can be obtained among different layers. A significant case is when the layers have alternate high and low acoustic impedances, thereby forming a Bragg reflector that acts as an acoustic mirror that isolates the film from the substrate, often termed *solidly mounted resonator* (SMR).

This section is focused on describing the R&D for development of an AlN–UNCD-based FBAR for biomedical applications. In this case, an FBAR piezoelectric biosensor is fabricated and characterized. Even though the analysis of the behavior of an FBAR piezoelectric mass sensor was carried out, it can be extended to new types of sensor devices.

8.4.2.1 Geometric Design and Fabrication of an FBAR

The simplest configuration of the FBAR membrane involves the acoustic resonance cavity formed by the air cavity underneath the resonating membrane by etching completely an Si substrate; the substrate was removed to acoustically isolate the resonance cavity, thus creating a freestanding membrane. The membrane was composed of a Pt–Ti–AlN–Pt–Ti–UNCD multilayer on a silicon substrate, as shown in the schematic of Figure 8.7a. Pt layers serve as the bottom and top electrodes to apply

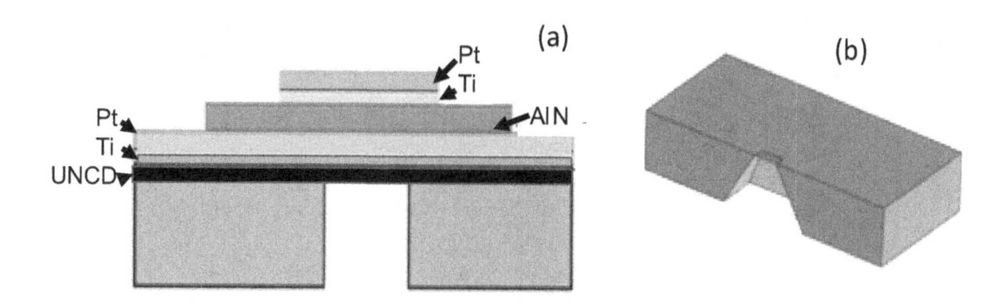

Figure 8.7 UNCD-based AlN FBAR. (a) Schematic of the heterostructure of Pt–Ti–AlN–Pt–Ti–UNCD–SiO_2–Si; (b) 3D longitudinal cut image of the freestanding FBAR heterostructure.

voltage to excite the piezoelectric effect on the AlN layer for actuation of the FBAR device. The top electrode (Pt–Ti) is similar to the bottom electrode, using a Ti layer on the UNCD surface as an adhesion layer for the Pt film.

Devices composed of multilayer structures, where the layers exhibit different thermal expansion coefficients, often suffer from intrinsic residual stresses that can be reduced using low-temperature processes. In addition, being a low-temperature process makes it promising for integrating with the IC technology.

A schematic of the FBAR device can be seen in Figure 8.7b; a longitudinal cut of the freestanding heterostructure membrane composed by a Pt–Ti–AlN–Pt–Ti–UNCD multilayer on the silicon substrate can be seen.

Mask Design and Fabrication

Three different masks for the entire device were designed and fabricated:

- mask 1: cavity on Si
- mask 2: piezoelectric AlN
- mask 3: top electrode

Figure 8.8 shows the mask for the cavity on the Si substrate. Four different sizes were used in this experimental step in order to compare their performances. Square membranes were designed for the FBARs.

For the cavity design, the substrate material and the used etching method have to be considered. In the case of silicon, a common etching method is based on wet anisotropic etching, produced by potassium hydroxide (KOH), which is a strong base capable of attacking the Si substrate in an anisotropic way. N-type Si (100) wafers with the surface polished to a mirror finish were used as the substrate, so the KOH etched the Si preferentially in the [100] plane.

The masks used have a multilayer structure with the following layers: PR–Cr_2O_3–Cr–glass. The photoresist (PR) is the last layer and the Cr_2O_3 serves as an antireflective film. The glass is highly flat and made of borosilicate. The fabrication of the masks was done using a laser pattern generator (LW405). The pattern was generated over the photoresist on the chrome layer and the glass (square shape of 125 mm) over

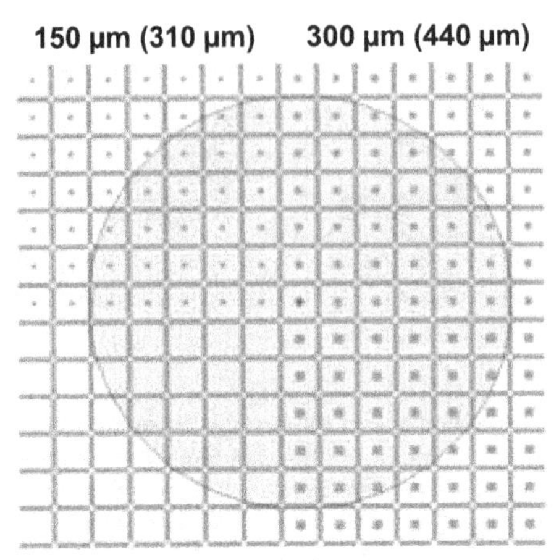

Figure 8.8 Mask layout for fabrication of the Si cavity. The four different sizes used in this experimental step can be observed, where the value in microns is the calculated length of the square, while in brackets is the real length due to undercuts.

24 h. After that, the photoresist was developed (developer 351) and a wet etching of the chrome film was done.

Fabrication of the FBAR Device

The freestanding membrane fabrication on the Si substrate involves several steps:

1. **UNCD film growth.** A UNCD film was grown on the front surface of the silicon wafer (N-type mirror-polished Si (100)) UNCD layers were grown using an MPCVD system (Lambda Technologies) powered by a 915 MHz magnetron generator at 2300 W. This generates a plasma from a mixture of Ar (99%)/CH_4 (1%) gases, producing C_2 dimers as the growth species at a pressure of 80 mTorr. This layer was used as a substrate for the AlN layer, and also serves as a stop layer for the anisotropic wet etching of the Si by using KOH.
2. **Si_3N_4 mask.** Silicon nitride (Si_3N_4) was used as a mask for the wet etching of the Si using KOH. Commercial, polished, low-stress Si_3N_4 residual (LSLPCVD, commercial grade, University Wafer, Inc.) was used. The pattern on the Si_3N_4 mask, to produce the cavities, was performed by photoresist spin coating (S1813), UV exposure for 5 s (mask aligner Karl Suss MA6), resist development (351 developer diluted in a 3:1 ratio in deionized water), and dry etching of the Si_3N_4 in a tetrafluoromethane (CF_4) plasma (Oxford CS-1701 RIE).
3. **Pt–Ti films growth.** The Pt layer was grown by magnetron sputter deposition on top of a Ti film deposited on the SiO_2 surface as an adhesion layer. The Pt bottom electrode was grown by sputter deposition (AJA ATC Orion thin film deposition system). The Ti film provides an adhesion layer for growing highly c-axis (002)-oriented AlN films. Films were deposited with a target RF power of 150 W,

pressure of 3 mTorr, Ar flow rate of 26 sccm, and base pressure of 1e–7 Torr. The thicknesses of the Pt and Ti were 150 nm (720 s deposition) and 10 nm (310 s deposition), respectively. In addition, this coating provides seeding capability and serves as a buffer layer, lowering the discrepancy in lattice parameters between substrate and film. It also helps to improve the adhesion of the AlN to the UNCD.

4. **AlN film growth.** AlN films for MEMS devices are most often synthesized in the Wurtzite hexagonal crystallographic structure with c-axis (002) orientation perpendicular to the surface (diffraction peak at 36°), which is the orientation that yields the highest piezoelectric constant in addition to a high acoustic wave velocity [26]. AlN (002)-oriented films were grown on the Pt layers at 500 °C at a pressure of 3 mTorr using reactive sputter deposition (AJA ATC Orion thin film deposition system) via sputtering material from an Al metallic target. Ar-plasma was used to produce Ar ions to sputter the Al material in an N_2 atmosphere to produce the AlN films. An aluminum target with a diameter of 4 inches and a purity of 99.999% was used. Several experiments were performed to reach the crystallographic orientation (002) of the AlN with the diffraction peak at 36°. AlN layers with thicknesses in the 260–420 nm range were synthesized on silicon oxide (SiO_2), and silicon nitride (Si_3N_4) was used as the intermediate layer grown on Si.
5. **AlN patterning.** The pattern was transferred to the AlN film by spin coating of the photoresist (ma-N 415 negative photoresist), UV exposure for 30 s (mask aligner Karl Suss MA6), resist development (533), and dry etching using anisotropic Cl_2-based reactive ion etching (Oxford CS-1701 RIE).
6. **Electrode patterning.** Patterning of Pt–Ti top electrode was done by applying lift-off techniques. First, a spin coating with S1813 on the frontside of the wafer, UV exposure, and resist development were done. Finally a Pt–Ti sputter deposition and posterior removal of the photoresist was performed.
7. **Cavity formation.** The cavity etching was carried out on the backside of the Si wafer, using KOH (30%). Hermetic protection to avoid frontside etching was used. Time for etching was estimated in 9 h at a temperature of 85 °C using 250 rpm stirring. The cavity prototype can be seen in Figure 8.9, using only UNCD as the membrane on the Si wafer substrate.

8.4.2.2 Characterization of the Materials and Structure of the FBAR Device

XRD Analysis

The XRD θ–2θ diffraction pattern taken from the XRD analysis of the AlN–Pt–Ti–UNCD–SiO_2–Si multilayer were scanned between $2\theta = 0°$ and $80°$. Figure 8.10 shows an XRD θ–2θ scan of the AlN–Pt–Ti–UNCD–SiO_2–Si layered film exhibiting high c-axis orientation for the AlN film. The presence of hexagonal AlN (002) diffraction peak at 36.05° can be noted, revealing the presence of mainly a (002) textured AlN film. Also, two Pt peaks, (111) at 40.05 and (002) at 46.5, are present. The weak peaks at 44.7 and at 65.1 correspond to Al (002) and Al (202), respectively. The strong narrow peak at 33.2 comes from the silicon wafer substrate and is the result of Si crystalline imperfections.

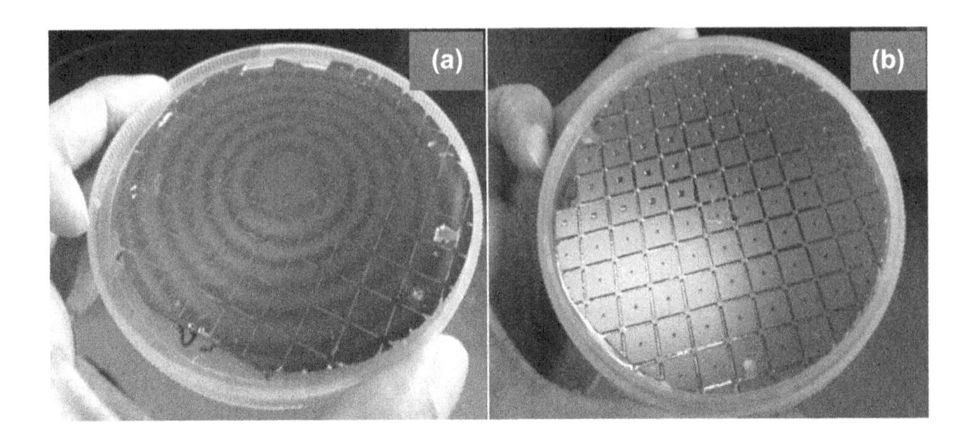

Figure 8.9 Optical pictures of fabricated cavities. (a) Frontside of the UNCD film; and (b) backside showing the different sizes of the cavities within dices.

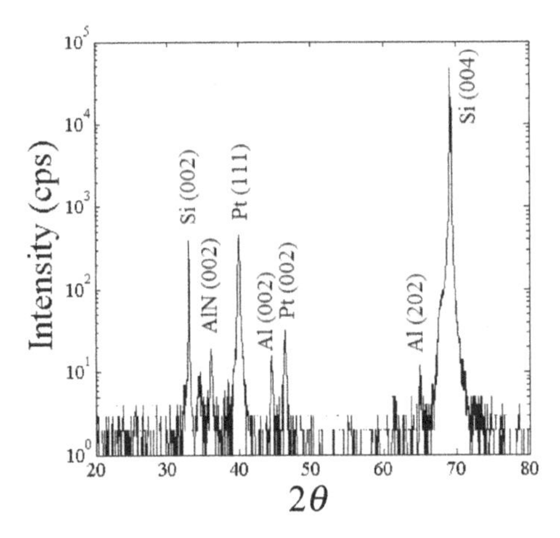

Figure 8.10 XRD spectrum of an AlN–Pt–Ti–UNCD–SiO_2–Si heterostructure showing the characteristic peaks of the Pt layers and Si substrate, in addition to the critical AlN (002) peak that reveals the high orientation of AlN, necessary to yield the high piezoelectric coefficient.

The X-ray wavelength was $\lambda = 1.540598$A (Kα1, long fine focus, Cu anode radiation). By using Bragg's law, it was calculated that the c-lattice constant for the AlN layer had a value of 4.9681 Å, which is in good agreement with the value of the AlN found in the literature [25].

Device Structure Imaging by Optical Microscopy

Optical microscopy images provided the first easy observation to evaluate the quality of the KOH etching in producing the cavities. The morphology of the membranes and cavities is shown in Figure 8.11. The difference in the square membranes between the frontside and backside of the Si wafer was caused by the classical pyramid-shaped Si

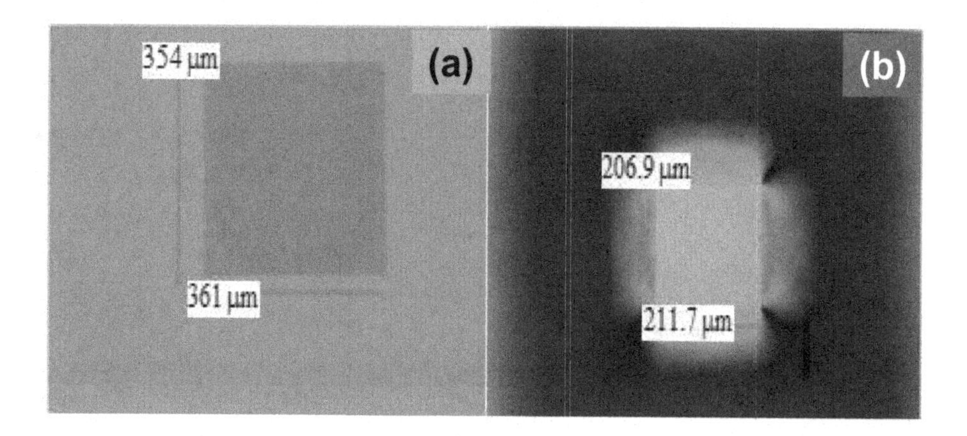

Figure 8.11 Square UNCD membranes. (a) Membrane (mask 300 μm) seen from the front; (b) membrane (mask 150 μm) seen from the back (cavity).

material; it obeys the preferential anisotropic etching direction (54.7°) of the (100) crystal plane of the Si wafer.

The size of each membrane is moderately higher than expected due to the generated undercuts below the membranes. Undercuts increase the square sizes in a range between 50 μm and 170 μm. The calculated etching rate for the Si using KOH was 1.16 μm/min.

SEM Studies of the Integrated FBAR Materials

Figure 8.12a shows a cross-section of the AlN–Pt–Ti–UNCD membrane (cantilevered). It exhibits a columnar microstructure. These columnar crystals are perpendicular to the substrate surface, in agreement with the XRD results where the AlN film is highly textured. The piezoelectric film reached a thickness of 417 nm after 5 h of growth, thus arriving at a deposition rate of 83.4 nm/h (Figure 8.12b). This rate was affected by the low Ar flow used, reducing the amount of heavy ions impacting on the target. The UNCD layer used was 420 nm in thickness.

A clarified view of the generated Si cavity can be appreciated in the SEM image of the freestanding AlN–Pt–Ti–UNCD membrane on the Si substrate (Figure 8.13). An in-focus image of the tilted Si walls (54.7°) can be seen where the big square has the dimensions of the mask and the smaller is the AlN–Pt–Ti–UNCD membrane. The white frame surrounding the big square is produced by the generated undercuts under the UNCD.

8.4.3 Materials Science and Fabrication of Thin Film–Based Drug Delivery Devices

Piezoelectric actuation could be used to break a membrane covering drug reservoirs, as a switch valve for opening the reservoirs, or by pushing the liquid outside of the reservoir. An increasing demand to fulfill unmet clinical needs in healthcare has

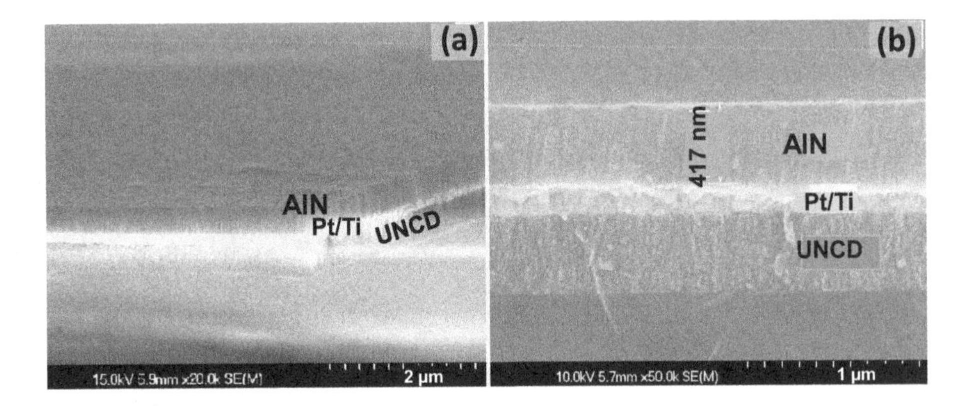

Figure 8.12 SEM cross-section of the film for the AlN–Pt–Ti–UNCD membrane. (a) Cantilevered membrane; (b) AlN film thickness measurement in the cross-section SEM image.

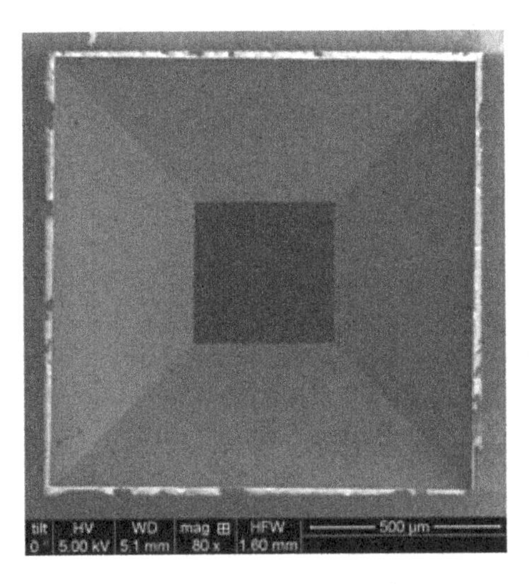

Figure 8.13 SEM image of the freestanding AlN–Pt–Ti–UNCD membrane on the Si substrate.

brought about a variety of new technologies to detect and treat a disease at an early stage and achieve a prompt therapeutic intervention. Early therapeutic interventions would result in higher survival rates, better quality of life for the patient, and reduced healthcare costs for the healthcare system. From a therapeutic standpoint, searching for a better performance in therapeutic systems represents better compliance with the therapeutic regimen, decreases potential for adverse reactions, and reduces on healthcare costs. As a consequence a medical technology paradigm shift toward more reliable, faster, actively controlled, and low-cost technologies are urgently needed. Fortunately, these kinds of technologies already exist in the market, known as bio-MEMS. This technology results from the integration of structural and functional microparts such as microchannels, microreservoirs, microsensors, and microactuators, and have recently been developed to overcome several limitations imposed by the current state of the art in pharmaceutical technology. In the case of drug delivery systems, the limitations include poor control over geometry, poor control over drug payload, stability constraints on the systems influenced by thermodynamic or kinetic parameters, poor reproducibility between sets of devices, and lack of active control over the drug release, among others. On the other hand, bio-MEMS for therapeutic applications are highly reproducible, could be integrated with electronics and actively controlled, their geometry could be tuned as desired, they are cost-effective when batch production is needed, and drugs could be contained in a precise amount in a sealed environment, keeping their properties intact. Drug delivery systems are important medical devices providing significant medical (improved pharmacokinetics decreasing drug dose and toxicity) and commercial (increasing product portfolios by adding new products and decreasing drug discovery costs by recycling old drugs) advantages.

The first part of this section describes the materials integration and fabrication and characterization of two different kinds of UNCD-based membranes for drug delivery devices. The first is a passive device where the drug diffuses through holes with controlled size, shape, and spacing between them. The second approach is an active system in which a piezoelectrically actuated valve allows a controlled drug release from a microreservoir by the application of an electric field.

8.4.3.1 Materials Processing and Device Fabrication

UNCD membranes were fabricated in a clean-room facility using thin-film deposition and microfabrication techniques, including: (1) UNCD thin films grown by MPCVD on Si substrates after a nanodiamond seeding process on the silicon surface; (2) photolithography, using a specially designed mask, used to define a window on the backside of the Si wafer; (3) reactive ion etching, used to pattern the Si_3N_4 mask to subsequently produce chemical etching of the Si substrate to produce the cavity sustaining the UNCD membrane on the front of the wafer; and (4) wet chemical etching, used to etch the Si wafer from the backside to create the cavity sustaining the UNCD membrane. Square membranes of 200–1000 μm with a thickness of 100–500 nm were fabricated and characterized by Raman spectroscopy, optical microscopy, SEM, and reflectometry.

Fabrication and Testing of a Passive Drug Delivery Device

The passive drug delivery device consists of a thin UNCD membrane in which an array of microholes were produced (see Figure 8.14). Size, shape, and spacing between holes control the amount of drug diffusing from the interior of the device reservoir to the outside. Thus, an accurate design permits a predictable release of the drugs to the media.

Fabrication of passive drug delivery devices was done using the focused ion beam (FIB) technique to produce holes with micron-size dimensions in the UNCD membranes to enable controlled drug diffusion through the latter (Figure 8.14).

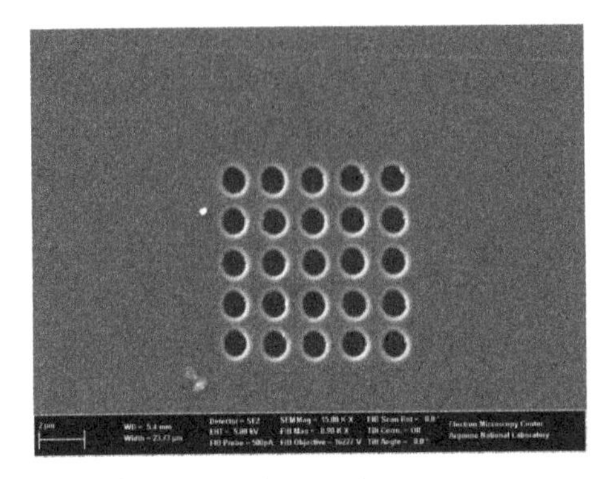

Figure 8.14 SEM micrograph of the passive UNCD drug delivery device.

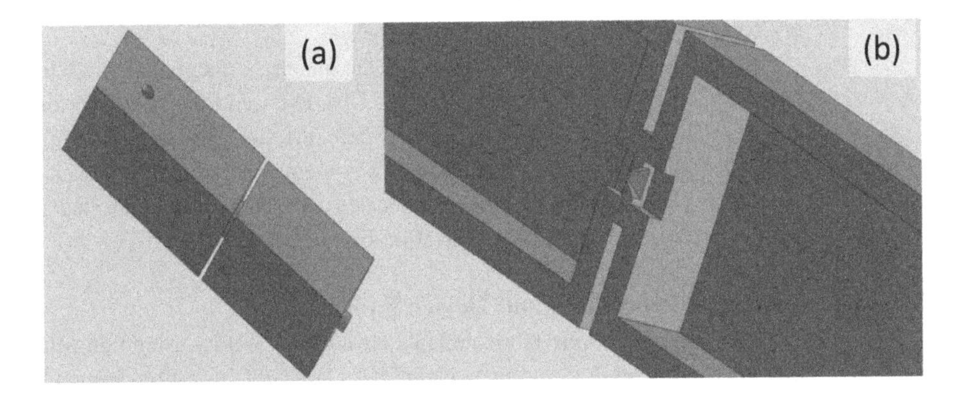

Figure 8.15 Drug delivery device with selective UNCD membrane and polymer reservoirs. (a) complete system; (b) membrane, with holes, attached to a polymer reservoir.

Figure 8.14 shows the structure of a passive drug delivery device, featuring an array of 5 × 5 circular holes with 1.217 μm diameter and 2.086 μm periodicity, with a magnification of 8000× and no tilting of the sample. Exposing the 5 × 5 array of 1 μm holes at a depth of ~500 nm took around 15 min.

The prototype drug delivery device is depicted in Figure 8.15, where the silicon substrate sustaining the membrane with the holes in is attached to a polymer reservoir containing the drug solution to be released.

A polymer polydimethylsiloxane (PDMS) was used to seal the bottom face of the Si substrate, where the integrated UNCD membrane–cavity structure was fabricated to create the reservoirs.

Fabrication and Testing of an Active Drug Delivery Device

For the active drug delivery device, based on a piezoelectrically actuated valve, a Pt–piezoelectric AlN–Pt layer heterostructure was grown on the UNCD membrane with a Ti adhesion layer, followed by FIB etching to define the valve aperture. The

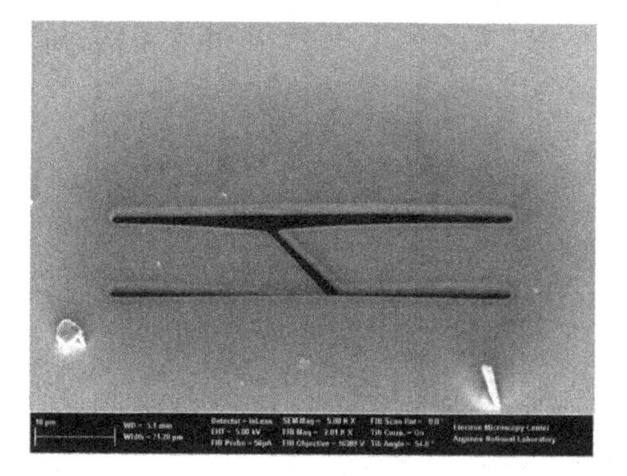

Figure 8.16 SEM image of a cantilever-based piezoelectrically actuated valve as part of an active drug delivery device.

UNCD cantilever valve, as the main part of the active drug delivery device, is depicted in Figure 8.16.

The active drug delivery device involves the actuation of a piezoelectric valve involving an AlN film as the biocompatible piezoelectric film encapsulated between two Pt electrodes and the whole structure grown on a UNCD film grown on a Si wafer. This device is capable of controlling drug delivery dosage with an applied electric field between the two Pt electrodes encapsulating the piezoelectric AlN layer, which actuates the lever valve, shown in Figure 8.16, inducing the flow of an aqueous media from a reservoir to the outside.

The morphology and operation of the piezoelectrically actuated cantilever is shown in the image from a numerical simulation (Figure 8.17), where the piezoelectric layer is grown on the top of the UNCD membrane, occupying a small part of the latter and thus increasing the strain of the membrane on the valve.

8.5 Discussion and Conclusions

A direct method for growing AlN films on an Si substrate, avoiding polishing steps and thus reducing fabrication times, has been demonstrated. It has been shown that thin AlN films with high (002) orientation can be produced on surfaces with rms roughness ≤ 1 nm, as is the case of atomically flat surfaces of SiO_2 films grown on Si wafers. The AlN films exhibit a highly textured surface with columnar structure, enabling the fabrication of SAW resonators and piezoelectric actuators. The Pt film has proven to be a good buffer layer, serving as the bottom and top electrodes.

The research provided valuable information for understanding the fundamental physical and chemical processes involved in the integration of piezoelectric AlN films with underlying UNCD, N-UNCD, and B-UNCD layers. AlN films deposited on as-grown UNCD or N-UNCD films exhibit relatively good (002) orientation due to the fact that these diamond films exhibit rms surface roughness of about 5–10 nm, which is the size of the grains. The columnar structure of the (002)-oriented AlN films grown

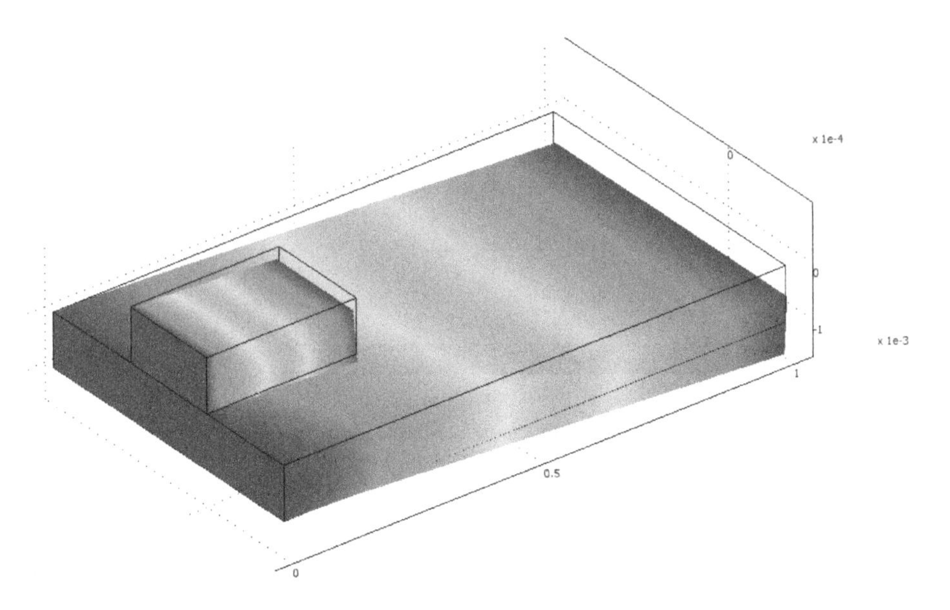

Figure 8.17 Computer simulation showing the morphology and operation of a piezoelectrically actuated Pt–AlN–Pt–Ti–UNCD cantilever valve for an active drug delivery device.

on the UNCD layers enable relatively high piezoelectric coefficient (~1.91 pm/V), but still require AlN film thickness $\geq\sim$ 400 nm to achieve the high (002) AlN orientation. A relatively high voltage (>10 V) applied between the bottom and top electrodes sandwiching the AlN layer is needed to produce the actuating piezoelectric effect in films as thick as 400 nm. It would be desirable to achieve high (002) AlN orientation with films $\leq$100 nm to produce actuation at much lower voltages ($\leq$2.5 V). In this sense, the CMP B-UNCD film, with rms surface roughness of about 0.2 nm, used as a substrate, yielded a highly oriented (002) AlN film even as thin as 80 nm. The 200 nm AlN on B-UNCD films exhibited a piezoelectric coefficient of about 5.3 pm/V, one of the highest for AlN film demonstrated today.

The feasibility of the fabrication of an FBAR AlN–diamond structure has been demonstrated. The AlN–diamond structure is a very promising device for FBAR applications, and it opens a huge field of biomedical applications.

The R&D demonstrated that the integration of two unique dissimilar materials, a piezoelectric one (AlN) and a diamond-based UNCD film, can produce a transformational new generation of external and implantable MEMS/NEMS sensors and drug delivery devices to improve the quality of life of people worldwide.

Acknowledgments

This work was supported by the US Department of Energy, Office of Science, under contract no. DE-AC02-06CH11357.

References

[1] V. Srinivasan, V. K. Pamula, and R. B. Fair, "An integrated digital microfluidic lab-on-a-chip for clinical diagnostics on human physiological fluids," *Lab Chip*, vol. 4 (4), p. 310, 2004.

[2] J. W. Grate, S. J. Martin, and R. M. White, "Acoustic wave microsensors: Part II," *Anal. Chem.*, vol. 65 (22), p. 987, 1993.

[3] M. Zalazar, *Mass Microsensors for Implantable MEMS*. Atlanta, GA: Scholars Press, 2014.

[4] M. Zalazar, "Design, simulation, fabrication and characterization of mass microsensors embeddable to an implantable microvalve for glaucoma treatment," Thesis, Universidad Nacional del Litoral, Facultad de Ingeniería y Ciencias Hídricas, Instituto de Desarrollo Tecnológico para la Industria Química, Santa Fe, Argentina, 2013.

[5] M. Zalazar and F. Guarnieri, "Quartz crystal microbalance: design and simulation," *Mecánica Computacional*, vol. 28, p. 2123, 2009.

[6] M. Zalazar and F. Guarnieri, "Analysis and evaluation of piezoelectric sensors behaviour," *Mecánica Computacional*, vol. 29, p. 6665, 2010.

[7] B. Rolfe, J. Mooney, B. Zhang, et al., "The fibrotic response to implanted biomaterials: implications for tissue engineering," in *Regenerative Medicine and Tissue Engineering: Cells and Biomaterials*, D. Eberli, Ed. London: IntechOpen, 2011.

[8] K. H. Chung, G. T. Liub, J. G. Duhb, and J. H. Wang, "Biocompatibility of a titanium–aluminum nitride coating on a dental alloy," *Surf. Coat. Technol.*, vol. 188, p. 745, 2004.

[9] C. C. Chen, C. T. Lin, S. Y. Lee, et al., "Biosensing of biophysical characterization by metal-aluminum nitride-metal capacitor," *Appl. Surf. Sci.*, vol. 253, p. 5173, 2007.

[10] C. H. Chou, Y. C. Lin, J. H. Huang, N. H. Tai, and I.-N. Lin, "Growth of high quality AlN thin films on diamond using TiN/Ti buffer layer," *Diam. Relat. Mater.*, vol. 15, p. 404, 2006.

[11] M. Dubois and P. Muralt, "Properties of aluminum nitride thin films for piezoelectric transducers and microwave filter applications,"*Appl. Phys. Lett.*, vol. 74, p. 3032, 1999.

[12] O. Auciello and A. V. Sumant, "Status review of the science and technology of ultra-nanocrystalline diamond (UNCD^{TM}) films and application to multifunctional devices," *Diam. Relat. Mater.*, vol. 19, p. 699, 2010.

[13] O. Auciello and B. Shi, "Science and technology of bio-inert thin films as hermetic-encapsulating coatings for implantable biomedical devices: application to implantable microchip in the eye for the artificial retina," in *Implantable Neural Prostheses 2: Techniques and Engineering Approaches*, D. D. Zhou and E. Greenbaum, Eds. New York: Springer, p. 63, 2016.

[14] P. Bajaj, D. Akin, A. Gupta, et al., "Ultrananocrystalline diamond film as an optimal cell interface for biomedical applications," *Biomed. Microdevices*, vol. 9, p. 787, 2007.

[15] O. Auciello, J. Birrell, J. A. Carlisle, et al., "Materials science and fabrication processes for a new MEMS technology based on ultrananocrystalline diamond thin films," *J. Phys. Condens. Matter*, vol. 16, p. 539, 2004.

[16] X. Xiao, X. Wang, J. Liu, et al., "*In Vitro* and *in vivo* evaluation of ultrananocrystalline diamond for coating of implantable retinal microchips," *J. Biomed. Mater. Res. B Appl. Biomater.*, vol. 77B, p. 273, 2006.

[17] D. D. Zhou and E. Greenbaum (Eds.), *Implantable Neural Prostheses: Techniques and Engineering Approaches*. New York: Springer, 2010.

[18] O. Auciello, P. Gurman, A. Berra, M. J. Saravia, and R. Zysler, "Ultrananocrystalline diamond (UNCD) films for ophthalmological applications," in *Diamond- Based Materials for Biomedical Applications*, R. Narayan, Ed. Cambridge: Woodhead Publishing, p. 151, 2013.

[19] M. Zalazar, P. Gurman, P. Park, et al., "Integration of piezoelectric aluminum nitride and ultranano-crystalline diamond films for implantable biomedical microelectromechanical devices," *Appl. Phys. Lett.*, vol. 102, p. 104101, 2013.

[20] M. B. Assouar,O. Elmazria, P. Kirsch, and P. Alnot, "High frequency surface acoustic wave devices based on AlN/diamond layered structure realized using e-beam lithography," *J. Appl. Phys.*, vol. 101, p. 114507, 2007.

[21] M. Ishihara, T. Nakamura, F. Kokai, and Y. Koga, "Preparation of AlN and LiNbO3 thin films on diamond substrates by sputtering method," *Diam. Relat. Mater.*, vol. 11, p. 408, 2002.

[22] A. Gruverman, O. Auciello, and H. Tokumoto, "Imaging and control of domain structures in ferroelectric thin films via scanning force microscopy," in *Ann. Rev. Mater. Sci.*, vol. 28, p. 101, 1998.

[23] O. Auciello, A. Gruverman, H. Tokumoto, et al."Studies of polarization phenomena in ferroelectric thin films via direct nanoscale scanning force imaging microscopy," *MRS Bull.*, vol. 23, (1), p. 37, 1998.

[24] A. Gruverman, H. Tokumoto, A. S. Prakash, et al., "Nanoscale imaging of domain dynamics and retention in ferroelectric thin films," *Appl. Phys. Lett.*, vol. 71, p. 3492, 1997

[25] A. Artieda, M. Barbieri, C. S. Sandu, and P. Muralt, "Effect of substrate roughness on *c*-oriented AlN thin films," *J. Appl. Phys.*, vol. 105, p. 024504, 2009.

[26] D. S. Ballantine, Jr., R. M. White, S. J. Martin, et al., *Acoustic Wave Sensors: Theory, Design, & Physico-Chemical Applications*. New York: Academic Press Inc., 1997.

9 Science and Technology of Integrated Multifunctional Piezoelectric Oxides/Ultrananocrystalline Diamond (UNCD™) Films for a New Generation of Biomedical MEMS Energy Generation, Drug Delivery, and Sensor Devices

Orlando Auciello and Geunhee Lee

9.1 Introduction

Microelectromechanical systems (MEMS) and nanoelectromechanical systems (NEMS) have become technological revolutions that are enabling many key devices, such as MEMS-based acoustic wave biosensors and MEMS-based energy generators, to power external and implantable medical devices.

A thin-film bulk acoustic resonator (FBAR or TFBAR) is a device consisting of a piezoelectric material produced in thin-film form encapsulated between two electrodes, also in thin-film form, and acoustically isolated from the surrounding medium. FBAR devices involve piezoelectric films with thicknesses in the range of hundreds to tens of micrometers, which can resonate in the frequency range 100 MHz to 20 GHz [1, 2]. Piezoelectric oxide-based materials like lead–zirconium–titanate ($PbZr_xTi_{1-x}O_3$ [PZT]), one of the most investigated [3–7] piezoelectric (ferroelectric) materials, synthesized in film form can act as an active material in an FBAR resonator. Aluminum nitride (AlN) and zinc oxide are the other two most studied piezoelectric materials for FBAR devices. Because of the compatibility with the silicon integrated circuit technology, AlN has become the most used material in commercial volume manufacturing of FBAR resonator-based products such as radio frequency filters, duplexers, and power amplifier modules. Doping or adding new materials like scandium (Sc) [8] are new directions to improve material properties of AlN for FBARs. Research of electrode materials focused on replacing one of the metal electrodes with very light materials like graphene [9] to minimize the loading of the resonator has been shown to yield better control of resonance frequency. FBAR resonators can be manufactured on ceramic (alumina), sapphire, glass, or silicon substrates. However, silicon is the most common substrate due to its scalability toward mass manufacturing and compatibility, with various manufacturing steps needed.

During early studies and the experimentation phase of thin-film resonators in 1967, cadmium sulfide (CdS) was evaporated on a resonant piece of bulk quartz crystal, which served as a transducer providing a quality factor of 5000 at the resonance frequency (279 MHz) [10]. This was an enabler for tighter frequency control for higher-frequency devices and utilization of FBAR resonators.

Most smartphones marketed in 2020 include at least one FBAR-based duplexer or filter. Some 4G or 5G products may even include 20–30 functionalities based on FBAR technology, mainly due to the increased complexity of the radio frequency front-end, electronics, and antenna system. Therefore, FBAR technology has become one of the critical enabling technologies in communication devices.

MEMS Biosensors. These devices detect changes in the resonant frequency of a mechanical resonator, like a cantilever-shaped (similar to a swimming pool diving board) structured device, where biomolecules, adsorbed on the surface of a biologically active membrane, can be detected. Since frequency change can be measured very precisely, very small mass changes can be measured. This leads to high sensitivity of the MEMS biosensors. Typical acoustic wave biosensors are bulk acoustic wave (BAW) and surface acoustic wave (SAW) sensors. Acoustic wave–based MEMS devices offer a promising technology platform for the development of sensitive, portable, real-time biosensors. MEMS fabrication of acoustic wave–based biosensors enables device miniaturization, power consumption reduction, and integration with electronic circuits. For biological sensing, the biosensors are integrated in a microfluidic system and the sensing area is coated with a bio-specific layer. When a bioanalyte interacts with the sensing layer, mass and viscosity variations of the bio-specific layer can be detected by monitoring changes in the acoustic wave properties such as velocity, attenuation, resonant frequency, and delay time. Few acoustic wave–type devices could be integrated in microfluidic systems without significant degradation of the quality factor. The acoustic wave–based MEMS devices reported in the literature (see, for example, the review in [11]), as biosensors, are based on FBAR, SAW resonators, and SAW delay lines.

Resonant microcantilever MEMS arrays are under development for label-free/real-time analyte monitoring and biomolecule detection. Research is being performed to develop MEMS cantilevers made of electroplated nickel, functionalized with hepatitis antibodies, to do biosensing of hepatitis molecules. Hepatitis A and C antigens at different concentrations are introduced in undiluted bovine serum to perform measurements in liquid, within a specifically designed flow cell, without drying the cantilevers throughout the experiment. Both actuation and sensing are done remotely, and therefore the MEMS cantilevers have no electrical connections, allowing for easily disposable sensor chips. Actuation is achieved using an electromagnet and the interferometric optical sensing is achieved using laser illumination and embedded diffraction gratings at the tip of each cantilever.

Resonant frequency of the cantilevers in dynamic motion is monitored using a self-sustaining closed-loop control circuit and a frequency counter. Specificity is demonstrated by detecting both hepatitis A and C antigens and their negative controls. Tiurdogan et al. [12] first reported hepatitis antigen detection by resonant cantilevers exposed to undiluted serum. A dynamic range in excess of 1000 and a minimum detectable concentration limit of 0.1 ng/ml (1.66 pM) was demonstrated for both hepatitis A and C. This result is comparable to labeled detection methods such as ELISA.

MEMS-Based Power Generation Techniques for Implantable Devices. Powering MEMS-based *in-vivo* devices is currently mostly done via conventional or thin-film battery systems. The problem is that the battery becomes a limiting factor in the lifespan and applicability of many biosensors and implantable medical devices requiring power. Although some biocompatible batteries may have long lifespans, the battery will eventually require replacement or recharging. For devices requiring short-term power, a battery may provide sufficient device lifetime, but for long-term or high duty cycle devices, alternative power sources are desirable to replacing dead batteries, especially for the case where the replacement/recharge procedure is invasive. For example, defibrillators/pacemakers are a common implantable medical device that requires an independent power source, operating completely autonomously from the outside world. The current standard for defibrillator/pacemaker operation is to utilize a high-life battery that supplies approximately 0.65–2.8 Ampere hours for 5.1–9 years [13]. Eventually, the battery for the defibrillator/pacemaker needs to be replaced, requiring additional surgery with extra cost and patient discomfort. Although a pacemaker is not necessarily a biosensing device, the power supply can be augmented by an electromagnetic-based MEMS generator. Roberts et al. [14] demonstrated a system by which an electromagnetic MEMS-based generator captures the vibrational energy produced by the heart muscle to generate power to supplement the pacemaker's internal battery. In initial clinical trials it was possible to produce up to 17% of the energy required to operate a conventional pacemaker [14]. Further development of this technology may be able to eliminate the costly and invasive surgeries required to maintain the pacemaker, both decreasing medical cost and improving the quality of life for the patient. MEMS-based power generation may also be a strategy for implementation in implantable biosensors. Many implantable biosensors could have their powering source replaced or augmented by MEMS-based power generators. The addition of MEMS-based generators to the conventional power systems of these sensors would allow for increased lifetime and the feasibility of adding components to the sensing platform, contributing to reduced costs. Additional hardware could also be inserted into the biosensor systems, enabling wireless communications and on-board computing to further increase the functionality and usefulness of MEMS-based implantable sensors.

Several types of MEMS-based power generation have been investigated and/or developed:

1. **Photovoltaic generator:** The operation of photoelectric generators based upon harvesting light energy from fiber optic cables may enable the use of photoelectric-based generators *in vivo*. These devices may have high efficiency, but the voltage available from these relatively small band gap diodes is too small for many switching and controlling applications.
2. **Thermoelectric generators:** Using the human body as a heat source has been explored by Leonov et al. [15]. Based on a wide range of tissues and fluids, each having their own unique material and thermal properties, it was found that the human body has an inherent nonuniform temperature distribution, which can be used for this type of MEMS power generation. The microgenerator itself is a microfabricated array of polysilicon–germanium (poly-SiGe) thermocouples, which are sandwiched between two silicon wafers and interconnected in series to form thermopiles.
3. **Microfuel cells:** These cells operate via harvesting electrons from controlled electrochemical reactions. Depending upon the fuel and oxidizing agents reacting in the microfuel cell, it can be considered either a regenerative or non-regenerative generation technique. If the electrochemical reactions are self-sustaining, such that the reactants are not irreversibly consumed, the fuel cell is regenerative. For example, glucose-based, self-contained fuel cells are completely regenerative, able to operate for extended periods of time without outside intervention [16].
4. **Electrostatic vibration-to-electricity conversion:** Electrostatic vibration-to-electricity energy harvesting involves using a comb drive to generate electricity from a vibrating base. Power is generated through a vibration-driven capacitance variance which causes charge transfer and current flow. The charge required for the power generation device to operate is supplied by a power source or passively through use of an electret layer [17] or a charge pump [18]. Based on an electret-driven microgenerator, an electret layer provides the polarization of the variable capacitor. The electrets are microfabricated from silicon wafers, with deposited layers of silicon oxide and silicon nitride. This MEMS power generation device needs to be packaged in a vacuum to avoid thin-film damping, which is a substantial limitation.
5. **Piezoelectric energy conversion:** Piezoelectric energy generation is a method of harvesting power from mechanical vibrations of piezoelectric materials. When electric fields are applied to a piezoelectric crystal structure, positive and negative ions are physically moved in opposite directions, distorting the microstructure of the crystal, and inducing a polarization in the material. In order to maintain electrical equilibrium within the crystal, the electrons become mobile, creating a current. This is referred to as the direct piezoelectric effect. Alternatively, the exact opposite phenomenon, the converse piezoelectric effect, can be produced, whereby vibration generation in a MEMS oscillating membrane or cantilever-type structure is translated into electricity, which is extracted to power external devices. Piezoelectric generation is frequency-dependent, maximized when the frequency at which the system is driven is at resonance [19]. The piezoelectric-based MEMS power generator involves using mainly a piezoelectric film to convert the

displacement and strain into electricity through the piezoelectric effect. Three materials have been explored as thin films for use in piezoelectric MEMS: PZT, which has the main drawback for application to implantable devices that it contains Pb, making it unacceptable; and zinc oxide (ZnO) and AlN, which to some extent are biocompatible. According to the literature, PZT is the dominant piezoelectric material used for power generation purposes, because it is the material with one of the highest piezoelectric coefficients, as shown in many publications. For example, Dufay et al. [20] measured piezoelectric coefficient d_{31} for $PbZr_{58}Ti_{42}O_3$ grown via the sol-gel process on Si with a bottom Al electrode coated with a 40–100 nm SrO_3 layer to induce an optimized interface to produce good stoichiometric PZT films. They demonstrate a coefficient of up to 33 pC/N (one of the highest shown in the literature).

In addition to piezoelectric membranes for power generation, a PZT microfiber generator has been demonstrated by Ishisaka et al. [21], in which the contractions of a heart muscle induce actuation of the piezoelectric microfiber generator, fabricated via growing PZT films onto a platinum wire and then plating the wire with nickel for use as the top electrode, to complete the electrical circuit. A polydimethylsiloxane (PDMS) membrane is then placed between the fiber and the heart muscle to provide biocompatibility. As the heart muscle contracts, the PDMS membrane is deflected, which in turn causes the embedded PZT microfiber to deflect mechanically. In experiments, cultured cardiomyocyte cells were used to actuate the piezoelectric generator at a frequency of 1.1 Hz, producing 40–80 mV for a single ~100 μm fiber. The relative strength of this microgenerator is provided by the biocompatibility of the PDMS membrane encapsulation, which prevents contact between the PZT and the cardiomyocytes. Besides being highly flexible, allowing for actuation, it is relatively biocompatible; the word "relative" is used here to account for the possibility that microcracks may be generated in the PDMS coating, exposing the heart to Pb.

Another fiber-based piezoelectric vibration power generation device uses ZnO nanowires to generate electricity on a microscale. The ZnO nanowires are grown on plastic substrates using a wet chemistry method that can be tailored to produce the desired material orientation and density [22]. Finally, the generator is encapsulated by a polymer layer. The ZnO-based power generator is capable of generating 1.26 V at a strain of 0.19%, potentially being capable of charging an AA battery [23], producing a peak power output of 2.7 mW/cm^3 [23]. The generator was able to generate more power from the heartbeat of a rat, generating 3 mV and 30 pA [24]. Although this experiment produced significantly less power than the previous studies that used purely mechanical stimulation, it is an important first step for ZnO nanowire-based *in-vivo* microgenerators.

A key problem with most MEMS-based biosensors and power generation devices, described above, is that the membranes or cantilevers are made of materials such as Si, metals, and polymers, which to a greater or less extent are not optimum biocompatible materials, and they generate undesirable reactions in the human body.

9.2 UNCD Thin Film: A New Best Biocompatible Paradigm Material for Implantable MEMS/NEMS Biosensors and Power Generators

9.2.1 Review of Properties of UNCD Films Critical for MEMS/NEMS Biosensors and Power Generation Devices

9.2.1.1 Bulk and Surface Properties

UNCD films exhibit a unique set of complementary mechanical and tribological properties [25] that make them extremely well suited for MEMS/NEMS devices in general. UNCD films exhibit hardness (98 GPa) and Young's modulus (980 GPa) [25] close to values for single crystal diamond (100 GPa and 1200 GPa, respectively). The Young's modulus is reduced to about 880 GPa when adding about 3% N_2 in the gas phase to the plasma chemistry in order to produce electrically conductive UNCD films [26, 27]. However, this value of Young's modulus is still superior to silicon and other MEMS/NEMS materials (see Table 9.1).

For RF MEMS resonators, a key figure of merit is acoustic velocity (AV). Diamond has the highest AV of any material (16,760 m/s), compared to 11,700 m/s for high-quality AlN and 8100 m/s for single crystal Si. The AV together with the high Young's modulus directly translates to higher frequency of operation for a given device geometry and allows for larger devices for a given frequency, facilitating manufacturing. Fixed–free UNCD cantilever-based resonators were fabricated and used to measure UNCD film's AV (15,400 m/s) and quality factor (Q; 10,000–13,000 [28]) using an atomic force microscope (AFM) in vacuum to avoid damping effects due to air [26, 29]. Here, it is relevant to mention that disc-shaped resonators based on nanocrystalline diamond (NCD) films exhibited some of the highest Qs [30,31]. Exposing the fixed–free beam surface to atomic hydrogen removes contaminants and nondiamond (amorphous) carbon phases, resulting in significant and stable increases in Q, even in air. This is in contrast to Si resonators, which exhibit increased Q upon oxide removal by vacuum annealing, but only temporarily, due to new oxide formation upon exposure to air or low partial pressures of oxygen.

Table 9.1 Comparison of mechanical and tribological properties between Si, SiC, and diamond

Property	Si	SiC	Diamond
Lattice constant (Å)	5.43	4.35	3.57
Cohesive energy (eV)	4.64	6.34	7.36
Young's modulus (GPa)	130	450	1,200
Shear modulus (GPa)	80	149	577
Hardness (kg/mm^2)	1,000	3,500	10,000
Fracture strength (GP)	1.0	5.2	5.3
Flexural strength (MPa)	127.6	670	2,944
Friction coefficient	0.4–0.6	0.2–0.5	0.01–0.04

Fracture strength is another important bulk parameter, particularly for MEMS/NEMS devices where moving components are subject to impact. Measurements by different groups revealed that the fracture strength of UNCD is about five times larger than for Si (see Table 9.1). For RF MEMS contact switches wear can seriously degrade the lifetime of the device. A thin coating of conductive UNCD could potentially be used to provide longer wear life for such contacting surfaces.

An important property relates to surface chemistry of MEMS/NEMS membranes or cantilevers. In relation to this property, diamond exhibits surface inertness, a very stable surface chemistry, and low stiction. UNCD has the lowest coefficient of friction (COF) (~0.01–0.05) [25] when compared to silicon (~0.4–0.6), conventional diamond-like carbon films (~0.2), and microcrystalline diamond (MCD) films (~0.4) [32, 33]. AFM studies provided quantitative information on interfacial adhesion and friction between AFM tips and UNCD film surfaces in ambient air as well as before and after H-plasma treatment. Silicon exhibits the highest work of adhesion, while the UNCD underside exhibits substantially lower work of adhesion before H-exposure than the "passivated" form of silicon (55 mJ/m^2 vs. 106 mJ/m^2, respectively) [34]. This value for UNCD is comparable to that of the untreated <111> diamond surface. The measured friction force for the as-released UNCD underside is comparable to that of the untreated diamond <111> surface. The very low nanoscale adhesion and friction revealed by the plain UNCD surfaces indicate that the underside of UNCD can significantly outperform Si in MEMS/NEMS devices where surface properties, due to potential contact of the moving membrane with the underlying substrate surface, may result in stiction and device failure.

Another important surface property is the rms roughness of the films. It has been shown that the surface roughness of UNCD and NCD films depend on the nucleation density and initial growth of diamond films [31]. A detailed analysis of the surface seeding and initial nucleation and growth of UNCD films is presented in Chapter 1.

The information presented above indicates that UNCD films exhibit superior mechanical and tribological properties over those of Si and other materials for key MEMS/NEMS devices, except for NCD films, particularly for high-frequency systems where AV and stable surface chemistries are key parameters to enable very high-frequency devices to the GHz range.

9.2.1.2 Properties of UNCD Films as a Biomaterial for MEMS/NEMS Devices Implantable in the Human Body

UNCD films were developed as a biocompatible/eye humor corrosion-resistant coating for encapsulation of a Si microchip implantable on the human retina as a key component of an artificial retina to restore sight to people blinded by genetically induced retina degeneration [35]. UNCD films are the only diamond films that can be grown on high-conductivity Si substrates at ~400 °C [25], which is within the thermal budget of Si microchips that will be destroyed if heated to ≥450 °C. Electrochemical tests of the plain UNCD coatings yielded relatively high leakage currents (~10^{-4} A/cm^2) that are not compatible with the needed functionality in the eye environment. Introducing

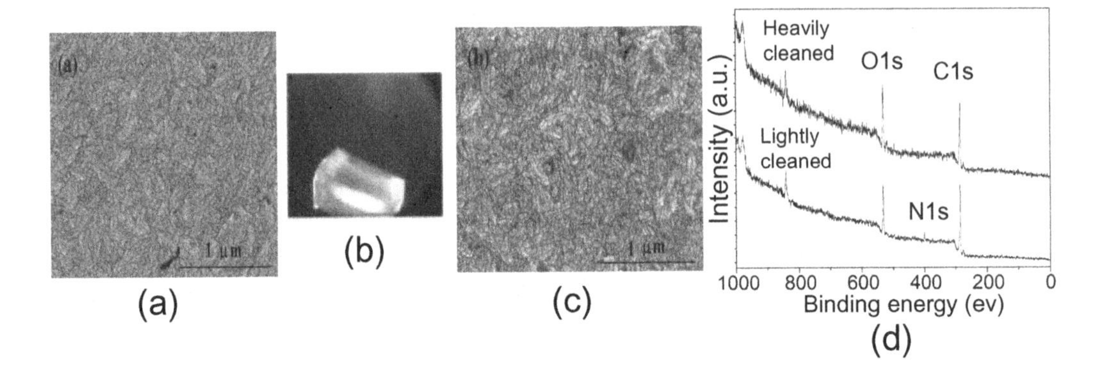

Figure 9.1 (a) SEM top view of the UNCD coating on an Si microchip before implantation in a rabbit's eye. (b) UNCD-coated Si microchip implanted in a rabbit's eye for ~3 years. (c) SEM top view of a UNCD coating on an Si microchip after implantation in a rabbit's eye for ~3 years. (d) XPS chemical analysis of the surface of the UNCD coating after implantation in a rabbit's eyes for three years, showing no chemical corrosion. (Reprinted from *MRS Bull.*, vol. 39 (7), 621–629, (2014) Fig. 3 in [37] with permission from Cambridge).

hydrogen into the Ar/CH_4 mixture resulted in the incorporation of hydrogen into the grain boundaries, which saturated the dangling bonds greatly decreased leakage current. Long-term (up to three years) *in-vivo* implantation of UNCD-coated Si chips in rabbit eyes demonstrated that UNCD is bioinert and biostable without any chemical attack (compare the scanning electron microscope [SEM] image of the surface of the UNCD coating on an Si microchip before implantation in rabbit eye (Figure 9.1a) and after three years in a rabbit's eye (Figure 9.1b,c), and X-ray photoelectron spectroscopy (XPS) chemical analysis of the surface of the UNCD coating after implantation in the rabbit's eye (Figure 9.1d).

The results described above provide preliminary validation of using low-temperature UNCD as a hermetic coating for encapsulation of microchips for an artificial retina and other biomedical implants [35, 37]. In addition, the *in-vivo* tests of UNCD demonstrate that UNCD can be used for the development of a new generation of implantable MEMS/NEMS, such as for drug delivery systems implantable inside the human body, since the UNCD coating is not attacked chemically by body fluids.

9.3 Materials Integration and Process Strategies for Fabrication of MEMS/NEMS Devices Based on Integrated Piezoelectric Oxide and UNCD Films

Two main fabrication processes to produce UNCD-based MEMS/NEMS were developed in the past by Auciello's group [25, 36] based on the processes described in the following subsections.

9.3.1 Selective Seeding and UNCD Film Growth Process for Fabrications of UNCD MEMS/NEMS Devices

In this process, the substrate is seeded by (1) using photoresist to prevent exposure of selected areas to the diamond powder; (2) using diamond-loaded photoresist to produce a patterned nucleation layer; or (3) seeding the substrate uniformly and then selectively etching portions of the surface to remove diamond-seeded areas. The feature resolution that can be achieved by this method is limited by the grain size. A schematic of the selective seeding plus UNCD growth process is shown in Figure 9.2a. Examples of MEMS structures produced by selective seeding and growth are shown in Figure 9.2b,d for UNCD-based structures and Figure 9.2c,e for MCD-based structures. It is clear that the resolution achieved by the UNCD-based process is an order of magnitude superior to that achieved with MCD films [25].

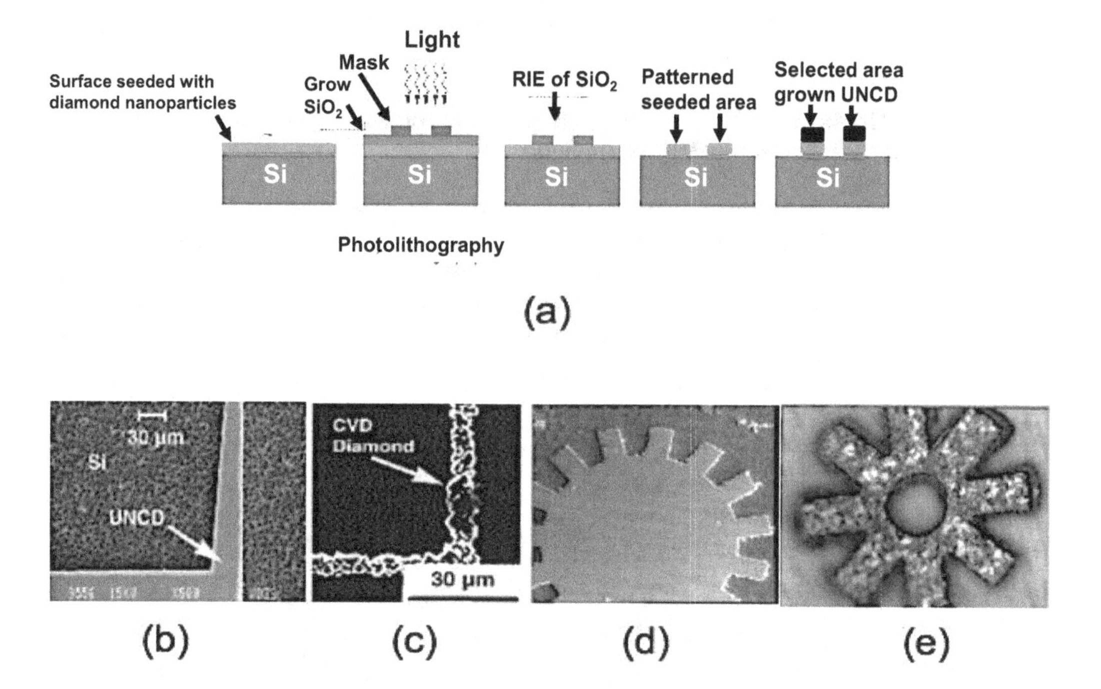

Figure 9.2 (a) Schematic showing the selective seeding via seeding of substrate uniformly, followed by photolithography and selectively etching of the substrate surface in the photoresist layer to remove diamond-seeded areas, and finally UNCD growth on selectively seeded areas. UNCD MEMS structures grown using the selective seeding plus growth, using the UNCD growth process (b) and (d) and the MCD growth process (c) and (e). It is clear that the UNCD growth process produces MEMS structures with much higher resolution and smoother surface morphology than the MCD process (reprinted from *Diam. Relat. Mater.*, vol. 19, p. 699, 2010 (Fig. 12) in [25] with permission from Elsevier Publisher).

9.3.2 Blank Seeding across Substrate Surface and UNCD Film Growth across Surface Plus RIE Process for Fabrication of UNCD MEMS/NEMS Device

In this fabrication process, blank UNCD layers are grown on a sacrificial release layer (SiO_2) using ultrasonic abrasion of the substrate surface exposed to a diamond powder suspension to produce a diamond seed layer. UNCD films grown using the Ar/CH_4 plasmas nucleate and grow directly on SiO_2 substrates with appropriate diamond seeding processes (see Chapter 1) [25]. The SiO_2 layer is used as a sacrificial layer for fabrication of UNCD film-based MEMS/NEMS structures, like cantilevers. Figure 9.3a shows a UNCD layer grown on a thermal SiO_2 layer, followed by a second SiO_2 layer grown on the UNCD layer, using a chemical vapor deposition (CVD) process. The photoresist is the hard SiO_2 mask, used to perform patterning reactive ion etching (RIE) with CF_4/CHF_3 gas mixture (2:1) plasma (Figure 9.3b,c). Oxygen plasma is then used to perform RIE etching of the UNCD layer, using the SiO_2 hard mask (Figure 9.3d). UNCD films can be etched at rates of about 1 μm/h. Finally, an HF wet and/or gas etch (Figure 9.3e) is used to remove the sacrificial oxide layer, leaving the UNCD film suspended above the Si substrate to form moving MEMS/NEMS structures such as cantilevers (Figure 9.3e).

9.3.3 UNCD Film Stress and Ohmic Contact of Electrode Layers on UNCD Films for Powering MEMS Devices

A critical property of UNCD films relevant to fabrication of high-performance MEMS/NEMS structures is that UNCD exhibits relatively low intrinsic tensile stress in the 100–400 MPa range (orders of magnitude smaller than for MCD films) [33]. This property may be due to attractive forces between the diamond grains. UNCD films also exhibit very low differential stress (down to 50 MPa/μm) due to both the microstructure, which does not vary with film thickness, and the use of

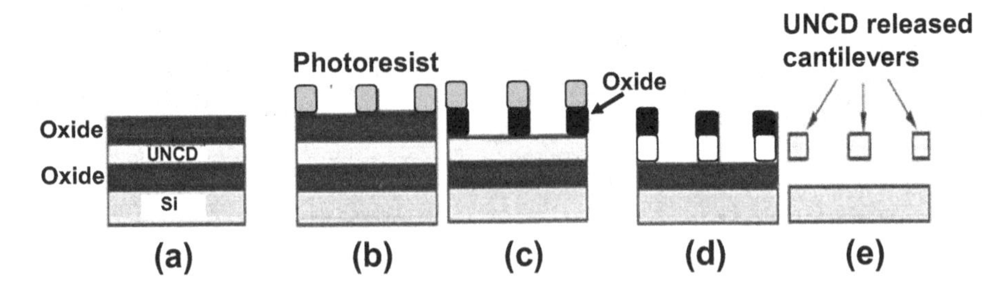

Figure 9.3 Schematic showing lithography-based process microfabrication of UNCD MEMS. (a) Growth of UNCD film on an SiO_2 sacrificial layer plus growth of a masking layer; (b) photoresist growth; (c) lithography process; (d) RIE in oxygen plasma to etch the UNCD layer to define a cantilever structure; (e) selective chemical etching of the SiO_2 sacrificial layer underneath the UNCD layer. This process works very well because UNCD is not etched by any acid used in the fabrication of Si microchips or Si-based MEMS (reprinted from *Diam. Relat. Mater.*, vol. 19, p. 699, 2010 (Fig. 13) in [25] with permission from Elsevier Publisher).

nanodiamond-based seeding techniques, which ensure uniform growth and density even in the first few layers of the film.

Another important property of UNCD film surfaces is that sputter-deposited Al, Au, Cr, Cu, Pt, and Ti layers, all about 200 nm thick, have been shown to establish Ohmic contact with the surface of UNCD films. This property of metallic layers/UNCD interfaces is critical for applications such as electrostatically actuated MEMS and NEMS devices, which require reliable electrical contacts.

9.3.4 Materials Integration for Hybrid Piezoelectric/UNCD-based MEMS/NEMS Devices

Most consumer-based mobile devices, including biosensors and power generation MEMS/NEMS devices, are evolving toward ever-decreasing power availability for such applications. These requirements induce severe strain in the current integration of electrostatic RF MEMS components with charge pumps needed to deliver 20–60 V. Such devices may become prohibitive in the future. Thus, alternative low-voltage actuation schemes are necessary and are being explored as described in this part of the chapter. In this respect, low-voltage RF MEMS/NEMS actuation can be achieved by integrating an electrode/piezoelectric/electrode layered heterostructure on top of a structural UNCD layer used for a MEMS/NEMS structure. Application of a voltage across the piezoelectric layer induces strain (mechanical deformation due to the piezoelectric effect), causing the structural material to deform in the same direction as the piezoelectric layer, thus inducing the MEMS/NEMS structure actuation.

A key aspect of integrating a piezoelectric PZT film on a underlying substrate material to fabricate optimum piezoelectrically actuated MEMS/NEMS biosensor and power generation devices is to determine which is the best substrate material. In this respect, calculations showed that the product resonance frequency/dynamic displacement at resonance is the largest for the PZT–UNCD hybrid compared to other PZT-based hybrids [38], including Si, nitrides, metals, and insulators (Figure 9.4). In addition, UNCD films provide the best support material for resonators due to the highest Young's modulus (Table 9.1).

Auciello's group was the first in demonstrating the integration of piezoelectric thin films with UNCD coatings [38] to enable the first generation of piezoelectrically actuated UNCD-based MEMS/NEMS devices. PZT was elected as the piezoelectric material, since it exhibits one of the highest piezoelectric coefficients (~33–40 pC/N) among piezoelectric materials. However, for implantable biosensors or power generation MEMS/NEMS devices, PZT is not the best piezoelectric material because it contains Pb. That is why Auciello's group is exploring alternative materials such as AlN (see Chapter 8), and most recently exploring $BiFeO_3$ in single stoichiometric composition [40], or as $BiFeO_3$–$SrTiO_3$ nanolaminates with higher piezoresponse performance over stoichiometric $BiFeO_3$ films [41]. Preliminary studies show that BFO is biocompatible, since it is formed by Bi, which is in the medicine Pepto Bismol, on the market today, and Fe and O are elements in the human body. BFO films exhibit a piezoelectric coefficient as high as PZT, and for nanolaminates even up to five times higher [41].

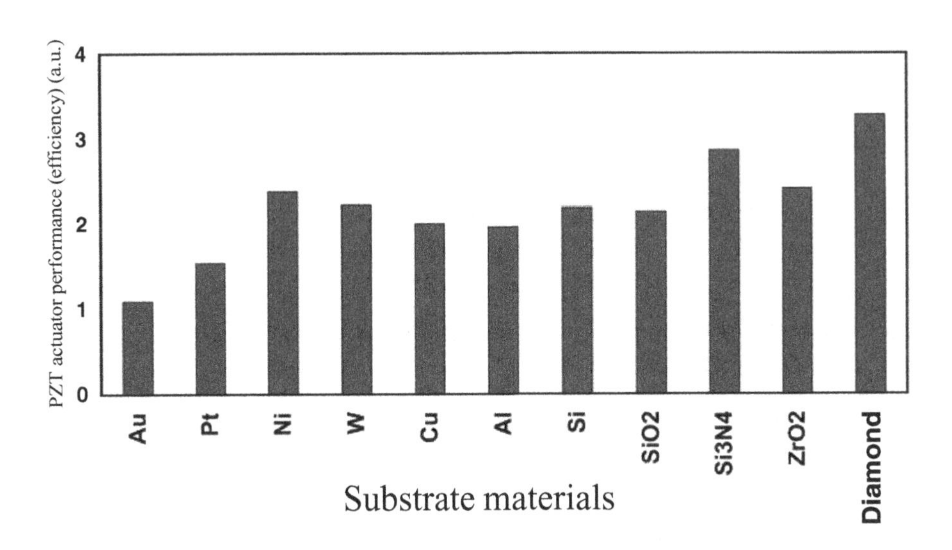

Figure 9.4 PZT film piezoelectric actuator performance vs. substrate material on which the PZT film is grown, showing that diamond is the best substrate.

In any case, Pt–Pb(Zr_xTi_{1-x})O_3(PZT)–Pt layered heterostructures, featuring PZT films with excellent piezoelectric and electromechanical coupling coefficients and high remnant polarization, were grown on UNCD structural layers to produce the first low-voltage piezoactuated UNCD MEMS cantilevers, and to obtain the information that will help to integrate the biocompatible AlN or BFO films with UNCD fims for implantable MEMS/NEMS medical sensors and power generators. Theoretical calculations performed by Auciello's group demonstrated that diamond is the best platform material for the proposed hybrid structures, including PZT as the piezoelectric actuation layer and UNCD as the structural mechanically robust biocompatible support layer.

The PZT–UNCD integration was achieved by using robust TiAl or TaAl (named TA) layers with dual functionality as an oxygen diffusion barrier, due to having the lowest energy of oxide formation compared to any other elements in the periodic table, thus stopping O atom diffusion through the bottom Pt electrode layer next to the C-based UNCD coating, which would be etched by the oxygen atoms at the 500–600 °C needed to grow the PZT layer [38] (Figure 9.5a). The TiAl or TaAl layers are also extremely good adhesion layers for the Pt bottom electrode layer on UNCD (Figure 9.5b). The novel TA oxygen diffusion barrier layer [38] enabled growing oxide piezoelectric films, such as PZT, on a carbon-based layer such as UNCD without chemical etching of the carbon by oxygen (Figure 9.5b–d).

9.3.4.1 Main Processes to Grow PZT Films on Different Substrates

Sol-Gel Growth Process for PZT Films

A suitable sol-gel solution to grow PZT films by the sol-gel process solution involves using Pb acetate tetrahydrate (Sigma-Aldrich 99%), Zr-propoxide (Sigma-Aldrich

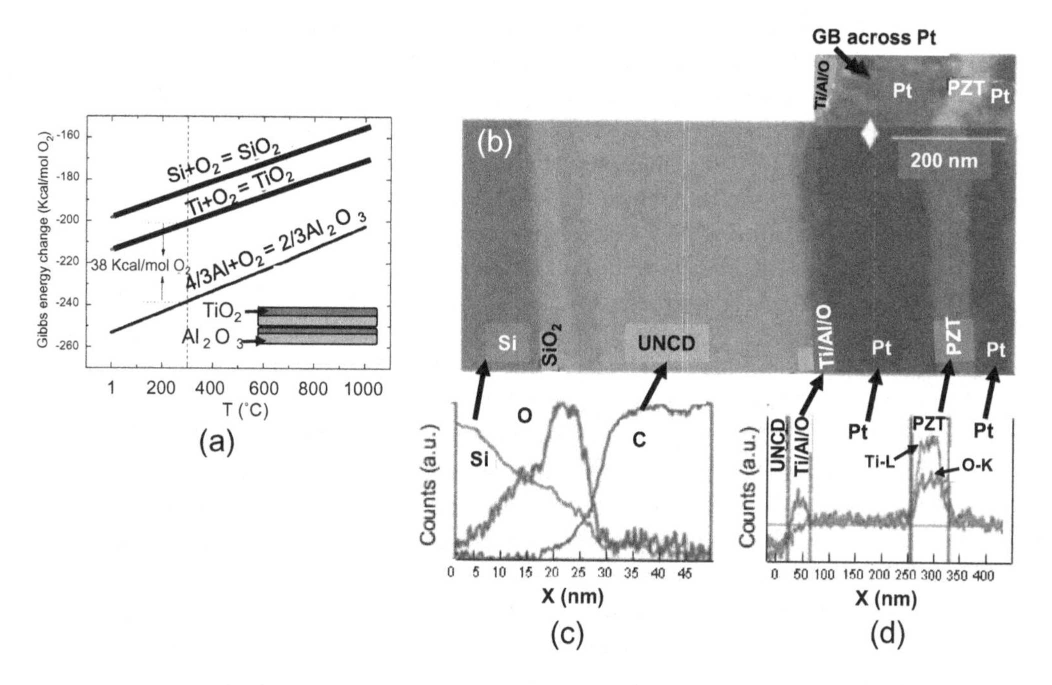

Figure 9.5 (a) Gibbs free energy of oxide formation vs. temperature, showing the two elements in the periodic table (Ti and Al) with the lowest energy of oxide formation of all elements, to stop O atom diffusion through grain boundaries of the Pt bottom electrode layer, thus inhibiting O-induced chemical etching of the C-based UNCD layer. (b) Cross-section transmission electron microscope (TEM) micrograph of a Pt–PZT–Pt–TiAl–UNCD heterostructure, showing the composition and microstructure of each layer and the interfaces. (c,d) EELS analysis of the different layers and interfaces shown in the top cross-section TEM images, showing the atom distribution in the layers and across interfaces (reprinted from *Appl. Phys. Lett.*, vol. 90, 134101, 2007 (Fig. 2) in [38] with permission from AIP Publisher).

97%), and Ti-isopropoxide as precursor materials and 2-methoxyethanol (2MOE) (Sigma-Aldrich 99.9 %) as a solvent. The molar ratio Zr–Ti is kept at Ti (52)/Zr (48), and 20% excess lead was used to compensate for Pb loss due to volatilization. Zr-propoxide powder was mixed with 2MOE in a flask by stirring. Ti-isopropoxide was then added to the above solution and stirred. Finally, the Pb (II) acetate trihydrate (Sigma-Aldrich 97%) was dissolved into the Zr–Ti solution by heating at 100 °C for 1 h, coupled with stirring. PZT layers (90 nm thick) were then grown on the LNO/ ITO coated glass substrate at a speed of 3000 rpm for 30 s. The wet film was pyrolized by heating in air at 450 °C for 5 min and then crystallized by annealing at 650 °C for 20 min in air. Details of the sol-gel process for growing PZT films can be found in [42–45]. This technique can produce stochiometric PZT films.

RF Planar Magnetron Sputtering-Deposition for Growth of PZT Films

RF planar magnetron sputtering-deposition [46] can produce PZT thin films via a relatively simple process, in which a plasma is created in front of a stochiometric PZT solid target, inducing bombardment by inert gas ions (mainly Ar ions, since Ar is the less expensive gas), which eject atoms (sputtering process) from the surface of the target, with the atoms flowing toward the substrate, where they grow the films. Several variations of RF sputtering have been developed and are still being optimized for the production of PZT thin films [46]. However, the main problem is that RF magnetron sputtering does not reproduce the stoichiometry of the target on the film due to preferential sputtering of O atoms, which change the target stoichiometry. In addition, there is loss of Pb atoms that may occur during deposition or post-deposition annealing. To account for this Pb loss, most groups use PbO-rich targets. But only a few reports were available in the literature on the RF-sputtered PZT films with stoichiometric PZT target without excess Pb. The exact amount of excess Pb required for obtaining the perovskite phase varies from report to report. It is reported that the stability of the Zr–Ti ratio depends on the composition of the target, but the Pb content depends largely on the substrate temperature and sputtering pressure [46]. The change in stoichiometry of the PZT films results in reduced polarization, and this affects the performance of the MEMS/NEMS devices due to reduced piezoelectric coefficient.

Pulsed Laser Ablation Deposition for Growth of PZT Films

Pulsed laser ablation deposition (PLD) of PTZ films involves an air evacuated chamber containing a target holder with a solid-state stoichiometric PZT target, and a substrate holder opposite the target at a distance of 10–15 cm. The substrate holder could be rotated to induce film growth with uniform thickness and composition on the area of growth. In general, a KrF excimer laser (wavelength 248 nm) is directed through a quartz SUPRASIL II window into the vacuum chamber, evacuated to ~10^{-7} Torr. Most commonly, solid targets with Pb $(Zr_{0.45}\ Ti_{0.55})\ O_3$ composition and 20 wt% excess of PbO are used, and their surfaces are polished with emery paper prior to each PLD process. The advantage of the PLD technique over the RF sputtering-deposition process is that the PLD process does not produce change in the stoichiometry of the target, since the laser beams induce evaporation of the material in a stoichiometric way [47–50].

Metalorganic CVD for Growth of PZT Films

PZT films have been grown on different substrates using metalorganic CVD (MOCVD) at temperatures ranging from 450° C to 525 °C, with a double-cocktail precursor solution liquid delivery system. The precursors used have been, generally, bis-tetramethylheptanedionato-Pb tetrakis1-methoxy-2-methyl-2-propoxy-Ti and tetrakis1-methoxy-2-methyl-2-propoxy-Zr for the Pb, Ti, and Zr, respectively. These new Zr and Ti precursors showed stable vaporization characteristics at 230 °C and allowed film deposition in the low-temperature range [51]. A self-regulation mechanism in cation composition control, where the ratio of the PZT components in the grown film is independent of the input ratio of the precursor solutions, was observed at temperatures as low as 475 °C. The self-regulation behavior became more evident with increasing temperature. However, the films grown at higher temperatures showed a larger leakage current and degraded ferroelectric performance because of the increased interfacial reaction. A 70 nm thick PZT film grown at 475 °C on an Ir electrode exhibited good ferroelectric performance, such as high remnant polarization at 4 V), small coercive voltage (~0.6 V at 4.5 V), and low-voltage (~4 V) saturation behavior post-annealing. The polarization fatigue test vs. the number of polarization switching cycles showed that the polarization was >90% of the initial value even with a simple Pt top electrode.

9.3.4.2 Growth and Characterization of PZT Films on UNCD Films for Fabrication of Piezoelectrically Actuated MEMS/NEMS Devices

The TA barriers were chosen based on thermodynamic arguments, which indicate that oxygen atoms react preferentially with Ti, Ta, and Al to form stable oxides due to having the lowest energy of oxide formation for these elements with respect to all other elements in the periodic table. The TA barrier inhibits oxygen-induced etching of the UNCD (carbon) layer during growth of PZT at relatively high temperature (450–500 °C) in oxygen. PZT films were grown by the PLD technique described above.

Measurement of the polarization vs. applied voltage between top and bottom Pt electrodes encapsulating the PZT film, inserted in cantilever-type MEMS/NEMS structure for biosensors and power generation devices, showed excellent polarization saturation (Figure 9.6). On the top right side of Figure 9.6, the unit cell of PZT shows the displacement of the Pb^{2+}, Zr^{4+}/Ti^{4+}, and O^{2-} ions in opposite directions, polarizing the PZT film in direction 1, while the bottom part of the polarization loop shows the same ions displaced in the opposite direction, related to polarization direction 0. The permittivity (ε_r), E_c (polarization switching from 0 to 1) electric field, and remnant polarization (P_r) values are one of the highest measured today [38].

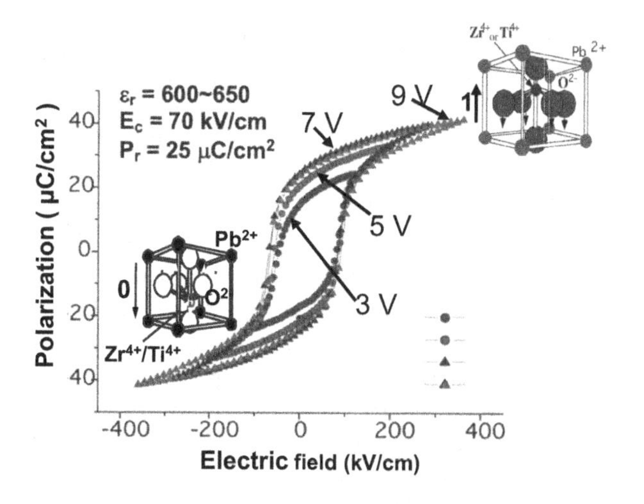

Figure 9.6 Polarization vs. electric field for polarization switching on a PZT film encapsulated between two Pt electrode layers integrated into a cantilever-type MEMS structure for biosensor and power generation (reprinted from *Appl. Phys. Lett.*, vol. 90, 134101, 2007 (Fig. 1) in [38] with permission from AIP Publisher).

The polarization performance of the piezoelectric PZT film shows values that were demonstrated to be very efficient at producing MEMS/NEMS cantilever displacement for piezoelectrically actuated biosensors/power generation devices, as shown in the next section.

9.4 MEMS and NEMS Biosensors and Power Generation Devices Based on Integrated Piezoelectric/UNCD FILMS

9.4.1 Fabrication of Integrated Pt–PZT–Pt–Interface Layer–UNCD Film MEMS/NEMS Cantilever-Based Biosensors and Power Generation Devices

The fabrication of the first integrated Pt–PZT–Pt–interface layer–UNCD film MEMS/NEMS cantilever-based biosensors and power generation devices involved the following steps [38]:

1. growth of ~1 μm thick UNCD layer on an Si (100) substrate coated with 1 μm sacrificial layer of SiO_2;
2. growth of ~10 nm thick TiAl barrier layer on the UNCD film to stop O atom diffusion during growth of PZT films in an O environment at ~600 °C, which would etch the C-based UNCD film;
3. growth of ~180 nm thick Pt layer (bottom electrode) on top of the TiAl barrier layer;
4. growth of a 60 nm thick $PbZr_{0.47}Ti_{0.53}O_3$ piezoelectric layer at ~600 °C in 100 mTorr of oxygen;
5. growth of the top 50 nm thick Pt layer (top electrode) to complete the capacitor-like structure to measure the polarization properties (Figure 9.6) of the PZT layers integrated with the UNCD films [38], and needed for piezoactuation via voltage application between the top and bottom Pt electrode layers in the Pt–PZT–Pt capacitor-type structure produced on the same UNCD film used for MEMS;
6. fabrication of UNCD (10–140 μm long × 1–10 μm wide) cantilevers, using a focused ion beam (FIB) system with a 30 keV Ga^+ ion beam to etch the integrated Pt–PZT–Pt–TiAl–UNCD films to define the cantilevers (Figure 9.7a);
7. chemical etching of the SiO_2 layer using HF acid, which etches only the SiO_2 layer, without any chemical attack on the UNCD, thus liberating the cantilever formed by the integrated Pt–PZT–Pt–TiAl–UNCD films (Figure 9.7b,c,e);
8. excitation in the range 1 Hz to 1 MHz (Figure 9.7d) and a PZT–UNCD horizontal nanoswitch actuated with 1 V in the range 1 Hz to 1 MHz (Figure 9.7e); both of the MEMS and NEMS cantilevers run up to one billion cycles without failure.

The integrated Pt–PZT–Pt–TiAl–UNCD film-based MEMS/NEMS cantilevers fabricated with the procedure described above were used to test piezoelectrically actuated MEMS biosensors and power generation MEMS/NEMS devices, as briefly described in the following subsection.

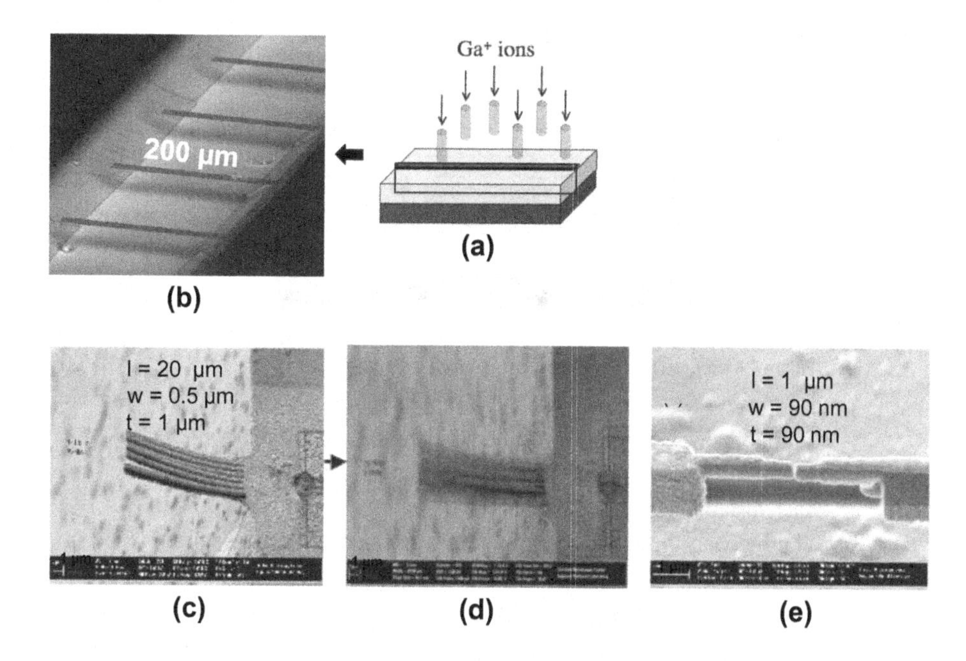

Figure 9.7 (a) Schematic showing fabrication of cantilever-type MEMS/NEMS structures for biosensors and power generation MEMS/NEMS devices, using FIB lithography, involving bombardment of integrated Pt–PZT–Pt–TiAL–UNCD grown on SiO_2–Si substrate to define the cantilevers. (b) SEM image of Pt–PZT–Pt–TiAL–UNCD released cantilevers, showing the straight cantilever due to the low strain of the UNCD layer. (c) SEM image of Pt–PZT–Pt–TiAL–UNCD released cantilevers before actuation. (d) Actuated Pt–PZT–Pt–TiAL–UNCD cantilevers by 3 V applied between top and bottom Pt electrodes, for up to one billion cycles without any actuation degradation. (e) Nanoscale Pt–PZT–Pt–TiAL–UNCD released cantilevers.

9.4.2 Test of Integrated Pt–PZT–Pt–TiAl–UNCD film MEMS/NEMS Cantilever-Based Biosensors

Figure 9.7d shows the integrated Pt–PZT–Pt–TiAl–UNCD film-based MEMS/NEMS at the second resonance (six resonances were demonstrated with oscillation amplitudes in the range 1–4 μm). The cantilevers were driven for one billion cycles, demonstrating the robustness of the PZT–UNCD hybrid structure. Although the PZT–UNCD cantilevers were fabricated using FIB, the same structures can be produced using industrial processes involving photolithography in conjunction with RIE in oxygen plasmas (Figures 9.2 and 9.3) to produce large arrays of PZT–UNCD structures for high-performance MEMS/NEMS piezoactuated biosensors.

One key aspect of the UNCD-based cantilever biosensor is that coupling of biological trapping molecules to the surface of the UNCD cantilever is orders of magnitude superior than to Si or other material-based cantilevers because of the unique biocompatibility of the C-terminated UNCD material. In this sense, implementation of other materials-based biosensors has been limited since the coupling strategies used to

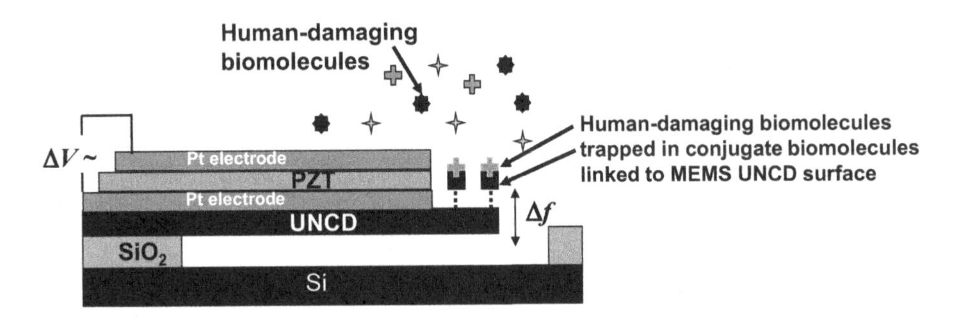

Figure 9.8 Schematic of a biosensor based on integrated Pt–PZT–Pt–TiAl–UNCD film.

immobilize biomolecules of interest (i.e., the bioinorganic interfaces) often suffer from long-term instability in aqueous or atmospheric environments. For one-time use devices, this instability may not be critical, but for biosensors/bioimplants that must work continuously in harsh operational environments, the robustness of the immobilization chemistry is of fundamental importance. Furthermore, the lack of biocompatibility of Si and other materials is problematic. The ideal platform material for biomedical devices or biosensors should possess two key properties: be bioinert and biofunctionalizable with conjugate biomolecules for chemically reacting with human body–damaging biomolecules for high-sensitivity detection (Figure 9.8).

MEMS/NEMS piezoelectrically actuated cantilevers in which the oscillating frequency changes when human-damaging molecules are chemically reacted with conjugate molecules grown on the UNCD-based cantilever thus provide information about the detected biological molecules [52].

The preliminary R&D on piezoelectrically based integrated Pt–PZT–Pt–TiAl–UNCD film cantilever-type MEMS/NEMS devices indicates that they can provide the transformational new generation of biosensors.

9.4.3 Test of Integrated Pt–PZT–Pt–TiAl–UNCD film MEMS/NEMS Cantilever-Based Biopower Generators

Cardiac cell–mediated actuation of piezoelectrically actuated UNCD-based biopower generation is based on cells of the myocardium, which exhibit compression and relaxation cycle beats in the range of 2–3 Hz autonomously when cultured in a proper environment. Cardiac cells from certain animals such as birds actuate at even higher frequencies. The actuation is driven by the energy supplied to the cells by nutrients in the fluid environment where the cells live.

Based on the discussion presented above, the opportunity exists for investigating the integration of myocardium cells with piezoelectric cantilevers to achieve biopower generation via cell beating–induced actuation of the cantilever and generation of electrical potential from the mechanical deformation of the piezoelectric layer through the top and bottom film electrodes encapsulating the piezoelectric layer.

Cardiac cells can be captured on the end of the hybrid UNCD–PZT (or $BiFeO_3$, a new alternative piezoelectric material being explored in the IMG and other laboratories,

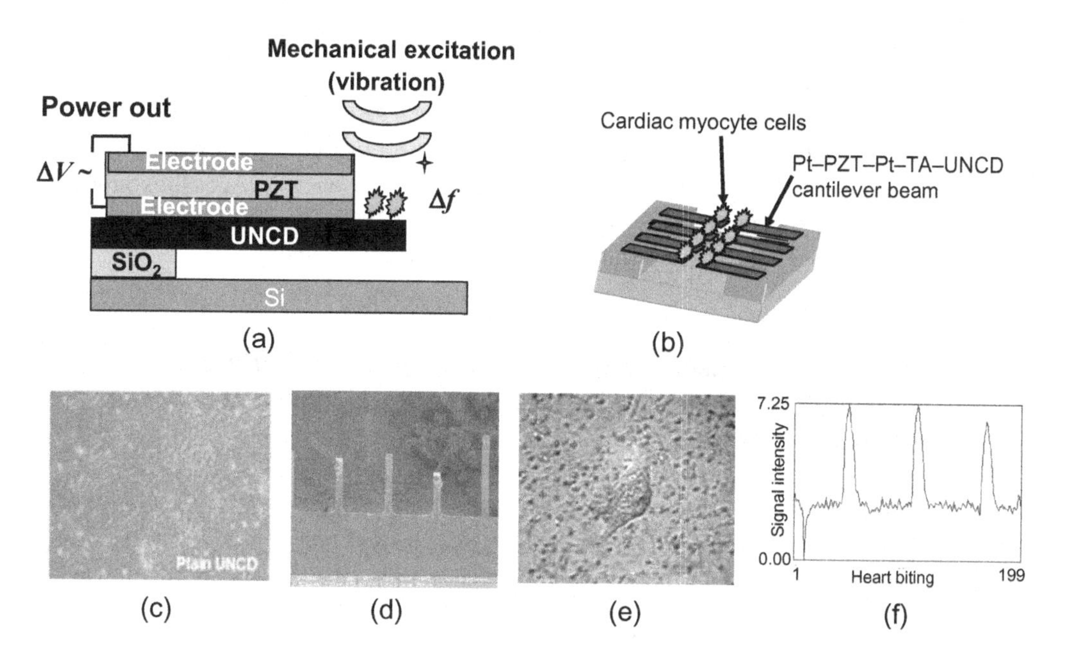

Figure 9.9 (a) Schematic of MEMS power generation device based on integrated Pt–PZT–Pt–TiAl–UNCD heterostructured film-based MEMS cantilever. (b) Schematic of the array of integrated Pt–PZT–Pt–TiAl–UNCD cantilevers, with heart cells grown on the exposed surface of the UNCD layer to induce the mechanical vibration, generating power (~3 mW already demonstrated by Auciello's group). (c) SEM image of cells of mouse heart grown on the exposed UNCD surface (see (a)) of the cantilever. (d) Low-magnification SEM imaging showing an array of Pt–PZT–Pt–TiAl–UNCD cantilevers, with mouse heart cells grown on the end. (e) Higher-magnification SEM image of a mouse heart cell grown on the surface of a UNCD cantilever. (f) Signal showing the biting pattern of the mouse heart cell shown in (e).

with higher polarization than PZT and environmentally friendly) (Figure 9.9). The cells are grown on the exposed UNCD end of the cantilever, which has been demonstrated to be a powerful platform for growing animal and human stem cells in recent publications [52–55]. The beating cardiac cell makes the cantilever vibrate, and the strain produced on the piezoelectric layer, induced by Pb^{2+}, Zr^{4+}, and Ti^{4+} ions displaced in one direction and the O^{2-} ion in the opposite direction (inverse piezoelectric effect), thus induces the PZT layer sandwiched between the bottom and top electrode layers to induce voltage pulses extracted between the electrodes. The part of the cantilever containing the cells can be immersed in blood cell fluid in the human heart, which provides the fluid with nutrients that energize the cardiac cells, enabling the piezoelectric actuated power generation MEMS device shown in Figure 9.9.

The integrated Pt–PZT–Pt–TiAl–UNCD MEMS power generation device powered by the heart may enable replacing the Li-ion batteries used today to power defibrillators/pacemakers, which need to be replaced in undesirable short periods of time (6–9 years from the time of first implantation), which requires high replacement costs and patient discomfort.

The information presented in this and the previous subsection shows that the integrated Pt–PZT–Pt–TiAl–UNCD MEMS–NEMS cantilever-based devices demonstrated the feasibility of producing a transformational new generation of integrated piezoelectric / UNCD films-based biosensors and biopower generation MEM/NEMS devices. However, it is critical to replace the PZT-based piezoelectric film by a biocompatible piezoelectric film, since Pb-based piezoelectric films are not suitable for insertion into the human body. This is a key reason why Auciello's group is currently performing R&D to develop $BiFeO_3$ (BFO) films for integration with UNCD, since first tests are showing that BFO is biocompatible.

9.5 Integrated Pt–"Biocompatible" Piezoelectric–Pt–TiAl–UNCD films for MEMS/NEMS Cantilever-Based Biosensors and Implantable Power Generation Devices

It is clear that integrated Pt–PZT–Pt–TiAl–UNCD MEMS/NEMS cantilever-based devices are not suitable for insertion into the human body because the Pb (lead) content. That is why the piezoelectric material shown in the sections above needs to be

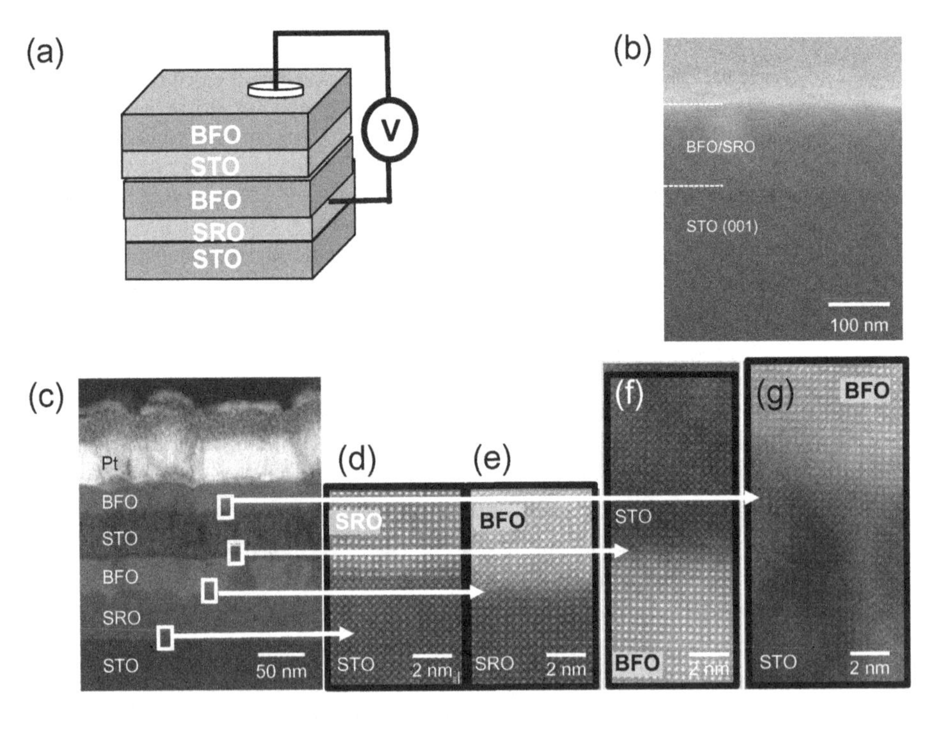

Figure 9.10 (a) Schematic of BFO–STO–BFO BSB-NLs (Pt (50 nm)/BFO (40 nm)/STO (0, 12, 24, and 48 nm)/BFO (40 nm)/SRO (55 nm)/STO (001 single crystalline substrate))-based capacitors used for electrical measurements. (b) SEM image of BFO (40 nm)/SRO (55 nm) initial layers on the (100) STO substrate. (c) Low-resolution TEM (LRTEM) image of BSB-NL with 48 nm thick STO layer. High resolution TEM (HRTEM) images at the interfaces of SRO–STO (d), BFO–SRO (e), STO–BFO (f), and BFO–STO (g), respectively (reprinted from *Appl. Phys. Lett.*, vol. 106, 022905, 2015 (Fig. 1) in [56] with permission from AIP Publisher).

replaced by a biocompatible piezoelectric material. A biocompatible piezoelectric material (AlN) is discussed in detail in Chapter 8. Another potential biocompatible piezoelectric material that can be used to replace the PZT layer is bismuth ferrite ($BiFeO_3$), which is being investigated for application as a piezoelectric and photovoltaic material, but no evidence of biocompatibility is found in the literature.

Preliminary R&D performed by Auciello's group [56] demonstrated that $BiFeO_3$ (BFO) and $SrTiO_3$ (STO) films integrated into novel BFO–STO–BFO nanolaminates (BSB-NLs), featuring nanometer-scale thickness of BFO and STO layers (Figure 9.10) can be tailored via introduction of an STO layer between two BFO layers, such that an optimized STO layer thickness can produce the highest piezoresponse value (Figure 9.11a) and lowest leakage current (Figure 9.11b) of BFO-based piezoelectric films for integration into new piezoelectrically actuated biosensors and power generation devices, with these nanolaminates providing biocompatible piezoelectrically based materials, since both BFO and STO layers are formed by potential biocompatible materials (Bi is inserted in the gastrointestinal relief medicine Pepto Bismol; Fe is in the human biological environment, strontium is also biocompatible, and oxygen is a component of the air breathed by humans).

The introduction of the STO layer between two BFO layers, all of them of nanometer thickness (Figure 9.11a), enables reducing the leakage current density (to 10^{-7} A/cm^2) by two orders of magnitude with respect to relatively high leakage currents (10^{-5} A/cm^2) of current single BFO layers (Figure 9.11b). The BSB-NL also shows very high piezoelectric response, which is about five times higher than that of the pure BFO with the same thickness (Figure 9.11a). The highly strained state of the BFO layers concurrently with the chemical/crystallographic state of the interfaces

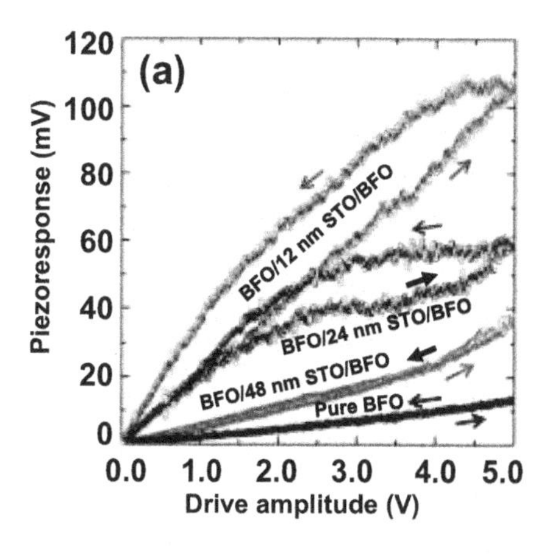

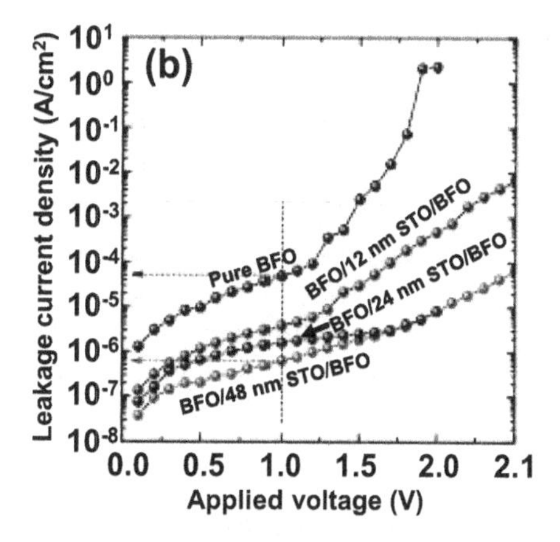

Figure 9.11 (a) Piezoresponse vs. drive voltage curves from piezoresponse force microscopy measured piezoelectric response of single BFO film and BSB-NLs, showing the substantially higher piezoelectric response of the nanolaminates. (b) Leakage current vs. applied voltage for a pure BFO film and BSB-NLs, confirming the hypothesis that the insulating STO layer contributes to reducing the leakage current (reprinted from *Appl. Phys. Lett.*, vol. 106, 022905, 2015 (Fig. 3a and 4a) in [56] with permission from AIP Publisher).

between the BFO and STO layers contributes to the very high values of piezoresponse and very low leakage current observed in the BSB-NLs [56].

The R&D described above on the piezoelectric BSB-NLs indicates that these are promising approaches to producing the piezoelectric materials with piezoresponse properties similar or even better than for PZT, but with the big expected advantage that the BFO–STO–BFO nanolaminates will be biocompatible (R&D is in progress at Auciello's laboratory to prove this hypothesis).

9.6 Conclusions

The R&D described in this chapter showed the materials science and technology development related to the use of UNCD films for fabricating MEMS and NEMS structures suitable for use in various devices for medical applications. Other R&D, performed by Auciello's group and cited in this chapter, indicates that the biocompatibility of UNCD demonstrated in other applications to medical devices and treatment of medical conditions, such as those described in Chapters 2, 4, 5, 6, and 8, provides the bases to envision that UNCD films are the appropriate base biocompatible material for the proposed integrated piezoelectric–UNCD heterostructured films to produce transformational piezoelectrically actuated biosensors and power generation devices.

The demonstrated integrated Pt–PZT–Pt–TiAl–UNCD films into piezoelectrically actuated MEMS/NEMS cantilever-based biosensors and power generation devices indicates that the technology is feasible, but that the Pb-based PZT piezoelectric material needs to be replaced by biocompatible piezoelectric materials. One possible such material is AlN, for which synthesis and properties were discussed in detail in Chapter 8. The other promising material is $BiFeO_3$, which shows very good competitive piezoelectric properties, as shown in Section 9.5, and which is potentially a good biocompatible material made of biocompatible elements like Bi, Fe, and O. Work in progress in a collaboration between Auciello's group and a group in Ireland is showing that BFO may be biocompatible.

Based on the scientific and technological discussion presented in this chapter, the feasibility of producing a new generation of piezoelectrically actuated MEMS/NEMS biosensors and power generation medical devices is promising.

Acknowledgments

O. Auciello acknowledges the contributions from several scientist, students, and postdocs through several years on the R&D that resulted in the science and technology development described in this chapter: J. A. Carlisle, R. W. Carpick, E. M. A. Fuentes-Fernandez, S. A. Getty, G. Lee, D. M. Gruen, J. Hiller, B. Kabius, R. S. Katiyar, A. R. Krauss, B. Shi, S. Sudarsan, and A. Sumant.

O. Auciello acknowledges support from several organizations while doing different parts of the R&D described in this chapter: the US Department of Energy and DARPA, both through several years of work at Argonne National Laboratory; and the Endowed Distinguished Professor Chair grant from 2012 to present from University of Texas-Dallas.

References

[1] K. M. Lakin and J. S. Wang, "Acoustic bulk wave composite resonators," *Appl. Phys. Lett.*, vol. 38, p. 125, 1981.

[2] K. Lakin, "A review of thin-film resonator technology," *IEEE Microwave Mag.* vol. 4 (4), p. 61, 2003.

[3] T. Matsushima, N. Yamauchi, T. Shirai, et al., "High performance 4 GHz FBAR prepared by Pb (Mn, Nb) O_3-Pb (Zr,Ti) O_3 sputtered thin film," in *IEEE International Frequency Control Symposium*, p. 248, 2010.

[4] C. A. Paz de Araujo, O. Auciello, and R. Ramesh (Eds), *Science and Technology of Integrated Ferroelectrics: Past Eleven Years of the International Symposium on Integrated Ferroelectrics Proceedings*. London: Gordon and Breach Publishers, 2000.

[5] O. Auciello, C. M. Foster, and R. Ramesh, "Processing technological for frroelectric thin films and heterostructures," *Ann. Rev. Mater. Sci.*, vol. 28, p. 501, 1998.

[6] A. Gruverman, O. Auciello, and H. Tokumoto, "Imaging and control of domain structures in ferroelectric thin films via scanning force microscopy," in *Ann. Rev. Mater. Sci.*, vol. 28, p. 101, 1998.

[7] O. Auciello, R. Dat, and R. Ramesh, "Pulsed laser ablation-deposition and characterization of ferroelectric thin films and heterostructures," in *Ferroelectric Thin Films: Synthesis and Basic Properties*, C. A. Paz de Araujo, J. F. Scott, and G. W. Taylor, Eds. London: Gordon and Breach Publishers, 1996, p. 525.

[8] M. Park, Z. Hao, G. Gyu Kim, et al., "A 10 GHz single-crystalline scandium-doped aluminum nitride lamb-wave resonator," in *20th International Conference on Solid-State Sensors, Actuators and Microsystems and Eurosensors XXXIII, Transducers & Eurosensors XXXIII*, p. 450, 2019.

[9] M. Knapp, R. Hoffmann, V. Lebedev, V. Cimalla, and O. Ambacher, "Graphene as an active virtually massless top electrode for RF solidly mounted bulk acoustic wave (SMR-BAW) resonators," *Nanotechnology*, vol. 29 (10), p. 10, 2018.

[10] T. R. Sliker and D. A.Roberts, "A thin-film CdS-quartz composite resonator," *J. Apply. Phys.*, vol. 38 (5), p. 2350, 1967.

[11] I. Voiculescu and A. N. Nordin, "Acoustic wave-based MEMS devices for biosensing applications (review)," *Biosens. Bioelectron.*, vol. 33 (1), p. 1, 2012.

[12] E. Tiurdogan, I. Erdem-Alaca, and U. Hakan, "MEMS biosensor for detection of Hepatitis A and C viruses in serum," *Biosens. Bioelectron.*, vol. 28 (1), p. 189, 2011

[13] E. Schroeppel and J. Lin, "Reliability and clinical assessment of pacemaker power sources," in *Handbook of Solid-State Batteries & Capacitors*, 2nd ed., M.Z.A. Munshi, Ed. London: World Scientific Publishing, 1999.

[14] P. Roberts, G. Stanley, and J. M. Morgan, "Abstract 2165: Harvesting the energy of cardiac motion to power a pacemaker," *Circulation*, vol. 118, p. S679, 2008.

[15] V. Leonov, P. Fiorini, S. Sedky, T. Torfs, and C. van Hoof, "Thermoelectric MEMS generators as a power supply for a body area network," in *Proceedings of the 13th International Conference on Solid-State Sensors, Actuators and Microsystems*, p. 291, 2005.

[16] S. Kerzenmacher, J. Ducree, R. Zengerle, and F. von Stetten, "Energy harvesting by implantable abiotically catalyzed glucose fuel cells," *J. Power Source*, vol. 182, p. 1, 2008.

[17] H. W. Lo, Y.-C. Tai, and H. T. Parylene, "Based electret rotor generator," in *Proceedings of IEEE 21st International Conference on Micro Electro-Mechanical-Systems*, p. 984, 2008.

[18] A. M. Paracha, P. Basset, D. Galayko, F. Marty, and T. A. Bourouina, "Silicon MEMS DC/DC converter for autonomous vibration-to-electrical-energy scavenger," *IEEE Electron. Dev. Lett.*, vol. 30, p. 481, 2009.

[19] F. Lu, H. P. Lee, and S. P. Lim, "Modeling and analysis of micro piezoelectric power generators for micro-electromechanical-systems applications," *Smart Mater. Struct.*, vol. 13, p. 57, 2004.

[20] T. Dufay, B. Guiffard, J.-C. Thomas, and R. Seveno, "Transverse piezoelectric coefficient measurement of flexible lead zirconate titanate thin films," *J. Appl. Phys.*, vol. 117 (20), p. 204101, 2015.

[21] T. Ishisaka, H. Sato, Y. Akiyama, Y. Furukawa, and K. Morishima. "Bio-actuated power generator using heart muscle cells on a PDMS membrane," in *Proceedings of International Solid-State Sensors, Actuators and Microsystems Conference*, p. 903, 2007.

[22] P. X. Gao, J. Song, J. Liu, and Z. L. Wang. "Nanowire piezoelectric nanogenerators on plastic substrates as flexible power sources for nanodevices," *Adv. Mater.*, vol. 19, p. 67, 2007.

[23] S. Xu, Y Qin, C. Xu, et al., "Self-powered nanowire devices," *Nat. Nano.*, vol. 5, p. 366, 2010.

[24] Z. Li, G. Zhu., R. Yang, A. C. Wang, and Z. L. Wang, "Muscle-driven *in vivo* nanogenerator," *Adv. Mater.*, vol. 22, p. 2534, 2010.

[25] O. Auciello and A. V. Sumant, "Status review of the science and technology of ultrananocrystalline diamond (UNCD^{TM}) films and application to multifunctional devices," *Diam. Relat. Mater.*, vol. 19, p. 699, 2010.

[26] D. M. Gruen, A. R. Krauss, O. Auciello, and J. A. Carlisle, "N-type doping of NCD films with nitrogen and electrodes made therefrom," US patent #6,793,849 B1, 2004.

[27] S. A. Getty, O. Auciello, A. V. Sumant, et al., "Characterization of nitrogen-incorporated ultrananocrystalline diamond as a robust cold cathode material," in *Micro-and Nanotechnology Sensors, Systems, and Applications-II*, T. George, S. Islam, and A. Dutta, Eds. Bellingham, WA: SPIE, p. 76791N-1, 2010.

[28] V. P. Adiga, A. V. Sumant, S. Suresh, et al., "Mechanical stiffness and dissipation in ultrananocrystalline diamond microresonators," *Phys. Rev. B*, vol. 79, p. 245403, 2009.

[29] A. V. Sumant, O. Auciello, H.-C. Yuan, et al., "Large area low temperature ultrananocrystalline diamond (UNCD) films and integration with CMOS devices for monolithically integrated diamond MEMS/NEMS-CMOS systems," *Proc. SPIE* vol. 7318, p. 7318171-7, 2009.

[30] J. E. Butler and H. Windischmann, "Developments in CVD-diamond synthesis during the past decade," *MRS Bull.*, vol. 23 (9), p. 22, 1998.

[31] A. V. Sumant, O. Auciello, R. W. Carpick, S. Srinivasan, and J. E. Butler, "Ultrananocrystalline and nanocrystalline diamond thin films for MEMS/NEMS applications," *MRS Bull.*, vol. 35, p. 1, 2010.

[32] A. R. Konicek, D. S. Grierson, P. U. P. A. Gilbert, et al., "Origin of ultralow friction and wear in ultrananocrystalline diamond," *Phys. Rev. Lett.*, vol. 100, p. 235502/1-4, 2008.

[33] O. Auciello, J. Birrell, J. A. Carlisle, et al., "Materials science and fabrication processes for a new MEMS technology based on ultrananocrystalline diamond thin films," *J. Phys. Condens. Matter*, vol. 16 (16), p. R539, 2004.

[34] A. V. Sumant, D. S. Grierson, J. E. Gerbi, et al., "Toward the ultimate tribological interface: surface chemistry and nanotribology of ultrananocrystalline diamond," *Adv. Mater.*, vol. 17, p, 1039, 2005.

[35] X. Xiao, J. Wang, J. A. Carlisle, et al. "*In Vitro* and *in vivo* evaluation of ultrananocrystalline diamond for coating of implantable retinal microchips," *J. Biomed. Mater.*, vol. 77B (2), p. 273, 2006.

[36] A. R. Krauss, O. Auciello, D. M. Gruen, et al., "Ultrananocrystalline diamond thin films for MEMS and moving mechanical assembly devices," *Diam. Relat. Mater.*, vol. 10, p. 1952, 2001.

[37] O. Auciello, P. Gurman, M. B. Guglielmotti, et al., "Biocompatible ultrananocrystalline diamond coatings for implantable medical devices," *MRS Bull.*, vol. 39 (7), p. 621, 2014.

[38] S. Srinivasan, J. Hiller, B. Kabius, and O. Auciello, "Piezoelectric/ultrananocrystalline diamond heterostructures for high-performance multifunctional micro/nanoelectromechanical systems," *Appl. Phys. Lett.*, vol. 90, p. 134101, 2007.

[39] J. E. Gerbi, O. Auciello, J. Birrell, D. M. Gruen, J. A. Carlisle, B. W. Alphenaar, "Electrical Contacts to Ultrananocrystalline Diamond," *Appl. Phys. Lett.*, vol. 83 (10), p. 2001, 2003.

[40] J. A. Klug, M. V. Holt, R. Nath Premnath, et al. "Elastic relaxation and correlation of local strain gradients with ferroelectric domains in (001) $BiFeO_3$ nanostructures," *Appl. Phys. Lett.*, vol 98, p. 052902, 2011.

[41] G. Lee, E. M. A. Fuentes-Fernandez, G. Lian, R. S. Katiyar, and O. Auciello, "Heteroepitaxial $BiFeO_3/SrTiO_3$ nanolaminates with higher piezoresponse performance over stoichiometric $BiFeO_3$ films," *Appl. Phys. Lett.*, vol. 106, p. 022905, 2015.

[42] Z. Liu, Q. Liu, H. Liu, and K. Yao, "Electrical properties of undoped PZT and Co-doped PCZT films deposited on ITO/glass substrates by a sol–gel method," *Phys. Stat. Sol. (a)*, vol. 202 (9), p. 1834, 2005.

[43] O. Auciello, "Science and technology of thin films and interfacial layers in ferroelectric and high-dielectric constant heterostructures and application to devices," *J. Appl. Phys.*, vol. 100, p. 051614, 2006.

[44] C. A. Paz de Araujo, O. Auciello, and R. Ramesh, "Science and technology of ferroelectric films and heterostructures for non-volatile ferroelectric memories: past eleven years and the future," in *Science and Technology of Integrated Ferroelectrics: Past Eleven Years of the International Symposium on Integrated Ferroelectrics Proceedings*. London: Gordon and Breach Publishers, vol. 11, p. xvii–lxxxi, 2000.

[45] O. Auciello, C. M. Foster, and R. Ramesh, "Processing technological for ferroelectric thin films and heterostructures," *Ann. Rev. Mater. Sci.*, vol. 28, p. 501, 1998.

[46] R. Thomas, S. Mochizuki, T. Mihara, and T. Ishid, "Preparation of Pb(Zr,Ti)O thin films by RF-magnetron sputtering with single stoichiometric target: structural and electrical properties," *Thin Solid Films*, vol. 413, p. 65, 2002.

[47] Z. Wang, H. Kokawa, and R. Maeda, *Epitaxial PZT films Deposited by Pulsed Laser Deposition for MEMS Applications*. Stresa: TIMA Editions, p. ISBN: 2-916187-03, 2006.

[48] O. Auciello, R. Dat, and R. Ramesh "Pulsed laser ablation-deposition and characterization of ferroelectric thin films and heterostructures," in *Ferroelectric Thin Films: Synthesis and Basic Properties*, C. A. Paz de Araujo, J. F. Scott, and G. W. Taylor, Eds. London: Gordon and Breach Publishers, p. 525, 1996.
[49] O. Auciello, "Pulsed laser ablation-deposition of multicomponent oxide thin films: basic laser ablation and deposition processes and influence on film characteristics," in *Handbook of Crystal Growth*, D. T. J. Hurle, Ed. Amsterdam: Elsevier vol. 3, p. 365, 1995.
[50] D. B. Chrisey and G. K. Hubler (Eds), *Pulsed Laser Deposition of Thin Films*. New York: Wiley-Interscience, 1994.
[51] J. S. Zhao, D.-Y. Park, M. J. Seo, et al., "Metallorganic CVD of high-quality PZT thin films at low temperature with New Zr and Ti precursors having MMP ligands," *J. Electrochem. Soc.*, vol. 151 (5), p. C283, 2004.
[52] B. Shi, Q. Jin, L. Chen, et al., "Cell growth on different types of ultrananocrystalline diamond thin films," *J. Funct. Biomater.*, vol. 3 (3), p. 588, 2012.
[53] B. Shi, Q. Jin, L. Chen, and O. Auciello, "Fundamentals of ultrananocrystalline diamond (UNCD) thin films as biomaterials for developmental biology: embryonic fibroblasts growth on the surface of (UNCD) films," *Diam. Relat. Mater.*, vol. 18 (2), p. 596, 2008.
[54] P. Bajaj, D. Akin, A. Gupta, et al., "Ultrananocrystalline diamond film as an optimal cell interface for biomedical applications," *Biomed. Microdevices*, vol. 9 (6), p. 787, 2007.
[55] W. Yang, O. Auciello, J. E. Butler, et al., "Preparation and electrochemical characterization of DNA-modified nanocrystalline diamond films," *Mater. Res. Soc.*, vol. 737, p. F4.4, 2002.
[56] G. Lee, E. M. A. Fuentes-Fernandez, G. Lian, R. S. Katiyar, and O. Auciello, "Heteroepitaxial $BiFeO_3/SrTiO_3$ nanolaminates with higher piezoresponse performance over stoichiometric $BiFeO_3$ films," *Appl. Phys. Lett.*, vol. 106, p. 022905, 2015.

10 Biomaterials and Multifunctional Biocompatible Ultrananocrystalline Diamond (UNCD™) Technologies Transfer Pathway

From the Laboratory to the Market for Medical Devices and Prostheses

Orlando Auciello

10.1 Introduction

The field of biomaterials has become a leader in materials science during recent years. Biomaterials represent a multidisciplinary field because of the many diverse technological applications. Biomaterials are defined as "materials used for production of new generations of implantable medical devices. prostheses to replace natural bone-based human body parts, and human body engineered tissue and artificial organs" This definition is supported by the fact that biomaterials are currently used in as many as 8000 medical devices [1]. Biomaterials are also being used in a variety of fields such as stem cell development and tissue engineering, gene therapy, and micro- and nanoelectromechanical systems (MEMS/NEMS), enabling a new generation of biosensors and implantable power generation devices to energize electronic medical devices [1, 2]. Technologies using biomaterials include cardiovascular and gastrointestinal stents, defibrillators, artificial hips, knees, dental implants, pacemakers, and much more, all of which are appearing in increasing numbers in the worldwide market. These products, which save the lives of millions of people in the world every year, must be manufactured with appropriate/functional biomaterials for biologically safe, optimum, robust, long-life performance when implanted in the human body.

Governments in countries around the world, including the USA, have created a number of agencies in charge of ensuring that new materials involved in the manufacture of new external and implantable medical devices, prostheses, and any other product will perform as expected, without causing harm to the bodies of people receiving those products.

The Food and Drug Administration (FDA) is the federal regulatory agency involved in the regulation of biomedical products in the USA, including products based on biomaterials. The regulation of biomaterials-based products is under the jurisdiction of one of the FDA divisions empowered to control their commercialization by enforcing manufacturers to comply with standards that ensure the safety and effectiveness of such products [3].

The objectives of this chapter are: (1) to summarize some regulatory challenges faced by medical devices containing biomaterials; (2) to provide a summary of regulatory process for readers not involved or familiar with this topic; (3) to describe new pathways being taken to obtain relatively fast approval for new medical devices and prostheses based on coating existing commercial metal-based devices (which fail due to body fluid–induced corrosion) with the unique biocompatible UNCD coating (made of C atoms – the element of life in human DNA, cells, and molecules, making UNCD probably the best biocompatible material for insertion in the human body).

The development of new technologies in the medical field requires the understanding of the science of biomaterials and their biological effects due to the intimate contact of materials with the human body. The history of development of biomaterials can be divided into three stages:

1. The first stage was characterized by the demand for materials capable of being implanted inside the body, with those materials taken from industrial raw materials and used in applications ranging from orthopedics to cardiovascular surgery and ophthalmology. However, they elicited unacceptable toxicity. Many materials were not biodegradable where this was desired. Classes of materials still used today include pure metals (e.g., iron, titanium, stainless steel), which elicited toxicity induced by corrosion due to exposure to biological media, polymers such as cellulose acetate (originally used for dialysis tubes), rubber, and ceramics such as zirconia.
2. The second stage in biomaterials development occurred during the 1970s, and provided materials that were not only biocompatible and biofunctional, but also biodegradable or bioactive. This second generation of biomaterials includes synthetic polymers such as absorbable sutures made of polylactic and polyglycolic acid or chitosan, bioactive ceramics (e.g., calcium phosphate), and biodegradable ceramics (e.g., hydroxyapatite), used around metallic prostheses to improve fixation to the bone.
3. The third stage features the development of biomaterials from 2000 to the present. These materials are both bioactive and biodegradable and take advantage of new microfabrication processes.

10.2 Ethics in Biomedical Research and Product Marketing

Ethical issues emerged related to the development and testing of new medical products, in many cases pushed forward by manufacturers looking for economic profits as the main goal. These activities resulted, in several cases, in catastrophic consequences for many patients, such as the case of Ti alloys used in dental implants, of which ~15% worldwide need to be replaced in the first 3–4 years due to destruction induced by chemical attack from oral fluids (Figure 10.1).

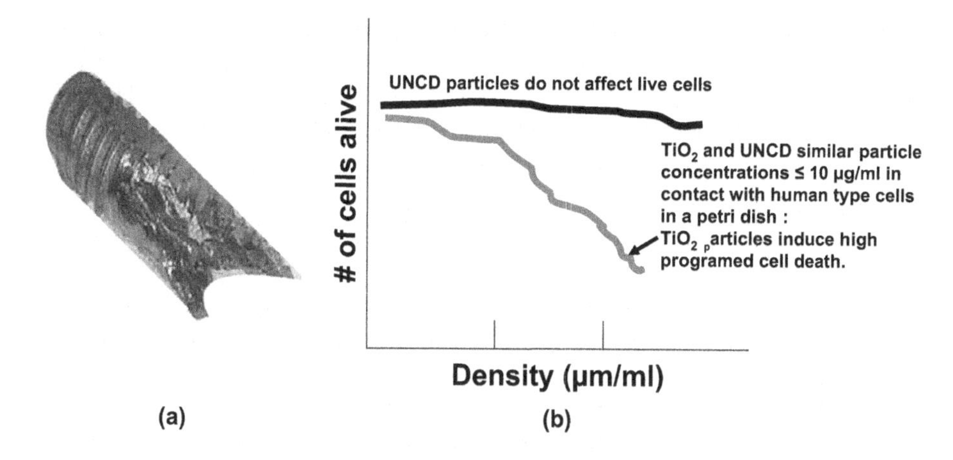

Figure 10.1 (a) Commercial Ti-6V-4Al dental implant extracted from a patient's mouth three years after implantation in the maxillary bone, showing extensive chemical corrosion induced by oral fluids, destroying the implant. (b) TiO_2 particles, dislodged from the oxidized surface of the Ti alloy dental implant, induce high programmed death of live cells, while UNCD particles do not affect live cells at all.

In many cases, product manufacturers' practices were implemented without careful consideration of the risks vs. the benefits for patients [4, 5]. It was not until the end of World War II, after the Nazi genocide, which included human experimentation, that the first code of conduct for research on humans was written and approved, in Nuremberg in 1947. The Nuremberg Code introduced three principles for conducting research on human subjects: voluntary/informed consent, favorable risk–benefit analysis, and patients' right to withdraw from the research or clinical trials at any time without further consequences [5]. Later, in 1964, the Helsinki Declaration was written by the World Medical Association, postulating two fundamental principles in addition to the Nuremberg principles: (1) the human involved in the research project would be provided with the best treatment available; and (2) care for the patient will be the first priority over the benefits for society. Then, in 1978, the Belmont Report was written, establishing three fundamental pillars for conducting biomedical research involving human subjects: (1) respect for the human subject; (2) benefits, and (3) justice [5]. Currently, the Helsinki Declaration (the latest amendment of which was approved in Seoul, Korea, in 2008) is being used as one of the main guidelines to conduct clinical research ethically [6].

10.3 Regulation of Medical Devices in the USA, Europe, and Japan

Several national agencies and worldwide private organizations are responsible for overseeing the manufacture and marketing of medical products. These organizations have different policies according to national requirements. In Japan, the

Pharmaceutical and Medical Devices Agency (PMDA) is in charge of approving new products, although final authority falls under the jurisdiction of the Japan Ministry of Health, Labor, and Welfare (MHLW). Within the MHLW, the Pharmaceuticals and Food Safety Bureau reviews the regulation of pharmaceuticals, food, and medical devices. Europe operates with a decentralized system based on notified bodies (NB), which are private organizations with the authority to bestow the CE mark, providing clearance for commercialization of medical device companies within European countries.

Manufacturers worldwide are audited by national agencies in each country, who regulate medical devices and ensure their safety and effectiveness for the general population [3, 7–10]. However, because the UNCD coating technology is being developed in the USA, it is relevant to discuss specifically the regulations stablished by the FDA; addressing also European and Asian countries would require more extensive discussion. In addition, the focus is mainly on the FDA because it is one of the most important regulatory agencies worldwide regulating pharmaceutical and medical device marketing. Finally, the FDA has launched new initiatives, such as the regulation of nanotechnology products and the Critical Path Initiative, which provides oversight of new initiatives in regulation of biomaterials and medical devices. However, since both European and Japanese regulatory systems are also key players in the global scenario of regulation of biomedical products, the reader is advised to read [10, 11], where they can find more comprehensive information.

10.4 A Brief History of the FDA

The FDA is the US federal agency within the US Department of Health and Human Services responsible for protecting public health by ensuring the safety, efficacy, and security of human and veterinary drugs, biological products, medical devices, food supplies, cosmetics, and products that emit radiation [3]. The FDA was created following several laws approved by the US Congress to protect public health. The first law was the Food and Drug Act of 1906, launched to monitor meat and food quality to protect people from possible food adulteration and to control medicines that were not properly labeled or had questionable compounds. Later, in 1938, following the deaths of 100 patients from exposure to diethylene glycol in sulfonamide products, created in the liquid form of this antibiotic, a comprehensive law, the Federal Food and Drug Cosmetic Act, was created in the USA to improve the safety of foods, drugs, and cosmetics. Following the tragedy of thalidomide in Europe, the FDA did not approve this drug for use in the USA.

In relation to the main topic of this book – the description of multifunctional medical devices and prostheses based on the unique biocompatible UNCD coating – a key regulatory development was the creation, in 1976, of the Medical Device Amendment Act, which was the first medical device regulation in the USA. This amendment called for quality control procedures and the registration, with the FDA, of new medical devices developed by manufacturers [12, 13].

10.5 Organization and Operation of the FDA

The FDA is organized into six product centers, one research center, and two offices:

Products centers:

1. The Center for Biologics and Evaluation Research (CBER) provides regulation for a large variety of products: gene therapy, blood/blood components, vaccines, tissues (e.g., biological heart valves), and tissue engineering products, for which regulations relate to licensing and safety of blood supplies and postmarketing surveillance of biological products to identify adverse events arising from the mentioned products introduction into the market.
2. The Center for Drug and Evaluation Research (CDER) evaluates all new drugs to be inserted into the US market.
3. The Center for Food Safety and Applied Nutrition (CFSAN) regulates labeling/safety of food and cosmetics.
4. The Center for Tobacco Products (CTP) evaluates and inspects tobacco products.
5. The Center for Veterinary Medicine (CVM) evaluates the safety and effectiveness of animal food, drugs, and devices.
6. The Center for Devices and Radiological Health (CDRH) evaluates medical devices, including biomaterials used for fabrication of those devices and their performance, including tracking of device malfunction and induced adverse reactions in human bodies. In addition, the CDRH reviews radiation safety performance standards, including products that emit X-rays, microwaves, radiofrequencies (RFs), and UV light. The relevance of carefully monitoring radiation-emitting products in biomaterials and medical devices relates to the extensive use of magnetic resonance imaging (MRI) in patients carrying pacemakers, where electromagnetic interactions between the pacemaker and the magnetic and RF fields produced by the MRI equipment may take place and interfere with the correct functioning of the pacemaker. Monitoring of the MRI process, in relation to new biomaterials, relates to the capability of the new materials to shield devices against unwanted magnetic fields. Another example of required regulation in this field relates to the use of superparamagnetic nanoparticles (Fe_2O_4 – magnetite) used as heating elements, via coupling of external magnetic fields, for cancer therapy, which being explored for a novel retina reattachment procedure (see Chapter 4). This center covers several areas, including biology, physics, and electrical engineering, all of which are related to the chemistry of new biomaterials.

Research center

7. The National Center for Toxicological Research (NCTR) addresses the toxicology of certain biomaterials, such as titanium oxide (TiO_2) and zinc oxide (ZnO) nanoparticles, involved in sunscreens and other products, and more recently demonstrated as released from oxidized surfaces of commercial Ti alloy–based dental implants (see Chapter 5), inducing deaths of live cells and inflammation in human mouth tissue.

Regulatory offices

8. The Office of Regulatory Affairs (ORA) is responsible for FDA inspections of manufacturers' facilities and products, as well as for control of imported products.
9. The Office of the Commissioner, created in 2002, is responsible for effective conduct of the FDA's mission. This includes addressing regulation of drugs delivery, such as drug-eluting stents, drug delivery polymer scaffolds, and antibiotic bone cements, which have been under development in recent years and are intended to obtain FDA clearance for commercialization. For example, for the case of drug delivery devices, where the drug is the main mode of action to produce the desired therapeutic effect, the CDER has primary jurisdiction over the product review, whereas if the device itself determines the primary mode of action to produce the desired therapeutic effect, then the CDRH will be the final reviewer of the documents submitted by the manufacturer [3].

Voluntary Standards and Organizations Providing Them. Many standards are developed by nongovernmental voluntary organizations, including the following:

1. The International Organization for Standardization (ISO), which is devoted to developing standards across a wide variety of disciplines, via committees, each providing guidance documents that become standards after being subjected to a general vote among all groups participating around the world. Examples of ISO committees are those focused on evaluation of biological performance of medical devices and, although still in a nascent stage, the ISO/TC 229 WG3 (Nanotechnologies in Health). Standards related to medical devices currently in use are ISO 10993, for the evaluation of biological performance of medical devices, and ISO 13485, a quality standard for medical devices [14].
2. The American Society for Testing and Materials (ASTM), which is internationally recognized, and sets standards for biomaterials and medical devices which are later adopted by the industry in trying to achieve competitive advantage and obtain permission from national organizations to commercialize their products. ASTM and other standards-developing organizations provide great help to industries because: (a) regulatory agencies do not have funding for covering the cost of human resources necessary to set standards for products or processes to be inserted in the market; and (b) voluntary standards allow industries to fulfill their objectives without being forced to comply with standards that are either not relevant or are not related to the industry's activities. Some key standards include corrosion and fatigue test for metallic implants, an *in-vitro* test for evaluation of fluid-induced degradation of polymers used in surgical implants, standard test for measuring magnetically induced torque in medical devices in a magnetic resonance environment, standard specifications for unalloyed titanium for surgical implant applications, and standard specifications for calcium phosphate coatings for

implantable materials. The F04 (Medical and Surgical Material and Device Committee) within the ASTM have set standards for evaluation of biomaterials and medical devices, which have been adopted by the FDA.

10.5.1 FDA Pathways for Approval of Medical Devices

Medical devices are used for a variety of purposes, including temperature-sensing bladder catheters, pulse oximeters (used to monitor oxygen in humans), drainage catheters, fracture fixation plates, implantable medical power and sense generation devices (pacemakers/defibrillators and microchips returning partial vision to blind people; see Chapter 2) and prostheses (dental implants, hips, knees, and much more), all used to improve the quality of life for people who have suffered the loss of natural human body components or functionality. The problem is that many materials used for fabrication of those devices and device geometry induce deleterious effects in human bodies (e.g., contaminated orthopedic fixation plates resulting in limb-threatening bone infection; misbehaving pacemakers causing cardiac arrest; the oxidized surface of Ti alloy dental implants inducing TiO_2 nanoparticles release and insertion into tissue, causing cells destruction and inflammation; and many more conditions induced by many biomaterials inserted in human bodies).

Based on this information, patients and/or their families and physicians need to be familiar with the materials and medical device designs in order to understand the potential deleterious effects when inserted into the human body, and, eventually, make plans for replacement, removal, or deactivation of the device or prostheses.

From the economical point of view, the manufacture and sale of medical devices and prostheses provides jobs for thousands of people. A large number of medical devices using many kinds of biomaterials are launched to the market every year. About 4% of the population in the USA uses at least one implantable medical device. Before the year 2000, the FDA had approved about 500,000 medical devices developed by 23,000 manufacturers [14]. In 2008, nearly 350,000 pacemakers, 140,000 defibrillators, and 1,230,000 stents were implanted in humans in the USA [15]. According to the Advanced Medical Technology Association, the medical devices market reached a value of $77 billion in 2002 [16], while a WHO report estimated a $260 billion market in 2006 [17]. Recent studies revealed that the costs of producing medical devices are $322–522 million [18].

10.5.2 Medical Device Regulation

The FDA has established three categories of medical devices based on risks to the patient: Class I (low risk, requiring low control level; e.g., examination gloves); Class II (intermediate risk, requiring higher control level than for Class I; e.g., CT scanners); and Class III (high risk, requiring high-level controls; e.g., defibrillators/pacemakers – that is, a device that could have life-threatening consequences if it fails). The requirements for each device class provide the "pathways" for obtaining FDA approval.

Premarket Notification (510k). This is the pathway for obtaining clearance from the FDA for most biomaterial-based medical devices to enter the US market. The 510k relates to existing devices, based on a level of equivalence of the new device for a particular usage. Evidence submitted for a 510k request is based on bench and sometimes pre-clinical data. Only 10–15% of the 510k notifications require clinical data. If a new device is equivalent to a preexisting device on the market, then a premarket notification is sufficient to market the product within the USA. If the device does not meet FDA criteria for equivalence to an existing device, a premarket approval is required.

Class I and Class II devices may be exempted from a premarket notification if they were on the market before 1976 and have not significantly changed since then, and/or if they are exempted by regulations. Examples of exempted Class I/II new devices from 510k are neurological devices such as two-point discriminators made of stainless steel, dental materials such as carboxymethylcellulose sodium denture adhesive, and a hydrogel wound dressing used in plastic surgery. Exemptions to the regulations occur when the FDA decides to refocus the resources used to assess 510k notifications onto public health issues of higher priority. A complete list of Class I and Class II devices exempted from 510k is available from the FDA website [20].

Premarket Approval (PMA). This is the way the FDA assess Class III devices. The PMA is required when it is not possible to prove substantial equivalence to an existing device, and therefore, in addition to bench and pre-clinical data, additional clinical data must be provided to demonstrate the effectiveness and safety of the device under investigation. This is the hardest and more expensive way of bringing medical devices to the market. High-risk, innovative devices are more likely to follow this pathway.

Two other important categories or pathways for new device insertion to the market are the **Investigational Device Exemption (IDE)** and the **Humanitarian Device Exemption (HDE).**

Investigational Device Exemption. An IDE must be submitted to the FDA if a new device is intended for use in a clinical trial by a manufacturer. An IDE could be required by the CDRH to support a 510k or PMA application when clinical data are needed. For a Class III device institutional review board (IRB) approval is also needed. IRB approval ensures that the principles behind the declarations will be followed, including protection of the rights of the patients enrolled in the clinical study [3, 6]. An IDE should also be submitted when the sponsor wants to submit a new use for an existing device.

Humanitarian Device Exemption. New medical devices projected for use to treat rare diseases affecting fewer than 4000 patients per year in the USA belong to the HDE category. Examples include fetal and small child bladder stents. The HDE category was developed to help manufacturers in commercializing medical devices that have small markets and therefore small financial incentives [13].

FDA Fast-Track Approvals. The FDA created this approval pathway to speed up insertion of certain products into the US market where they fulfill an unmet medical need and are not available to the general population. These devices are exempted from FDA review and approval, including registry, standard conformity, or pre-market approval. Examples of such devices include dental and orthopedic implants [12, 20].

10.6 The Helsinki Declaration for Current Clinical Research

The Helsinki Declaration is very important in relation to conducting clinical trials ethically. Key parts of the Helsinki Declaration are:

1. ***Informed consent.*** Clinical trials, involving new biomaterials, require informing patients about the risks involved in the trial so they can decide voluntarily whether or not to be involved in the research.
2. ***Operator experience.*** A clinical trial related to exploration of a new implantable medical device requires the participation of an expert trained to perform any procedure, such as the implantation of a pacemaker.
3. ***Well-being of the patient.*** During a clinical trial, exploring implantation of a new medical device takes precedent over any other interest, implying that any procedure that threatens the patient's well-being is not justified even if it is proving to be beneficial for people.
4. ***Scientific principles.*** Any clinical trial on devices using new biomaterials must be based on comprehensive scientific and technological knowledge validated by the international scientific community through the literature, such as peer review journals or other relevant publications, and by IRBs responsible for ensuring compliance with these principles.

10.7 Examples of Faster Pathways for Insertion of New Biomaterials, Such as UNCD Coatings and UNCD-Coated Devices, into the US Market, Overcoming the Current Long Pathway Induced by FDA Regulations

10.7.1 UNCD-Coated Silicon-Based Microchip Implantable in the Eye to Restore Partial Vision to People Blinded by Retinitis Pigmentosa

The Argus II Retinal Prosthesis (see detailed description in Chapter 2) involves an Si-based chip that in the final rendition would be implanted inside the eye on the ganglion cell layer (Figure 10.2), receive an image from a camera on glasses and inject processed electrical charges to the ganglion cells through a large electrode array, finally transmitted to the brain via the ganglion cell axons bundle (optic nerve) [23].

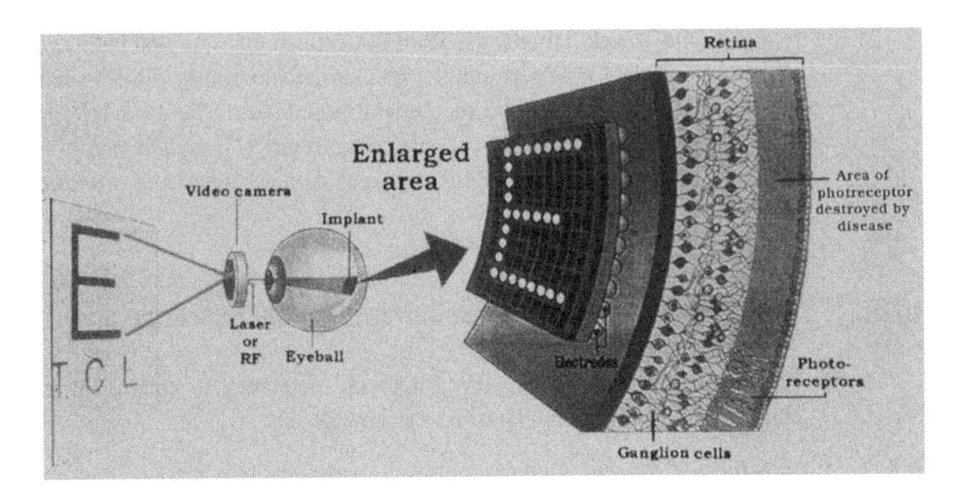

Figure 10.2 Schematic showing a charge-coupled device (CCD) camera mounted on glasses outside the eye, capturing images and sending wireless electromagnetic signals, with image information, to an Si-based microchip implanted inside the eye, as desirable in its final future rendition, with a large array of electrodes connected to the ganglion cells. The Si microchip should be coated with a biocompatible, hermetic, corrosion-resistant coating (UNCD is probably the best coating, as demonstrated in 10 years of animal studies – see the detailed description in Chapter 2) (reprinted from *MRS Bull.*, vol. 39, p. 621, 2014 (Fig. 3) in [36] in Ch. 2 with permission from Cambridge Publisher).

The Argus II device was developed by a team of researchers from universities, national laboratories, and Second Sight (the company currently commercializing the device) during a 10 year (2000–2010) US Department of Energy-funded project. The Argus II is currently the most advanced artificial retina device, and the only one currently implanted commercially in the USA and Europe to restore partial vision to people blinded by retinitis pigmentosa. The Argus II device was named one of the top 25 inventions for 2013 by *TIME Magazine*.

After 10 years of approved clinical trials in humans in the USA, México, the UK, France, and Switzerland, including 31 people who each received an Argus II device implant, Second Sight, with the consensus of the large team who developed the implantable device, requested approval from the FDA to implant the Argus II device (based on the Si microchip inserted in a sealed metallic box located outside the eye and connected to the retina's ganglion cells via Pt wires encapsulated in a biocompatible polymer) in the eyes of blind people in the USA. After one year (January 2010 to December 2010) of waiting for a response, the Argus II device development team discussed an alternative pathway for approval and commercialization of the Argus II device. The team decide to request approval from the European Community. Thus, Second Sight requested approval (January 2011) to implant the Argus II device in Europe. It received approval in six months (July 2011), and the first Argus II device was implanted in a person blinded by retinitis pigmentosa in October 2011, with the entire process taking only nine months. Subsequently, Second Sight, with consensus from the Argus II device team, requested approval, again, from the FDA, to implant the Argus II device in the USA, showing that the device was being implanted commercially in Europe. The FDA approved the commercial implantation of the Argus II device in the USA in six months (see a detailed description of the process in [24]).

The example of the Argus II device indicates that getting approval from the FDA for new medical devices implantable in humans can be speeded up by getting approval first in countries outside the USA, where agencies provide a faster – but still strongly based on medical safety – approach. However, there is still not FDA approval for implanting UNCD-coated Si microchips inside the human eye.

10.7.2 UNCD-Coated Dental Implants

Original Biomedical Implants (OBI-USA and OBI-México), companies co-founded by Auciello and colleagues, have been performing R&D for developing UNCD-coated dental implants (Figure 10.3), with the UNCD coating providing the best biocompatible material because it is made of C atoms and is extremely resistant to chemical corrosion by oral fluids, as demonstrated by five years (2014–2019) of animal studies and three years (2018 to present) of clinical trials in humans, involving 20 UNCD-coated dental implants inserted in patients in the world-class clinic of Dr. Gilberto López-Chávez in Querétaro-México (see Chapter 5). The animal studies and clinical trials in humans demonstrated that UNCD-coated commercial Ti alloy dental implants provide a new transformational technology for this type of prosthesis, orders of magnitude superior to current commercial Ti alloy–based dental implants, since the UNCD coating provides the best biocompatible/oral fluids corrosion-resistant biomaterial to be in contact with the human mouth.

The reason why the clinical trials are being undertaken in México is because faster approval for clinical trials can be obtained, compared with the USA. In addition, extensive market research indicates that inserting UNCD-coated dental implants in the market in México would be a good commercial approach, since currently thousands of

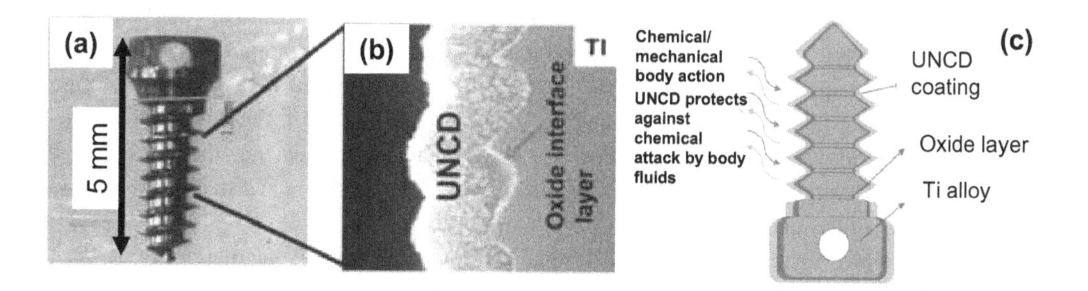

Figure 10.3 (a) A commercial Ti alloy dental implant. (b) Cross-section scanning electron microscope (SEM) image of a UNCD-coated commercial Ti alloy dental implant. (c) A schematic of a UNCD-coated Ti alloy dental implant showing key component and mechanical/chemical actions.

people living in the USA go to México to get dental implants because getting a dental implant costs $800–900 in México vs, $4000–5000 in the USA (US dollars). In addition, the UNCD-coated dental implants are projected to get approval for insertion into markets in Latin America, Europe, and Asia much sooner than from the FDA. Then, the same approach used for getting approval for the artificial retina, described above, will be followed for getting approval from the FDA for marketing of UNCD-coated dental implants and other prostheses (e.g., hips, knees, and more) in the USA.

10.7.3 Integrated Injection of Superparamagnetic Nanoparticles Inside the Human Eye, Attracted by a Magnetic Field Produced by an External UNCD-Coated Magnet to Reattach a Detached Human Retina

Chapter 4 describes the R&D done through a collaboration between researchers in Argentina and the USA, focused on developing a transformational new treatment to reattach a detached retina in the human eye, to avoid blindness. The new treatment involves injecting a fluid solution containing superparamagnetic nanoparticles, made of Fe_2O_4 (approved by the FDA for other medical applications), which are attracted by a magnetic field produced by a UNCD-coated magnet positioned outside the eye (see details in Chapter 4), to push the retina back onto the eye's inner wall for reattachment. Although the superparamagnetic particles have been approved by the FDA, the UNCD coatings have not yet been approved. The team decided to conduct clinical trials in Argentina, because the regulatory agency (ANMAT) provides approval in less than a year, as opposed to the time frame, measured in years, to get approval from the FDA. Thus, this is another example on how UNCD-coated new medical devices can get approved faster for clinical trials, and eventually for marketing, outside the USA. Two patients were treated with the new retina reattachment procedure, and they recovered full vision.

10.8 Conclusions

It is expected that insertion of new medical devices in the worldwide market will be supported by fewer reports of adverse events and early device failure, thus improving the quality and extension of human life. It is expected that new micro-/nano-biomaterials and micro-/nanotechnologies will be part of the insertion of new generations of implantable medical devices and prostheses into the market, with new functionalities and orders of magnitude better performance than current products in the market.

However, the current status of medical device technology regulation favors technological improvements that are incremental in nature (i.e., improvements of devices already on the market) rather than disruptive technologies. The reason for such regulatory device approval status in the USA is that breakthrough medical devices are subjected to a much more comprehensive review process by the FDA, favoring the development of devices with less innovation. Therefore, transformational innovation in medical devices and prostheses will be possible only by modernization of regulatory science. This goal may be achieved by interdisciplinary groups, including scientists and industries, working jointly with the FDA to bring new medical devices/prostheses faster and more safely to the population. It is expected that initiatives like the Critical Path Initiative may impact the regulatory science and policy by adopting new approaches in clinical trials for testing new medical devices in people which incorporate new technologies such as computer simulations for studying implantation performance and other advanced technological tools that are now available.

From the ethical point of view, reducing or eliminating potential conflicts of interest may need the insertion of independent committee assessment groups, collaborating with FDA experts, in monitoring the strength of the evidence from clinical trials for new medical devices and prostheses technologies, to make the regulatory process more transparent and less based on economic bias.

Another important action needed to reduce the high barriers for FDA approval of new medical devices and prostheses is that better control of the often uncontrolled behavior of lawyers using lawsuits to generate big profits. These actions are most probably the strongest basis for the high FDA barriers for approving clinical trials and insertion of new medical devices and prostheses to the US market, favoring more rapid insertion in international markets outside the USA, as proven for the Argus II device and UNCD-coated dental implants described in this book.

Acknowledgments

O. Auciello acknowledges the support from the University of Texas-Dallas through his Distinguished Endowed Chair Professor position. O. Auciello acknowledges also the great contributions of collaborators in the development of UNCD-based medical devices, described in Chapters 2–5, and the performance of animal studies and clinical trials in humans, namely: M. Humayun (USA) and other colleagues, A. Berra (Argentina), G. López-Chávez (México), P. Gurman (Argentina), K. Kang (USA), E. Lima (Argentina), D. Olmedo (Argentina), M. J. Saravia (Argentina), D. Tasat (Argentina), and R. Zysler (Argentina).

References

[1] R. Woo, D. D. Jenkins, and R. S. Greco, "Biomaterials, and historical overview and current directions," in *Nanotechnology in Nanoscale Technology in Biological Systems*, R. Greco, Ed. Boca Raton, FL: Taylor and Francis, 2005.

[2] D. F. Williams, "On the nature of biomaterials," *Biomaterials*, vol. 30, p. 5897, 2009.

[3] FDA. Homepage. www.fda.gov.

[4] B. Constantz, "Crossing the chasm: adoption of new medical device nanotechnology," in *Nanoscale Technology in Biological Systems*, R. Greco, Ed. Boca Raton, FL: CRC Press, 2005.

[5] T. W. Rice, "The historical, ethical and legal background of human subjects' research," *Respir Care*, vol. 53 (10), p. 1325, 2008.
[6] World Medical Association, Helsinki Declaration. www.wma.net.
[7] European Commission Enterprise and Industry. Homepage. http://ec. europa.eu/atoz_en .htm.
[8] Ministry of Health Labor and Welfare, Japan. Homepage. www.mhlw.go.jp/english.
[9] Pharmaceutical and Medical Device Agency, Japan. Homepage. www.pmda.go.jp/eng lish/index.html.
[10] Medicines and Health Care Products Regulatory Agency, UK. Homepage. www.mhra.gov .uk/index.htm.
[11] P. Gurman, O. Rabinovitz-Harison, and T. B. Hunter, "Regulatory challenges in biomaterials: focus on medical devices," in *Biomaterials Science: A Clinical and Engineering Approach*. Boca Raton, FL: CRC Press, 2012.
[12] FDA. *United States FDA Medical Devices Control and Regulation Handbook*, 4th ed. Washington, DC: International Publisher Publications, 2010.
[13] FDA, "Overview of device regulation." www.fda.gov/MedicalDevices/DeviceRegulation andGuidance/Overview/default.htm.
[14] ISO. Homepage. www.iso.org.
[15] M. D. Feldman, A. Petersen, L. Karliner, et al., "Who is responsible for evaluating the safety and effectiveness of medical devices? The role of independent technology assessment," *J. Gen. Inter. Med.*, vol. 23 (suppl. 1), p. 57, 2008.
[16] S. S. Dhruva, L. A. Bero, and R. F. Redberg, "Strength of the evidence examined by the FDA in premarket approval of cardiovascular devices," *JAMA*, vol. 302 (24), p. 2679, 2009.
[17] S. L. Brown, R. A. Bright, and D. R. Travis, "Medical device epidemiology and surveillance: patient safety is the bottom line," *Expert Rev. Med. Dev*. vol. 1 (1), p. 1, 2004.
[18] Medical Device Regulations, "Global overview and guiding principles: World Health Organization Report," 2003. www.who.int/whr/2003/en/whr03_en.pdf.
[19] FDA, "Medical device exemptions 510(k) and GMP requirement." www.accessdata.fda .gov/scripts/cdrh/cfdocs/cfpcd/315.cfm?GMPPart=878#start.
[20] FDA, www.accessdata.fda.gov/scripts/cdrh/cfdocs/cfpcd/315.cfm?GMPPart=878#start.
[21] M. Humayun, "Interim results from the international trial of Second Sight's visual prosthesis," *Ophthalmology*, vol. 119, p. 779, 2012.
[22] Second Sight. Homepage. www.secondsight.com.

Index

Page numbers in **bold** indicate figures; and ***bold italics*** indicate to Tables

Printed in the United States
by Baker & Taylor Publisher Services